Cyclic AMP, Cell Growth,
and the Immune Response

WERNER BRAUN *(1914–1972)*

Cyclic AMP, Cell Growth, and the Immune Response

*Proceedings of the Symposium
Held at Marco Island, Florida
January 8-10, 1973*

Werner Braun

Lawrence M. Lichtenstein

Charles W. Parker

Editors

SPRINGER-VERLAG NEW YORK · HEIDELBERG · BERLIN

1974

Library of Congress Cataloging in Publication Data

Cyclic AMP, cell growth, and the immune response.

　　1. Immunopathology—Congresses. 2. Cyclic
adenylic acid—Congresses. 3. Cell proliferation—
Congresses. I. Braun, Werner, 1914-1972, ed.
II. Lichtenstein, Lawrence M., ed. 1934–　　. III. Parker,
Charles Ward, 1930–　　, ed. [DNLM: 1. Adenosine
cyclic 3′, 5′ monophosphate—Congresses. 2. Cell divi-
sion—Congresses. 3. Immunology—Congresses. 4.
Neoplasms, Experimental—Congresses. QW504 C995
1973]
QR188.6.C9　　　574.2′9　　　74-1278
ISBN 0-387-06654-3

Printed in the United States of America

ISBN 0-397-06654-3 Springer-Verlag New York · Heidelberg · Berlin
ISBN 3-540-06654-3 Springer-Verlag Berlin · Heidelberg · New York

PREFACE

The brilliant research of Dr. Earl Sutherland and his colleagues has had a broad impact on many areas of biology. Among the fields influenced rather late by the insights arising from this work were immunology and oncology. Although research relating cyclic AMP metabolism to the development and manifestations of the immune response and the control of mammalian cell growth is relatively recent, the growth of knowledge in these areas has been rapid and there is already a considerable amount of empirical information. This conference provided an opportunity to collate and begin to interpret that information. A deliberate attempt was made to bring together investigators nominally involved in immunology, biochemistry, pharmacology, or cellular biology for in many instances parallel observations are being obtained in these fields. For example, the immunologist studying the transformation of lymphocytes by antigens or mitogens is carrying out experiments that are very close to those of the biologist studying the growth of cells in culture; in both cases, the phenomena they observe are modulated by changes in the intracellular level of cyclic nucleotides. Many other examples of closely analogous experiments in different fields could be cited, but perhaps the point is clear.

The meeting was convened for two major reasons: to reap the benefits of cross-fertilization which attend the interaction between different disciplines interested in common phenomena, and to permit interested investigators in several fields to formulate preliminary concepts of cyclic nucleotide control of inflammation, immunologic differentiation, and tumor growth. We hoped by this process to discard nonviable ideas and to project the directions in which future progress might be made.

The meeting was primarily the brainchild of Dr. Werner Braun; it was he who obtained funds from the National Institutes of Health and contacted most of the participants. In November 1972, shortly after meeting with us to finalize plans for the meeting, Dr. Braun died suddenly. His death was a great loss to science and to us personally; we feel privileged to have had the opportunity to organize the meeting with him.

There are many without whose assistance the meeting could not have taken place. Primary thanks, of course, must go to the National Institute of Allergy and Infectious Diseases for providing funds for the meeting. Dr. Dorland J. Davis, Dr. Maurice Landy, and Dr. Bill Gay of that institute were of particular assistance in many aspects of organizing the meeting, as was Mrs. Gwen Northcutt. Sincere thanks are due as well to the Schering Corporation, Merck, Sharpe and Dohme, and The Nelson Research and Development Corporation for additional funds. We wish also to express our gratitude to Miss Lee Stein, Dr. Braun's secretary, who was able to bring together the arrangements he had made for the meeting, and to Mrs. Anne Sobotka, who helped organize both the meeting and the book.

    We owe special thanks to Dr. Otto Plescia, Dr. Braun's long-time associate at Rutgers, who was able to reconstruct from Dr. Braun's notes and papers in progress an up-to-date account of his current research.

L. M. Lichtenstein
C. W. Parker

CONTRIBUTORS

°A STEN, K. FRANK, Harvard Medical School, Robert B. Brigham Hospital, Department of Medicine, Parker Hill Avenue, Boston, Massachusetts 02120.

AUSTIN, KATHRYN E., Department of Medicine, University of California, San Francisco Medical Center, San Francisco, California 94122.

BECK, N. P., Departments of Pediatrics and Pathology, University of Minnesota Medical School, Minneapolis, Minnesota 55455.

BECKER, B., Institut für Therapeutische Biochemie der Universität Frankfurt, Frankfurt (Main), Germany.

BLANCUZZI, V., Research Department, CIBA-Geigy Corp., Ardsley, New York 10502.

°BLOOM, BARRY R., Department of Microbiology and Immunology, Albert Einstein College of Medicine, Bronx, New York 10461.

°BOURNE, HENRY R., Division of Clinical Pharmacology, Departments of Medicine and Pharmacology, University of California Medical Center, San Francisco, California 94122.

BRAUN, WERNER (1914–1972).

CARPENTER, C. B., Department of Medicine, Peter Bent Brigham Hospital, Boston, Massachusetts 02115.

°CHANDRA, PRAKASH, Institut für Therapeutische Biochemie der Universität Frankfurt, and Farbwerke Hoechst AG, Frankfurt-Höchst, Germany.

°DEISSEROTH, A., Building 10, Room 7 D 04, National Institutes of Health, Bethesda, Maryland 20014.

ESTENSEN, R., Departments of Pharmacology and Pathology, University of Minnesota Medical School, Minneapolis, Minnesota 55455.

°Denotes conference participant.

FUDENBERG, H. HUGH, Department of Medicine, University of California, San Francisco Medical Center, San Francisco, California 94122.

°GERICKE, DEITMAR, Institut für Therapeutische Biochemie der Universität Frankfurt, and Farbwerke Hoechst AG, Frankfurt-Höchst, Germany.

°GILLESPIE, ELIZABETH, Division of Clinical Immunology, The Johns Hopkins University School of Medicine at The Good Samaritan Hospital, Baltimore, Maryland 21239.

°GOLDBERG, NELSON D., Department of Pharmacology, University of Minnesota, Minneapolis, Minnesota 55455.

GOOD, ROBERT A., 410 East 68th St., New York, New York 10021.

HADDEN, ELGA, Departments of Pharmacology and Pathology, University of Minnesota Medical School, Minneapolis, Minnesota 55455.

°HADDEN, JOHN W., Departments of Pathology and Pediatrics, University of Minnesota, Minneapolis, Minnesota 55455.

HADDOX, M. K., Departments of Pharmacology and Pathology, University of Minnesota Medical School, Minneapolis, Minnesota 55455.

HAN, IHN H., Department of Microbiology, The University of Michigan Medical School, Ann Arbor, Michigan 48104.

°HARDMAN, JOEL G., Department of Physiology, Vanderbilt University School of Medicine, Nashville, Tennessee 37232.

°HENNEY, CHRISTOPHER S., The Johns Hopkins University School of Medicine at The Good Samaritan Hospital, O'Neill Memorial Laboratories, Baltimore, Maryland 21239.

°HIRSCHHORN, ROCHELLE, Department of Medicine, New York University School of Medicine, New York, New York 10016.

HOFFSTEIN, SYLVIA, Department of Biology, Baruch College, City University of New York, New York, New York 10016.

°JOHNSON, ARTHUR G., Department of Microbiology, The University of Michigan, Ann Arbor, Michigan 48104.

°JOLLEY, WELDON, Surgical Research Laboratory, Loma Linda University School of Medicine, Loma Linda, California 92354.

°JONES, CAROL, Department of Biophysics and Genetics, University of Colorado Medical Center, Denver, Colorado 80220.

°KALINER, MICHAEL, Department of Medicine, Harvard Medical School, Robert B. Brigham Hospital, Parker Hill Avenue, Boston, Massachusetts 02120.

°LICHTENSTEIN, LAWRENCE M., Department of Medicine, The Johns Hopkins University School of Medicine at The Good Samaritan Hospital, Baltimore, Maryland 21239.

LOPEZ, C., Departments of Pharmacology and Pathology, University of Minnesota Medical School, Minneapolis, Minnesota 55455.

MCCLURG, JAMES E., University of Nebraska, College of Medicine, Omaha, Nebraska 68105.

MCCRERY, JERRY, University of Nebraska Medical Center, Omaha, Nebraska 68105.

MCCURDY, FRED, University of Nebraska Medical Center, Omaha, Nebraska 68105.

°MACMANUS, JOHN, Division of Biological Sciences, National Research Council of Canada, Ottawa, Canada K1A OR6.

°MELMON, KENNETH L., Division of Clinical Pharmacology, Departments of Medicine and Pharmacology, University of California Medical Center, San Francisco, California 94122.

MERRILL, J. P., Department of Medicine, Peter Bent Brigham Hospital, Boston, Massachusetts 02115.

MORGANROTH, JOEL, Building 10, Room 7 M 220, National Institutes of Health, Bethesda, Maryland 20014.

°MORSE, STEPHEN I., Departments of Microbiology and Immunology, Downstate Medical Center, State University of New York, Brooklyn, New York 11203.

°NELSON, ERIC, Nelson Research and Development Company, Irvine, California 92664.

ORONSKY, A., Research Department, CIBA-Geigy Corp. Ardsley, New York 10502.

°PARK, BEN H., Departments of Pediatrics and Pathology, University of Minnesota Medical School, Minneapolis, Minnesota 55455.

°PARKER, CHARLES W., Division of Allergy and Immunology, Department of Medicine, Washington University School of Medicine, St. Louis, Missouri 63110.

PERLMAN, JANICE, Department of Medicine, University of California, San Francisco Medical Center, San Francisco, California 94122.

°PERPER, ROBERT J., Immunology Research, CIBA-Geigy Corp., Ardsley, New York 10502.

°PLESCIA, OTTO, Institute of Microbiology, Rutgers University, The State University of New Jersey, New Brunswick, New Jersey 08903.

PUCK, THEODORE, T., Department of Biophysics and Genetics, University of Colorado Medical Center, Denver, Colorado 80220.

°RIGBY, PERRY G., University of Nebraska Medical Center, Omaha, Nebraska 68105.

RIXON, R. H., Division of Biological Sciences, National Research Council of Canada, Ottawa, Canada K1A OR6.

°ROBISON, G. ALAN, Department of Pharmacology, Vanderbilt University School of Medicine, Nashville, Tennessee 37222.

°RYAN, WAYNE L., College of Medicine, University of Nebraska, Omaha, Nebraska 68105.

SCHULTZ, GÜNTER, Department of Physiology, Vanderbilt University School of Medicine, Nashville, Tennessee 37232.

°SHEARER, GENE M., Immunology Branch, National Cancer Institute, National Institutes of Health, Bethesda, Maryland 20014.

°SHEPPARD, J. R., University of Colorado Medical School, Denver, Colorado 80220.

SHIOZAWA, CHIAKI, Institute of Microbiology, Rutgers University, The State University of New Jersey, New Brunswick, New Jersey 08903.

°SMITH, SIDNEY, Department of Physiology, Schering Corporation, Bloomfield, New Jersey 07003.

SORKIN, ERNST, Biochemical and Biological Section, Medical Department, Schweizerisches Forschungsinstitut, CH-7270 Davos-Platz, Switzerland.

STECHER, VERA J., Biochemical and Biological Section, Medical Department, Schweizerisches Forschungsinstitut, CH-7270 Davos-Platz, Switzerland.

°STECHSCHULTE, DANIEL, Department of Medicine, Harvard Medical School, Robert B. Brigham Hospital, Boston, Massachusetts 02120.

STITES, DANIEL P., Department of Medicine, University of California, San Francisco Medical Center, San Francisco, California 94122.

°STROM, TERRY, Department of Medicine, Peter Bent Brigham Hospital, Boston, Massachusetts 02115.

°SULLIVAN, TIMOTHY, Department of Medicine, Washington University School of Medicine, St. Louis, Missouri 63110.

Sutherland, Earl W., Department of Physiology, Vanderbilt University School of Medicine, Nashville, Tennessee 37232.

°Tytell, Alfred A., Merck, Sharpe and Dohme, Research Laboratories, West Point, Pennsylvania 19486.

°Webb, David R., Jr., Department of Medicine, University of California, San Francisco Medical Center, San Francisco, California 94122.

°Wedner, H. James, Department of Medicine, Washington University School of Medicine, St. Louis, Missouri 63110.

Weinstein, Yacob, Division of Clinical Pharmacology, Department of Medicine and Pharmacology, University of California Medical Center, San Francisco, California 94122.

°Weissmann, Gerald, Department of Medicine, New York University School of Medicine, New York, New York 10016.

White, J. G., Departments of Pharmacology and Pathology, University of Minnesota Medical School, Minneapolis, Minnesota 55455.

Whitfield, J. F., Division of Biological Sciences, National Research Council of Canada, Ottawa, Canada K1A OR6.

°Willingham, Mark, National Cancer Institute, National Institutes of Health, Building 37, Room 4B22, Bethesda, Maryland 20014.

°Winchurch, Richard A., Department of Pharmacology, Smith Kline and French Laboratories, Philadelphia, Pennsylvania 19101.

°Wissler, Josef, H., Biochemical and Biological Section, Medical Department, Schweizerisches Forschungsinstitut, CH-7270 Davos-Platz, Switzerland.

Zurier, Robert B., Department of Medicine, New York University School of Medicine, New York, New York 10016.

CONTENTS

A TRIBUTE TO WERNER BRAUN

It is with heavy heart that I stand before you today on this occasion. It seems only yesterday that Werner Braun was describing, in his usual enthusiastic way, his plans for this meeting. We were in his room at the Campus Inn in Ann Arbor, where we were attending the Cyclic AMP-Prostaglandin Conference last November. Besides the plans for this meeting, we talked about his extensive travel and seminar activities, his council responsibilities, his family, and his particular concern for the health of his wife, Barbara, and then he asked me to read over two rather large manuscripts he had just written. I remember commenting that I wondered how he managed to do so much. He was indeed a dynamo. Three nights later in Bethesda, he suffered a fatal heart attack.

Werner was born in Berlin on November 16, 1914. During his high school days, he did research on problems of genetics as a guest in the Kaiser-Wilhelm Institut for Biology. He later entered the University of Goettingen, where he received his doctorate in biology.

He had completed three years of medical school at the time he left Germany in 1936. He was a guest investigator in genetics at the University of Michigan for a year before becoming a research associate of Dr. R. Goldschmidt at the University of California in 1937, conducting research on problems of physiological genetics. In 1942, he joined the Department of Veterinary Science at the University of California where he studied problems of bacterial genetics, immunology, and biochemistry.

In 1948, Werner became chief of the microbial genetics branch of the Biological Laboratories, Chemical Corps, Camp Detrick, Maryland, and served in that position until he joined the Rutgers Institute of Microbiology as professor of microbiology in 1955. He remained at Rutgers until his death in 1972.

By any standards Werner's scientific accomplishments were impressive. He was a renowned microbial geneticist who, during recent years, applied microbiological approaches with great originality to the study of cellular immunology and cancer. His studies on microbial population changes, especially in the case of *Brucella*, are classical and led him to investigations of the nature and regulation of immune response with possible application to the control of tumor growth. He is the author of over 200 publications in scientific journals, received the Superior Accomplishment Award of the U.S. Chemical Corps and the Barnett Cohen Award from the American Society for Microbiology in 1954.

Werner was a member and leader of many scientific organizations. He was past chairman of the American Academy of Microbiology, former chairman of the Board of Scientific Counselors of the National Institute of Allergy and Infectious Disease, and a member of the National Advisory Allergy and Infectious Disease Council.

He served as a member of the editorial board of the *Journal of Bacteriology, the American Journal of Hygiene, the Journal of the Reticuloendothelial Society, and Infection and Immunity.* He was past president of the Theobald Smith Society.

He was a member of Sigma Xi, Phi Sigma, the New York Academy of Science, the American Association of Immunologists, the American Society for Microbiology, the Genetics Society of America, the Society for Experimental Biology and Medicine, and the Society for General Microbiology.

Werner also was the editor of several books, a contributor to numerous textbooks and encyclopedias, and the author of *Bacterial Genetics* published by W. B. Saunders in 1953 and reedited in 1965. Several of these books are recognized as classics in their field.

Apart from his many scientific accomplishments, Werner Braun was a remarkable human being. He was a rare person given to deep concern for his fellowman and a great love for science. He had an unusual appreciation of life and enjoyed it fully.

Werner was from the beginning a multidisciplinary scientist preferring to look on life and living things as the sum of all integrated activities. He blended his background of genetics, microbiology, biochemistry, pharmacology, and immunology with an astute ability to conceptualize what was really happening in a living system. Though he read a great deal, he preferred to be in direct contact with people in laboratories everywhere—seeking, questioning, and then synthesizing what he had learned into testable hypotheses.

He was an architect, a visionary often ahead of his contemporaries. He once told me, "The world is full of bricklayers, but there are very few architects." It is sad that there are now even fewer.

In his incessant travels to meetings and laboratories, he sought out those who, like himself, had the ability to perceive, to generate ideas, or who had special talents. Often these were students—no one was more encouraging, no one a better teacher and stimulus to the young. Ideas were the important thing, and Werner recognized that young people abound in ideas. He was recently very concerned about the trend toward reduction in research support for numerous young investigators in favor of increased support to relatively few established investigators.

Life is to be lived; this was Werner's philosophy. He wanted to be where the action was, to experience things personally and most of all to know the people who made life happen.

He was a collector of art, and he had a knack for discovering new artists. He had an interesting collection of Israeli paintings and numerous French works. Always, though, he wanted to know the artist personally, to talk with him, and to know the individual behind the work.

He carried a camera not to photograph scenery, but to photograph people. Some of his character shots, close-ups of faces, are classic. His knowledge of food, fine wines, and restaurants the world over was unusual, and I never knew him to be without his Gude Michelin and the small black book in which he noted superb dishes, wines, hotels, restaurants, and the names of the people who make them so—each a personal friend.

Like many of you here, I could continue to recount the marvelous experiences shared, the meals, the fishing trips, the late night story sessions, the vigorous scientific discussions, but time does not permit.

Let it be remembered that this meeting, like so many others, was conceived by Werner Braun to advance our knowledge and to stimulate ideas. Most importantly for Werner, were the people; he was the catalyst bringing bright people together in a friendly atmosphere with enthusiasm and dedication to generate new understanding. His presence will be felt here and in meetings and laboratories yet to be conceived.

I owe him a great deal, and I will always remember him, as I'm sure you will, a true friend.

ERIC L. NELSON

REGULATORY FACTORS IN THE IMMUNE RESPONSE: ANALYSIS AND PERSPECTIVE

WERNER BRAUN (1914-1972)

I. Introduction

Cellular immunology has made rapid advances in elucidating, on a broad scale, the nature of cells involved in immune responses and the diversity of their functions and products, with appropriate recent emphasis on the differences between cell-mediated responses and humoral antibody formation (1). We have learned much in recent years about the usual need for cell cooperation in the activation of antibody-forming cells (2–6), the synthesis and nature of products that can be released from such cells (7–9), and we have begun to learn something about the nature of mediators of cell-mediated immune responses (10,-11). We also have become increasingly aware of the importance of genetic and physiological factors that can influence the magnitude of immune responses (12,13). Yet, until recently there remained a paucity of information regarding the nature of molecular events that are responsible for the activation of immunocompetent cells and for the regulation of the magnitude of their response. While adjuvant effects and tolerogenic effects have been familiar and widely investigated phenomena for a long time and are known to involve the same cells (14), we have only just begun to understand some of the molecular events and cellular factors that appear to be critically involved in regulating the intensity of immune responses in general, and the functions of specific cell populations in particular (e.g., those responsible for antibody formation versus those involved in cell-mediated immunity).

Our own contribution to this subject has been based on attempts to elucidate basic mechanisms of inter- and intracellular events in immune responses by studying the effects of chemically well-defined modifiers of immune responses. Our overall objective has been the identification of normal regulatory factors controlling the activation and performance of cells involved in immune responses, and the utilization of this knowledge for a more effective manipulation of such responses.

---

*Compiled by Dr. Otto Plescia from excerpts of Dr. Braun's recent writings.

In the course of studies during the past three years, we have made significant advances in meeting this objective by identifying cyclic AMP-mediated intracellular events as an important, perhaps critical component in the regulation of responses by cells involved in antibody formation. Available evidence indicates that endogenous levels of cAMP play a role in controlling to what extent an activated immunocompetent cell is "turned on" or is "turned off," and preliminary data suggest that genetic and physiologic differences in responses also may reside, in part, at this level. Furthermore, cell cooperation in the immune response can be interpreted as involving a cAMP-mediated amplification system. It now appears that cAMP plays a central role in the regulation of immune responses. We need to learn more about the way in which cAMP and the intracellular enzymes that control its formation and persistence control cellular activities, and to utilize this new information for a more deliberate and effective manipulation of immune responses.

Heretofore, we concentrated on antibody formation as a convenient criterion of the immune response, but it now becomes important to focus attention on ways to alter selectively the performance of cells involved in cell-mediated immunity, as opposed to those participating in humoral antibody formation. It is well established that immunogenic stimuli give rise to qualitatively different types of immune responses, of which cell-mediated immune responses and humoral antibody formation are two major examples. Although these two types of responses share specificity for antigenic determinants, they have different roles and importance in the spectrum of the host's responses to foreign materials. To activate preferentially one type of response and not the other has become a major challenge in attempts to improve our ability to manipulate immune responses to infectious agents, to control tumor and transplantation immunity, as well as autoimmune diseases, and to repair genetic and acquired immunodeficiencies, including those occurring in disease and with aging. There are indications that one effective way of achieving this goal might be via differential effects of modifiers of endogenous adenyl cyclase activity and cAMP levels on different classes of cells involved in different types of immune responses.

II. Recognition of Immunoenhancing Effects of Oligo- and Polynucleotides

Our present concentration on the role of cAMP-mediated events in immune responses is the direct consequence of our earlier discovery and analysis of the effects of oligo- and polynucleotides on immune responses. It has now become apparent that these nucleotides and many other regulators of immune responses may operate via effects on endogenous cAMP levels, and it is, therefore, appropriate to review the history of our involvement with the regulatory effects of oligo- and polynucleotides.

It all began with studies that we conducted many years ago on the influence of nucleic acid breakdown products on population changes in bacteria (18, 19). These early studies clearly indicated a capacity of deoxyribose-containing oligo- and polynucleotides, regardless of their source, to stimulate the rate of macromolecular syntheses and cellular proliferation in a relatively nonspecific manner. This finding was in sharp contrast to the usual association of nucleic acids with functions that are specifically informative, and we, therefore, explored the regulatory effects of nucleic acids, which were independent of base sequences, in a variety of microbial systems. Subsequently, in attempts to determine the possible nonspecific effects of nucleic acid breakdown products on multiplying eukaryotic cells, we tested the effects of heterologous DNA and DNA breakdown products on antibody formation and observed a stimulation both in terms of early appearance of antibody titers as well as in terms of numbers of antibody-forming spleen cells (18–23); stimulatory effects on phagocytic cells also were detected (24). Independently, and at approximately the same time, Johnson and associates found stimulatory effects of oligoribonucleotides and oligodeoxyribonucleotides on antibody formation (25). There followed a period during which both our group and Johnson's group explored various immunological responses that could be affected, related these stimulatory effects to the action of other known adjuvants (19, 25–30) and to naturally released stimulators of antibody formation (31), and began attempts to define the mode of action of the immunoenhancing oligo- and polynucleotides (29, 32). The latter effort was disappointingly unsuccessful, which, in retrospect, is not surprising inasmuch as we were then completely unaware of cAMP and its role in regulating the performance of cells involved in the immune response.

A. *Immunoenhancing Effects of Synthetic Polynucleotides*

A critical turning point was the discovery that double-standed synthetic polynucleotides, including poly A:U, poly G:C, and poly I:C, were even better and more consistent stimulators than the natural nucleic acids previously employed (33, 34).

Our report on this effect led to the subsequent demonstration by Hilleman and co-workers that poly I:C is a potent inducer of interferon (35). It soon became apparent that only double-stranded polynucleotides were active in vivo, whereas both single- as well as double-stranded polynucleotides were active in vitro, the difference being due to the too rapid depolymerization of the single-stranded molecules in vivo, whereas double-stranded molecules are more resistant to nuclease activities. Also, poly I:C proved to be toxic and pyrogenic (36), whereas poly A:U, which is almost equally effective in stimulating immune responses, does not have these detrimental properties (presumably because it is degraded more rapidly). For this reason, we concentrated on the use of poly A:U in all of our subsequent studies, but confirmed from time to time that the findings with poly A:U were equally applicable to the other double-stranded polynucleotides, with the one notable difference that poly I:C is a good interferon inducer, whereas poly A:U is a very poor one.

A detailed recitation of the subsequent work on various poly A:U effects on normal and deficient immune responses will not be included here, since much of it is summarized in reviews (18, 19, 25, 37). Briefly, poly A:U is now known to stimulate:

(1) phagocytosis (19, 25, 38, 46);

(2) antibody formation (19S, 7S, primary and secondary responses) (19, 25, 30);

(3) cell-mediated immune responses (delayed-type hypersensitivity, graft-versus-host reactions) (19); lymphocyte transformation (39);

(4) protective effects of immunization against toxins, bacterial and viral agents (*Brucella* and rabies, G. Renoux, personal communication; tetanus toxoid, L.L. Csizmas, personal communication; influenza, Woodhour et al. (40); cholera toxin, W. Braun and M. J. Rega, unpublished data);

(5) host resistance to syngeneic tumors, where tested (41);

(6) high levels of antibody in certain types of genetic low responders (42);

(7) premature antibody formation in newborns (43);

(8) in the absence of readministered antigens, anamnestic-type responses in primed animals (30);

(9) recovery from tolerance (P. Liacopoulos, personal communication).

And poly A:U restores:

(1) antibody formation in aged C57 BL mice (44);

(2) Responses to T cell–dependent antigens in thymectomized mice (45);

(3) responses to T cell–dependent antigens in T cell–deficient spleen cell cultures in vitro (D. Webb and W. Braun, preliminary observations);

(4) near-normal responses in certain cases of "antigenic competition."

Evidence has been presented (19, 25, 47) that poly A:U can stimulate functions of all the major cell types involved in immune responses, i.e., macrophages, T and B lymphocytes. In some isolated instances, poly A:U effects can be inhibitory (46), and we shall come back to this point and a possible explanation for this apparent paradox, when we discuss dose-response curves in cAMP-mediated systems. Also, since at least a partial explanation for the various above listed effects of poly A:U can be based on our more recent findings regarding the mode of action of poly A:U, we shall defer additional discussion on this point to a subsequent section.

We have demonstrated, with the aid of radioactively labeled poly A:U, that the double-stranded polyanions complex extremely rapidly with cell surfaces. We suspect that, like other polyanions (48), these materials produce   conformational changes in the membrane, but whereas other polyanions eventually damage the membrane (resulting in a release of pyrogenic and toxic factors), poly A:U appears to be depolymerized into membrane-inactive mononucleotides before it has a chance to cause undesirable membrane damage. We suspect that it is this "mild," transient effect of poly A:U on membranes that triggers the modification of responses by cells that are interacting with specific antigens. It should be noted that, except for the alteration of macrophage functions (24, 46), the stimulatory polynucleotides are only active in conjunction with a specific antigen (19). Thus, they are principally modifiers of specific activations and not activators per se.

III. Cyclic AMP and Immune Responses

A. *Polynucleotides as Stimulators of Endogenous cAMP*

Part of the mystery of how poly A:U, and probably many other adjuvants, affect immune responses disappeared when it was discovered, during the past two years, that (1) the response of immunocompetent cells to extracellular activating signals is, like most hormone-dependent activations of cells, mediated and modified by endogenous cAMP levels (49–69), and (2) poly A:U and other oligo- and polynucleotides enhance the activity of adenyl cyclase, the enzyme responsible for cAMP formation (39, and unpublished data by C. Shiozawa and W. Braun).

The role of endogenous cAMP as a regulator of intercellular interactions (70–72), and particularly its apparent significance for influencing the activities of enzymes involved in nucleic acid syntheses (73–76), has become a prominent topic in recent times and should not require extensive review here. Suffice it to state that even though the "second-messenger" activites of cAMP have been known to endocrinologists for some time (70), the involvement of cAMP in the regulation of a multitude of other intercellular regulations, including not only those of secretory cells but also muscle function, nerve function, behavioral activities, vision, etc. (70, 71, 77), has been recognized only recently. The initial concern of some investigators that cAMP is doing too many things and, therefore, its willful regulation for the control of activities of specific cells may be difficult or impossible to accomplish appears to become less acute as the result of the recent discovery that there may be a degree of specificity, since the functions of cAMP in different cells and tissues may be selectively regulated through cofactors (e.g., GTP) and by a whole family of enzymes (phosphodiesterases) that are responsible for the conversion of cAMP into biologically inactive AMP (77).

In studies with hormone-dependent cells, it was established that following the interaction of a hormone with appropriate membrane-associated receptor sites, adenyl cyclase, a membrane-associated enzyme, is activated and will convert ATP into cAMP. Cyclic AMP, in turn, will combine with a specific intracellular cAMP-binding protein, thereby activating a variety of kinases, mostly phosphorylating enzymes, that can influence macromolecular biosyntheses, including transcription and translation, as well as the activation of preexisting enzymes (70). Cyclic AMP, once formed, is quite rapidly converted by phosphodiesterases into inactive AMP. Among the stimulators of adenyl cyclase activity, first uncovered in studies by endocrinologists, are the catecholamines, including epinephrine, norepinephrine, and isoproterenol; among the inhibitors of phosphodiesterase activity, which thus stabilize endogenous cAMP levels, are theophylline, caffeine, and papaverine.

We demonstrated (49–54), in recent studies (1) that cAMP or dibutyryl cAMP enhances specific antibody formation in vivo and in vitro and (2) that the stimulatory effects of poly A:U can be greatly potentiated, in vivo and in vitro, by the concurrent administration of theophylline or caffeine. These findings indicated that the immunoenhancing effects of poly A:U and other polynucleotides may be the consequence of a stimulation of adenyl cyclase activity, a conclusion that was supported by our earlier findings that epinephrine, a known adenyl cyclase enhancer (70), can stimulate antibody-forming cells (30), and by the subsequent observation (78) that another potent adenyl cyclase stimulator, isoproterenol, also stimulates antibody formation. As in the case of hormone-dependent systems (70), the stimulatory effects of isoproterenol on antibody formation in vivo and in vitro are antagonized by propranolol. In addition, the involvement of cAMP in antibody formation has been supported by the finding that materials derived from *Vibrio cholera,* known to stimulate adenyl cyclase activity (81), also exert potent adjuvant effects. Whereas such evidence was somewhat indirect in nature, we (39) were able to obtain direct evidence for the capacity of poly A:U and poly I:C to stimulate adenyl cyclase activity in spleen cells by assaying the activity of this enzyme in mouse spleen cells that had been exposed to either poly A:U or its breakdown products in the absence or presence of PHA. These studies also demonstrated that exposure of mouse spleen cells to PHA increased adenyl cyclase activity.

The publication of our results on the influence of PHA on adenyl cyclase activity of murine lymphocytes was preceded by a report from C.

Parker's group (56) on such effects of PHA on cyclase activity in human peripheral lymphocytes, and by reports of others (55, 82) on the capacity of cAMP to *inhibit*, rather than to stimulate, lymphocyte transformation and antibody formation. However, such apparently conflicting reports are no longer surprising in view of our subsequent observations that, depending on the degree and duration of stimulaton of endogenous cAMP levels, one can obtain either enhancing or inhibitory effects. Similar biphasic responses have been observed in various studies on cAMP-mediated responses of hormone-dependent cells (70, 83). These observations are relevant to mechanisms of immune paralysis, specific tolerance, and the ability of excessive stimulation of cAMP-mediated events to "turn off" immunocompetent cells.

Up to this point, all direct measurements of an elevation of adenyl cyclase activity in activated immunocompetent cells employed mitogens or, in some of our recent tests, interactions between allogeneic lymphocytes. One recent study (84), however, succeeded in demonstrating that such an elevation also occurs following contact of sensitized lymphocytes with the specific antigen. An elevation of cAMP levels during phagocytosis also has been reported (67), and we participated in a study (85) which showed that clearance in vivo can be stimulated by modest amounts of cAMP and reduced by higher amounts.

Recent studies of others on immediate and delayed-type hypersensitivity (60–62, 64, 65, 86) have shown that an elevation of endogenous cAMP levels can "turn off" the activities of cells involved in these responses, but it is not yet clear whether such cells can only be "turned off" or might be stimulated at low levels of cAMP elevation; the available evidence would suggest that they can only be turned off. This would not be surprising in view of observations with neoplastic cells. In fact, there are increasing indications that cells in the process of activation can be either stimulated or inhibited by an elevation of endogenous cAMP levels, the direction of the response depending on the extent of elevation of endogenous cAMP. On the other hand, cells that are already activated and then are exposed to external stimuli that influence their performance, e.g., exposure of sensitized lymphocytes to allergens, may only be inhibited by an elevation of endogenous cAMP levels. There are also some indications, in studies conducted by others (I. Pastan and associates, personal communication), that the "piling up" of endogenous cAMP may inhibit cell activities, whereas increased adenyl cyclase activity without "piling up" of cAMP may enhance cell responses.

In Table 1 we have summarized some of the available data regarding cAMP-mediated stimulations and inhibitions of a number of systems related to immune responses.

An additional aspect of potential importance to the understanding and manipulation of factors influencing the magnitude of immune responses has developed with the discovery that certain prostaglandins

Table 1. Some cAMP-Mediated Effects

System	Stimulation	Inhibition	Reference
Carbon clearance in vivo	+	+	(85)
Phagocytosis in vitro	−	+	(61)
Release of lysosomal enzymes from polymorphonuclear leucocytes	−	+	(68)
Candidacidal activity of neutrophils	−	+	(61)
Lymphocyte transformation	l	+	(89), (39), (90)
Cytolytic activity of sensitized lymphocytes	−	+	(92)
Antibody formation	+	+	(54)
Allergic, IgE-mediated histamine release from basophils and mast cells	−	+	(86)
Delayed hypersensitivity	+	+	St.: (19); In.: (86)
Mitosis of thymocytes	+	+	(66)

(87), notably PGE_1, PGE_2, and PGA_1, natural materials with activity at exceedingly small concentrations, can stimulate cAMP production in mixed populations of human leukocytes (61), in purified human lymphocyte populations (57), in guinea pig granulocytes (88), and in mouse macrophages (61). Except in the case of human lymphocytes, where both stimulation and inhibiton of transformation have been reported, physiological functions of the other systems have been inhibited by such elevations (90).

Obviously, the regulatory effects produced by cAMP, and more indirectly by prostaglandins, are exceedingly complex, and this complexity may reflect an involvement of more than one receptor site per cell and the use, in all of the studies so far published, of highly heterogeneous cell populations (90).

B. *Natural Role of Intracellular cAMP Levels in Immune Responses*

Lymphocyte populations participating in immune responses are heterogeneous in their mode of maturation and in their functions. Although all of them originate from the bone marrow, it is known (2–6) that antibody formation to many antigens requires the cooperation of relevant members of a lymphocyte population that migrates from the bone marrow to lymphoid organs via the thymus (T cells) and of a directly bone marrow (or bursa-equivalent)-derived lymphocyte population (B cells). The basis for the requirement of T-B cell cooperation in activating B cell performance (antibody formation) to so-called T cell–dependent antigen is poorly understood. However, recent observations suggest that B cell performance may be dependent on both a specific signal by antigen and an adequate supportive stimulation of other cellular functions that we suspect involve cAMP-mediated events.

Such considerations are strongly supported by the very recent data of Kreth and Williamson (92a), who also conclude that the activation of B cells requires both an antigenic and an auxiliary signal.

In this connection, we may cite the findings of Johnson et al. (45) that poly A:U, now known to stimulate adenyl cyclase activity, can restore antibody formation to sRBC in neonatally theymectomized mice. Also, quite preliminary tests in our laboratory indicate that antibody formation to sRBC in vitro, which fails to occur following treatment of spleen cell populations with anti-τ + complement, may be restored by poly A:U. Similarly, we have observed that in the presence of stimulatory events that can result from interactions among allogeneic lymphocytes, T cell–deficient° populations of mouse spleen cells will produce normal or even enhanced antibody responses.

Stimulatory effects as a consequence of allogeneic interactions in vivo or in vitro also have been reported, for example, by Katz et al. (93), Mensah and Kennedy (94), Bonmassar et al. (95), Schimpla and Wecker (96), Janis and Bach (97), and are reflected in earlier data of Dutton (98). In contrast, Hirano and Uyeki (99) noted principally inhibitory effects of allogeneic spleen cell interactions on antibody formation, and we also observed such inhibitory effects in certain allogeneic combinations. Similarly, graft-versus-host (GVH) reactions in vivo may result either in stimulation of subsequent antibody responses to unrelated antigens or produce a temporary interference with such responses ("antigenic competition"). Stimulators of endogenous cAMP levels can either enhance interference effects or alleviate them, the result depending on whether the stimulators are employed at the time of immunization with the first antigen or at the time of subsequent immunization with a different antigen.

It is an intriguing question whether such opposite results in allogeneic interactions in vitro, and following GVH reactions in vivo, i.e., either stimulation or inhibition of antibody formation, are merely an expression of sufficient *versus* excessive stimulation of cAMP-dependent intracellular events. We already have a clue that this may be so. Mixing spleen cell populations from CBA and C57 BL mice, in a ratio of 3:1, results in a stimulation of antibody formation to sRBC in vitro (and also elevates adenyl cyclase activity), whereas similar mixtures of C57 BL cells and DBA cells causes an inhibition of responses to sRBC. However, when the proportion of DBA cells in the latter mixture is decreased to 2%, stimulation of antibody formation to sRBC is observed (S. Adler and W. Braun, unpublished data). Thus, there is a distinct probability that many opposite results in the immunological literature, stimulation on the

---

°It probably would be better to refer to these populations as T cell deprived, because we have no assurance at all that these populations are completely devoid of T cells, yet we know that they contain an insufficient number of T cells for normal responses.

one hand, inhibiton on the other, may turn out to be different expressions of cell responses triggered by the same initiating event, the critical variable being the quantity of the trigger.

The release of nonspecific stimulatory factors from previously sensitized spleen cells on exposure to antigen has been reported by us as a result of in vitro (19) and in vivo (31, 37) studies, and cell-released nonspecific stimulators have also been described by Dutton et al. (100), Smith (91), Gordon and MacLean (101), Radovich and Talmage (103), and many others. As yet, however, we do not know if any of these factors operates through effects on endogenous cyclase activity or cAMP levels.

The role of the carrier in antibody formation to haptenic determinants has been beautifully analyzed by Mitchison (14), Benacerraf et al. (107), Nossal (14), Rajewsky (108), Plescia (104), and others, and it has been recognized that a principal function of the carrier is to bring specific T cells (helper cells) into contact with hapten-responsive B cells. Tests employing "educated T cells" have been particularly instructive in this regard (14, 107). Although these events are very specific, we shall pose the question whether they might not lead also to nonspecific amplification events, involving cAMP levels, in specific hapten-activated B cells. We have attempted to emphasize for some time (31, 37) that apart from the specific aspects of priming with the carrier, reexposure of sensitized cells to carrier antigen may result in the release of nonspecific stimulators of B cells, and probably also of T cells, and this, in fact, may account for some of the nonspecific stimulations associated with hypersensitivity reactions, including the effect of Freund's adjuvant and endotoxin. Recently, there has been more general interest in these nonspecific aspects associated with specific events, and a number of additional more precise substantiations of such effects have been published (107).

Cell-released nonspecific inhibitory factors also have been recognized in a number of studies, including those by Gershon and Kondon (109) and Baker et al. (110), among others. Again the question arises to what extent such factors may operate via cAMP-mediated events.

Regulatory effects of antibodies (14, 111), of antigen-antibody complexes, particularly in "low zone tolerance" (112), as well as of antigen dosages (14) are familiar phenomena. In connection with the possible involvement of altered endogenous cAMP levels of lymphocytes in such events, it is noteworthy that the biphasic dose-response curve for stimulatory and inhibitory effects of antigen-antibody complexes on antibody formation in vitro (112) resembles that obtained for known stimulators of adenyl cyclase activity (53). A conceivably related phenomenon is the apparent ability of complexes of antibody and antigenic fragments to interfere with specific cell-mediated immune responses in many tumor-bearing animals (113).

A number of antigens, e.g., endotoxin and polymerized flagellin-DNP (but not monomeric flagellin-DNP), are apparently T cell independent as far as antibody formation is concerned (112). Others, particularly Nossal (114), have considered the density of antigenic determinants on a molecule ("epitope density") as a critical factor in achieving T cell–independent responses by B cells. In relation to our present interest in the apparent need for cAMP-mediated nonspecific amplifications of relatively weak specific antigenic signals to relevant B cells, one wonders whether high epitope density may result in a simultaneous specific triggering and amplification of the specific signal in a given B cell, where low epitope density may provide only the signal and require amplification by cell cooperation or other nonspecific stimulatory factors.

It is now well established that the nature of receptors for portions of an immunogen is quite different in the case of T and B cells and also that both affinity of T cells for an antigen and their response to antigenic stimulation differ considerably from those typical for B cells (91). These differences are expressed in the influence of different dosages of antigen, its molecular properties (115) on the activation of humoral antibody formation *versus* cell-mediated immune responses and immunological memory. The selective activation and regulation of T and B cell populations, and of their subpopulations, are central issues of immunology; some recent approaches can be found in the studies of Parish (115), Goodman (personal communication), and Freedman (116). In our studies with modifiers of endogenous cAMP levels (54), we noted a reduction of antibody formation with a concomitant increase in memory following the administration of high levels of theophylline + poly A:U. If we assume that this memory resided in activated T cells, the interesting implication arises that there may be differences in the responses of T and B cells to the inhibitory effects of high concentrations of agents that influence endogenous cAMP levels, and that this difference may be utilized for manipulations of the quality of the immune response. However, since memory can reside either in T cell populations or in partially activated B cell populations (117), it remains to be established that the effects we observed really involved T cell memory.

Studies that we conducted in association with Mozes et al. (42) demonstrated that some genetic deficiencies in specific immune responses (in this case antibody formation to (T, G)-Pro-L in a low responder strain of mice) can be overcome, and normal responses can be obtained by administering the antigen in the presence of poly A:U. Similar restorative effects by poly A:U, and an even better stimulation by poly A:U + theophylline, have been observed by G. Biozzi (personal communication) in the case of a line of mice selected for low responses to sRBC. Where analyzed, such effects appear to overcome a paucity of relevant B cells (42). It should be noted, however, that there are also a number of low-responder situations that have not been repairable by poly A:U (McDevitt, personal communication).

Inherently low responses to specific antigens are usually attributed to a deficit in relevant T or B cells or to a change or paucity of receptors on these cells (1, 12, 91), and there is evidence for linkage between H2 and HL-A loci and some loci responsible for differences in specific immune responses (1, 91). Since poly A:U is known to stimulate proliferative responses by immunocompetent cells (19), it is, of course, feasible to assume that its effect on the response of genetic low responders is principally that of expanding the population of specifically activated cells in the presence of a paucity of relevant stem cells. However, there is also the possibility that general differences in immune responsiveness by different strains and individuals might be attributable to inherent and physiological differences in the responsiveness of the amplification system. Immunogenic and responsive differences among lymphocytes with different H2 alloantigens are known (95, 118, 119), and we have begun to analyze differences among various allogeneic spleen cell combinations in their capacity to modify antibody formation. In the course of these studies, we noted the existence of differences in the time pattern and extent of adenyl cyclase activation of PHA in spleen cells from mouse strains differing at the H2 locus (and also at some other loci). Figure 1 il-

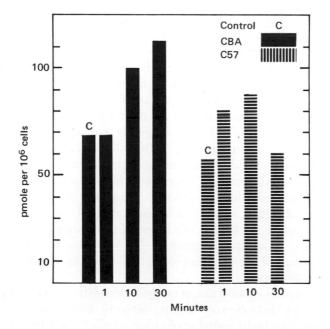

Fig. 1. Changes in adenyl cyclase activity, measured in terms of cAMP formed, in CBA and C57 Bl spleen cells after exposure to PHA in vitro.

lustrates such differences, showing the time pattern of increases in adenyl cyclase activity in CBA and C57 BL spleen cell population after exposure to PHA in vitro. It can be seen that, in comparison to untreated controls, C57 BL cells show an increase in cyclase activity in samples removed for assay almost immediately after addition of PHA ("1 minute"), reach maximum activity some time between 10 and 30 min, and return to normal cyclase activity after 30 min. In contrast, CBA cells show a slower response, and their cyclase activity is still increasing 30 min after the beginning of PHA exposure. Since we suspect that the non-specific amplification of specific immune responses is related to endogenous adenyl cyclase activity and cAMP levels, the question now arises whether some genetic and physiological differences in immune responsiveness might also reside, or might be reflected, at this level.

IV. Cyclic AMP Levels and Tumor Development

Poly I:C and poly A:U possess a degree of antitumor activity, which has been attributed to a stimulation of immune responses and/or interferon induction. In recent studies with several syngeneic mouse tumor systems (including a methylcholanthrene-induced tumor and a Rauscher leukemia virus-induced tumor), we observed that the growth of tumor cell populations is usually retarded more effectively by poly A:U + theophylline than by poly A:U alone. However, it has turned out that such effects can also be produced (1) in animals that had received sufficient irradiation to depress immune responses and (2) also after pretreatment of the tumor cells in vitro prior to their implantation into syngeneic hosts. Thus, it became apparent that a major portion of the effect was directly on the tumor cells, which appear to be uniquely susceptible to agents that have the potential for elevating endogenous cAMP levels. This susceptibility may be related to the fact that some tumor cell lines have unusually high adenyl cyclase activity (120–121) and, therefore, any superimposition of additional stimulation of cyclase activity may lead to inhibition as in the case of excessive stimulation of lymphocytes. Others have reported inhibitory cAMP effects on the growth of tumor cells in vitro (122–124), as well as a restoration of normal phenotype after cAMP exposure (124–125); inhibition of proliferation of tumor cells by prostaglandins have also been reported (126). These findings are cited as another instance of the apparent influence of endogenous cAMP levels on cellular proliferation.

In recent studies carried out in cooperation with Dr. H. Levy, it was observed that interferon preparations contain an adenyl cyclase–stimulating factor capable of enhancing antibody formation. Accordingly, it is now suspected that the antitumor effect of interferon is attributable to this adenyl cyclase–enhancing factor.

V. Perspective

Development of methods to achieve a better quantitative as well as qualitative modification of immune responses will have a vast area of potential applicability. More specifically, approaches to the control of the nature of the immune response via modifiers of endogenous cAMP levels would permit, if successful, a more meaningful intervention in tumor immunology, where cell-mediated immune responses have been recognized as beneficial to the host, whereas antibody formation can interfere and result in tumor enhancement. On the other hand, in the case of tissue or organ transplantation, one wants to suppress cell-mediated immune responses without sacrificing antibody formation, which is considered of value to the transplant recipient's ability to cope with many infectious agents. There is now the distinct prospect that many allergic conditions may be controllable via cAMP-mediated systems, since it is already known that high levels of endogenous cAMP can turn off the release of effector substances from allergen-exposed sensitized cells. While others have studied such effects with cell populations from already sensitized donors, our investigations have been concerned principally with the initiation of such cell responses and as such may lead to means for a better prevention of sensitization to allergens.

The capacity to enhance immune responses quantitatively by agents free of toxicity and pyrogenicity is of value for prophylactic immunizations, since many potentially useful vaccines require a potentiation of the protective immune response that presently is difficult to achieve with agents free of undesirable side effects. Also, a large number of vaccines fail to achieve significant protection because they are apparently unable to evoke the proper kind of immune response, or they elicit the latter while also eliciting interfering responses by other activated immunocompetent cells. The induction of host resistance to agents like tubercle bacilli, staphylococci, streptococci, and brucellae appears to belong in this category. A better understanding of how to manipulate the character of the immune response should permit a better control over such economically still very important infective diseases.

There is also the large area of immunodeficiency diseases and the alteration of immunologic competence with aging, which would benefit from procedures that permit a normal immune response despite the paucity in numbers or the abnormality of certain relevant immunocompetent cells. Data already collected support the validity of this conclusion.

It can also be anticipated that the results obtained from study of cAMP and the immune response will help us to gain a better understanding of relationships between hormone activities and immunological responsiveness. Clinically valuable diagnostic procedures, such as the mixed lymphocyte reaction, also may benefit, since available evidence suggests that the elevation of endogenous adenyl cyclase activity may consititute a uniquely advantageous, rapid indicator of genetic relationship (in regard to histocompatability antigens) between cell donors. Also, if there exists a relationship between general immunological competence and pattern of alteration of endogenous cAMP levels or cyclase activity in lymphocytes (preferably from the circulation), it may become possible to utilize the latter test for a prognosis of immune responsiveness and for appropriate adjustments of thereapeutic treatment with modifiers of endogenous cAMP.

Apart from these pragmatic considerations, analysis of the role of cAMP as a regulator of the immune response should aid our basic understanding of cellular factors responsible for the activation and regulation of immune responses; it may elucidate the molecular basis for the phenomenon of cell cooperation and, by assisting in the development of immunopharmacology, contribute to a better understanding of problems of cellular differentiation and dedifferentiation.

References

1. Amos, B., ed. *Progress in Immunology.* New York: Academic Press (1971).
2. Claman, H. N., and Chaperon, E. A. *Transplant. Rev.* 1:92 (1969).
3. Davies, A. J. S. *Transplant. Rev.* 1:43 (1969).
4. Taylor, R. B. *Transplant Rev.* 1:114 (1969).
5. Miller, J. F. A. P., Basten, A., Sprant, J., and Cheers, C. *Cell. Immunol.* 2:469 (1971).
6. Mitchell, G. F., Mishell, R. I., and Herzenberg, L. A. In *Progress in Immunology*, B. Amos, ed. New York: Academic Press (1971).
7. Hood, L., and Talmadge, D. W. *Science* 168:225 (1970).
8. Scharff, M. D., and Laskov, R. *Progr. in Allergy* 14:37 (1970).
9. Kochwa, S., and Kumkel, H. G., eds. "*Immunolglobulins,*" *Ann. N.Y. Acad. Sci.* 183 (1971).
10. Lawrence, H. S., and Landy, M. *Mediators of Cellular Immunity.* New York: Academic Press (1969).
11. Revillard, J. P., ed. *Cell-Mediated Immunity—in Vitro Correlates.* Baltimore, Md.: University Park Press (1971).
12. McDevitt, H. O., and Benacerraf, B. *Advan. in Immunol.* 11:31 (1969)
13. Sigel, M., ed. *Aging and Autoimmunity.* Springfield, Ill.: C. C. Thomas (1971).
14. Landy, M., and Braun, W., eds. *Immunological Tolerance.* New York: Academic Press (1969).
15. Braun, W. *J. Cell. Comp. Physiol.* 52 (Suppl. 1):337 (1958).
16. Makinodan, T., Hoppe, I., Sado, T., Capalbo, E. E., and Leonard, R. M. *J. Immunol.* 95:466 (1968).
17. Coffino, P., and Scharff, M. D. *Proc. Nat. Acad. Sci.* (U.S.A.) 68:219 (1971).
18. Braun, W., and Firshein, W. *Bacteriol. Rev.* 31:83 (1967).
19. Braun, W., Ishizuka, M., Yajima, Y., Webb, D., and Winchurch, R. In *Biological Effects of Polynucleotides*, R. F. Beers and W. Braun, eds. New York: Springer-Verlag (1971).

20. Braun, W., and Nakano, M. *Proc. Soc. Exp. Biol. Med.* **119**:701 (1965).
21. Braun, W. In *Molecular and Cellular Basis of Antibody Formation,* J. Sterzl, ed. Prague: Publ. House of the Czech. Academy of Sciences, p. 525 (1965).
22. Braun, W., and Nakano, M. In *Ontogeny of Immunity.* Gainsville, Fla.: Univ. of Florida Press (1967).
23. Braun, W., and Nakano, M. In "Adjuvants of Immunity," *Symp. Ser. Immunobiol. Standard* **6**:227, Basel: Karger (1967).
24. Freedman, H. H., and Braun, W. *Proc. Soc. Exp. Biol. Med.* **120**:222 (1965).
25. Johnson, A. G., Cone, R. E., Friedman, H. M., Han, I. H., Johnson, I.G., Schmidtke, J. R., and Stout, R. D. In *Biological Effects of Polynucleotides,* R. F. Beers and W. Braun, eds. New York: Springer-Verlag (1971).
26. Braun, W., and Kessel, R. W. I. In *Bacterial Endotoxins,* M. Landy and W. Braun, eds. New Brunswick, N.J.: Rutgers Univ. Press, p. 397 (1964).
27. Kessel, R. W. I., Braun, W., and Plescia, O. J. *Proc. Soc. Exp. Biol. Med.* **121**:449 (1966).
28. Kessel, R. W. I., and Braun, W. *Nature* **211**:1001 (1966).
29. Braun, W., and Kessel, R. W. I. In *Nucleic Acids in Immunology,* O. J. Plescia and W. Braun, eds. New York: Springer-Verlag, p. 397 (1968).
30. Braun, W., Yajima, Y., Jimenez, L., and Winchurch, R. In *Developmental Aspects of Antibody Formation and Structure,* J. Sterzl, ed. Prague: Academia Publ. House, p. 799 (1970).
31. Nakano, M., and Braun, W. *J. Immunol.* **99**:570 (1967).
32. Johnson, A. G., Schmidtke, J., Merritt, K., and Han, I. In *Nucleic Acids in Immunology,* O. J. Plescia and W. Braun, eds. New York: Springer-Verlag, p. 379 (1968).
33. Braun, W., and Nakano, M. *Science* **157**:819 (1967).
34. Turner, W., Chan, S. P., and Chrigos, M. A. *Proc. Soc. Exp. Biol. Med.* **133**:334 (1970).
35. Hilleman, M., and Tytell, A. A. *Sci. Amer.* **225**:26 (1971).
36. Braun, W. *Nature* **224**:1024 (1969).
37. Braun, W., Nakano, M., Jaraskova, L., Yajima, Y., and Jimenez, L. In *Nucleic Acids in Immunology,* O. J. Plescia and W. Braun, eds. New York: Springer-Verlag, p. 347 (1968).
38. Winchurch, R. Thesis. Rutgers University (1970).
39. Winchurch R., Ishizuka, M., Webb, D., and Braun, W. *J. Immunol.* **106**:1399 (1971).
40. Woodhour, A. F., Friedman, A., Tytell, A. A., and Hilleman, M. R. *Proc. Soc. Exp. Biol. Med* **131**:809 (1969).
41. Braun, W., Plescia, O. J., Raskova, J., and Webb, D. *Israel J. Med. Sci.* **7**:72 (1971).
42. Mozes, E., Shearer, G. M., Sela, M., and Braun, W. In *Biological Effects of Polynucleotides,* R. F. Beers and W. Braun, eds. New York: Springer-Verlag, p. 197 (1971).
43. Winchurch, R., and Braun, W. *Nature* **223**:843 (1969).
44. Braun, W., Yajima, Y., and Ishizuka, M. *J. Reticuloendothel. Soc.* **7**:418 (1970).
45. Cone, R. E., and Johnson, A. G. *J. Exp. Med.* **133**:665 (1971).
46. Johnson, H. G., and Johnson, A. G. *J. Exp. Med.* **133**:649 (1971).

47. Campbell, P. A., and Kind, P. *J. Immunol.* **107**:1419 (1971).
48. Braun, W., Regelson, W., Yajima, Y., and Ishizuka, M. *Proc. Soc. Exp. Biol. Med.* **133**:171 (1970).
49. Ishizuka, M., Gafni, M., and Braun, W. *Proc. Soc. Exp. Biol. Med.* **134**:963 (1970).
50. Braun, W., Ishizuka, M., Winchurch, R., and Webb, D. *Ann. N.Y. Acad. Sci.* **181**:289 (1971).
51. Braun, W., Ishizuka, M., Winchurch, R. and Webb, D. *Ann. N.Y. Acad. Sci.* **185**:417 (1971).
52. Braun, W., and Ishizuka, M. *Proc. Nat. Acad. Sci.* (U.S.A.) **68**:1114 (1971).
53. Ishizuka, M., Braun, W., and Matsumoto, T. *J. Immunol.* **107**:1027 (1971).
54. Braun, W., and Ishizuka, M. *J. Immunol.* **107**:1036 (1971).
55. Hirschhorn, R., Grossman, J., and Weissmann, G. *Proc. Soc. Exp. Biol. Med.* **133**:1361 (1970).
56. Smith, J. W., Steiner, A. L., and Parker, C. W. *Fed. Proc.* **29**:369 (1970).
57. Smith, J. W., Steiner, A. L., Newberry, W. M., and Parker, C. W. *J. Clin. Invest.* **50**:432 (1971).
58. Smith, J. W., Steiner, A. L., and Parker, C. W. *J. Clin. Invest.* **50**:442 (1971).
59. Gallo, R. C., and Whang-Peng, J. *J. Nat. Cancer Inst.* **47**:91 (1971).
60. Bourne, H. R., and Melmon, K. L. *J. Pharmacol. Exp. Ther.* **178**:1 (1971).
61. Bourne, H. R., Lehrer, R. I., Cline, M. J., and Melmon, K. L. *J. Clin. Invest.* **50**:920 (1971).
62. Bourne, H. R., Melmon, K. L., and Lichtenstein, L. M. *Science* **173**:743 (1971).
63. Cross, M. E., and Ord, M. G. *Biochem. J.* **120**:21 (1970).
64. Ishizaka, T., Ishizaka, K., Orange, R. P., and Austen, K. F. *J. Immunol.* **106**:1267 (1971).
65. Lichtenstein, L. M., and Margolis, S. *Sci.* **161**:902 (1968).
66. McManus, J. P., Whitfield, J. F., and Yandale, T. *J. Cell. Physiol.* **77**:103 (1971).
67. Park, B. H., Good, R. A., Beck, N. P., and Davis, B. B. *Nature New Biol.* **229**:27 (1971).
68. Weissmann, G., Dukor, P., and Zurier, R. B. *Nature (New Biol.)* **231**:131 (1971).
69. Watts, H. G. *Transplant.* **12**:229 (1971).
70. Robison, G. A., Butcher, R. W., and Sutherland, E. W. *Cyclic AMP.* New York: Academic Press (1971).
71. Robison, G. A., Nahas, G. G., and Triner, L. "Cyclic AMP and Cell Function," *Ann. N.Y. Acad. Sci.* **185**:1–556 (1971).
72. Rasmussen, H. *Science* **180**:404 (1970).
73. Varmus, H. E., Perlman, R. L., and Pastan, I. *J. Biol. Chem.* **245**:2259 (1970).
74. Zubay, G., Schwartz, D., and Beckwith, J. *Proc. Nat. Acad. Sci.* (U.S.A.) **66**:104 (1970).
75. Pastan, I., and Perlman, R. L. *Science* **169**:339 (1970).
76. Dokas, L. A., and Kleinsmith, L. J. *Science* **172**:1237 (1971).
77. Greengard, P., Paoletti, R., and Robison, G. A., eds. *Advances in Cyclic Nucleotide Research*, vol. 1. New York: Raven Press (1972).

78. Braun, W., and Rega, M. J. *Red. Proc.* (in press).
79. Winchurch, R. (submitted for publication).
80. Northrup, R. S., and Fauci, A. S. Personal communication.
81. Kimberg, D. V., Field, M., Johnson, J., Henderson, A., and Gushon, E. *J. Clin. Invest.* **50**:1218 (1971).
82. Gericke, D., Chandra P., Haenzel, I., and Wacker, A. *Hoppe-Seyler's Z. Physiol. Chem.* **351**:305 (1970).
83. Fassina, G. *Life Sci.* **6**:825 (1967).
84. Jolley, W. B., and Epstein, G. Abstracts of the Conference on Physiology and Pharmacology of cAMP (Milan, 1971), p. 101.
85. Bolis, L., Luly, P., and Braun, W. Abstracts of the Conference on Physiology and Pharmacology of cAMP (Milan, 1971), P. 72.
86. Lichtenstein, L. M., and Henney, C. In *Prostaglandins in Cellular Biology and the Inflammatory Process*, P. Ramwell, ed. (in press).
87. Ramwell, P., and Shaw, J. E. "Prostaglandins," *Ann. N.Y. Acad. Sci.* **180**:1–568 (1971).
88. Stossel, T. P., Polland, T. D., Mason, R. J., and Vaughan, M. *J. Clin. Invest.* **50**:1745 (1971).
89. Parker, C. W. In *Prostaglandins in Cellular Biology and the Inflammatory Process*, P. Ramwell, ed. (in press).
90. Bourne, H. R. In *Prostaglandins in Cellular Biology and the Inflammatory Process*, P. Ramwell, ed. (in press).
91. Uhr, J., and Landy, M. *Immunologic Intervention*. New York: Academic Press (1971).
92. Henney, C. S., and Lichtenstein, L. M. *J. Immunol.* (in press).
92a. Kreth, H. W., and Williamson, A. R. *Nature* **234**:454 (1971).
93. Katz, D. H., Paul, W. E., and Benacerraf, B. *J. Immunol.* **107**:1319 (1971).
94. Ekpahah-Mensah, A., and Kennedy, J. C. *Nature* **233**:174 (1971).
95. Bonmassar, E., Goldin, A., and Cudkowicz, G. *Proc. Soc. Exp. Biol. Med.* **137**:1486 (1971).
96. Schimpl, A., and Wecker, E. *Eur. J. Immunol.* (in press).
97. Janis, M., and Bach, F. *Nature* **255**:238 (1970).
98. Dutton, R. W. *J. Exp. Med.* **122**:759 (1965).
99. Hirano, S., and Uyeki, E. M. *J. Immunol.* **106**:619 (1971).
100. Dutton, R. W., Falkoff, R., Hirst, J. A., Hoffman, M., Lesley, J. F., and Vann. D. In *Progress in Immunology*, B. Amos, ed. New York: Academic Press (1971).
101. Gordon, J., and MacLean, L. D. *Nature* **208**:795 (1965).
102. Trainin, N., Burger, M., and Kaye, A. M. *Biochem. Pharmacol.* **16**: 711 (1967).
103. Radovick, J., and Talmage, D. *Science* **185**:512 (1967).
104. Plescia, O. J. *Curr. Top. Microbiol. Immunol.* **50**:78 (1969).
105. Wigzell, H., and Anderson, B. *J. Exp. Med.* **129**:23 (1969).
106. Wofsy, L., Kimura, J., and Truffa-Bachi, P. *J. Immunol.* **107**:725 (1971).
107. Paul, W. E., Katz, D. H., Goidl, E. A., and Benacerraf, B. *J. Exp. Med.* **132**:283 (1970).
108. Rajewsky, K., Schirrmacher, S., Nase, S., and Jerne, N. K. *J. Exp. Med.* **129**:131 (1969).

109. Gershon, R. K., and Kondo, K. *J. Immunol.* **106**:1524, 1531 (1971).
110. Baker, P., Prescott, B., Barth, R. F., Stashak, P. W., and Amsbaugh, D. F. *Ann. N.Y. Acad. Sci.* **181**:34 (1971).
111. Dixon, F. J., Jacot-Guillarmod, H., and McConahey, P. J. *J. Exp. Med.* **125**:1119 (1967).
112. Diener, E., Feldmann, M., and Armstrong, W. D. *Ann. N.Y. Acad. Sci.* **181**:119 (1971).
113. Hellstrom, K. E. Hellstrom, I., Sjogren, H. O., and Warner, G. A. In *Progress in Immunology*, B. Amos, ed. New York: Academic Press (1971).
114. Nossal, G. J. V., and Ada, G. L. *Antigens, Lymphnodes and the Immune Response.* New York: Academic Press (1971).
115. Parish, C. R. *Ann. N.Y. Acad. Sci.* **181**:108 (1971).
116. Freedman, H. H. *Proc. Soc. Exp. Biol. Med.* (in press).
117. Thorbecke, G. J., Takahashi, T., and McArthur, W. P. In *Morphological and Functional Aspects of Immunity*, Alm and Hanna Lindahl-Kiessling, eds. New York: Plenum Publ. Co. (1971).
118. Lengerova, A., and Matousek, V. *Folia Biol.* **15**:263 (1969).
119. Demant, O. *Folia Biol.* **16**:273 (1970).
120. Brown, H. D., Chattopadhyay, S. K., Morris, H. P., and Pennington, S. N. *Cancer Res.* **30**:123 (1970).
121. Brown, H. D., Chattopadhyay, S. K., Spjut, H. J., Spratt, J. S., and Pennington, S. N. *Biochim. Biophys. Acta* **192**:372 (1969).
122. Heidrick, M., and Ryan, W. L. *Cancer Res.* **30**:376 (1970).
123. Johnson, L. D., and Abell, C. W. *Cancer Res.* **30**:2718 (1970).
124. Johnson, G. S., Friedman, R. M., and Pastan, I. *Proc. Nat. Acad. Sci.* (U.S.A.) **68**:425 (1971).
125. Hsie, A. H., and Puck, T. T. *Proc. Nat. Acad. Sci.* (U.S.A.) **68**:358 (1971).
126. Pastan, I., and Johnson, G. In *Prostaglandins in Cellular Biology and the Inflammatory Process*, P. Ramwell, ed. (in press).
127. Stjernsward, J., and Levin, A. *Cancer* (Sept. 1971).
128. Leuchars, E., Wallis, V. J., and Davies, A. J. S. *Nature* **219**:1325 (1968).
129. Bert, G. , Forrester, J. A., and Davies, A. J. S. *Nature (New Biol.)* **234**:86 (1971).
130. Sterzl, J. *Nature* **209**:416 (1966).
131. Adler, W. H., Takiguchi, T., and Smith, R. T. *J. Immunol.* **107**:1357 (1971).
132. Pasanen, V. J., and Makela, O. *Immunol.* **16**:399 (1969).
133. Baker, P. J., Prescott, B., Stashak, P. W., and Amsbaugh, D. F. *J. Immunol.* **107**:719 (1971).
134. Hartmann, K.-U. *J. Exp. Med.* **132**:1267 (1970).
135. Andersson, B., and Blomgren, H. *Cell. Immunol* **1**:32 (1971).
136. Cohen, J. J., Fishbach, M., and Claman, H. N. *J. Immunol.* **105**:1146 (1970).
137. Edelman, G. M., Rutishauser, U., and Milette, C. F. *Proc. Nat. Acad. Sci.* (U.S.A.) **68**:2153 (1971).
138. Takasugi, M., and Klein, E. *Transplant.* **9**:219 (1970).
139. Hellstrom, I., Hellstrom, K. E., and Sjorgren, H. O. *Cell. Immunol.* **1**:18 (1970).
140. Canty, T. G., and Wunderlich, J. R. *J. Nat. Cancer Inst.* **45**:761 (1970).

141. Bach, J.-F. In *Cell-Mediated Immunity—in Vitro Correlates*, J. P. Revillard, ed. Baltimore: University Park Press (1971).
142. Wolstencroft, R. A. In *Cell-Mediated Immunity—in Vitro Correlates*, J. P. Revillard, ed. Baltimore: University Park Press (1971).
143. Brunner, K. T., Mariel, J., Cerottini, J. C., and Chapins, B. *Immunol.* **14**:181 (1968).
144. Coe, J. E., and Salvin, S. B. *J. Immunol.* **93**:495 (1964).
145. Shortman, K., and Palmer, J. *Cell. Immunol.* **2**:399 (1971).
146. Kindred, B. *J. Immunol.* **107**:1291 (1971).
147. Krishna, G., Weiss, B., and Brodie, B. B. *J. Pharmacol. Exp. Ther.* **163**:379 (1968).
148. Gilman, A. G. *Proc. Nat. Acad. Sci.* (U.S.A.) **67**:305 (1970).
149. Moller, G. *Immunol.* **20**:597 (1971).

THE CELLULAR BASIS OF THE IMMUNE RESPONSE

Barry R. Bloom

The basic fascination that the immune system holds for biologists is its extraordinary variability and specificity. It is possible to generate antibodies and cellular immune responses to enormous numbers of antigenic determinants—proteins, polysaccharides, organic molecules, and so on. Yet antibodies recognize their antigens with extraordinary specificity, being able to discriminate a single nitro group on an aromatic hapten, D or L-amino acids, and α- or β-linked sugars. The principal known agents for recognition are antibodies, which are classified chemically as immunoglobulins. Figure 1 illustrates a four-chain model for the IgG immunoglobulin molecule, with two heavy chains and two light chains. Antibodies have two antigen-combining sites, each involving the two chains and occupying approximately one-third of the molecular weight of the molecule, and these are designated as Fab fragments. The remainder, known as the Fc fragment, is responsible for many of the biological pro-

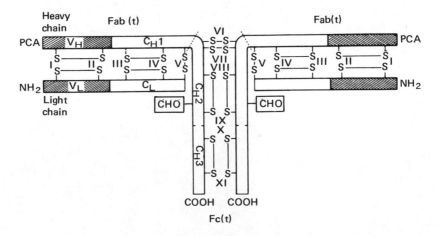

Fig. 1. Structure of human IgG immunoglobulin [Cunningham et al. *Progr. in Immunol.* 1:10 (1972)].

perties of the immunoglobulin molecule, such as ability to cross the placenta, secretion into extravascular compartments, and fixation to cell surfaces.

In addition to IgG, there are four other classes of immunoglobulins, which share common light chains but differ from one another in the nature of the heavy chains and the Fc fragment. When cells of the lymphoid series are surveyed, it becomes clear that some lymphocytes have immunoglobulins on their surface and during the process of immunization enlarge, proliferate, and ultimately differentiate into plasma cells (1,2). It is these cells and their immediate precursors that are primarily responsible for the production of antibodies.

Cells of this line arise from the bone marrow and, under appropriate conditions, can differentiate into the antibody-forming plasma cells and are termed B cells or B lymphocytes. However, the production of antibodies to most common protein bacterial and viral antigens requires, in addition to the B cells, the participation of a second population of lymphocytes, which originates in the bone marrow but passes through the thymus. These thymus-dependent or thymus-derived "T cells" are unable to secrete antibodies themselves, apparently recognize different determinants on antigens than the B lymphocytes, and carry immunological memory. In addition, T cells are responsible for the cell-mediated immune responses, such as rejection of allografts and tumors, chemical allergies, and resistance to many intracellular parasites such as tubercle and lepra bacilli and some viruses.

From studies in many in vitro systems on the production of antibodies by B cells or activation of T lymphocytes, the participation of a third cell, the macrophage, is often obligatory. Macrophages do not have immunologic specificity or produce antibodies. The mechanism by which they participate in "cell cooperation" with B and T lymphocytes is not clear. It has been suggested that they may fix antigen or their membranes and present it in an effective manner for lymphocyte activation, or they may secrete certain "conditioning" factors, which enhance the survival and activation of lymphocytes. The principal characteristics for distinguishing these populations of lymphoid cells are summarized in Table 1. They differ in surface markers, the B cells, for example, possessing surface immunoglobulins, and receptors both for the activated third component of complement and for the Fc portion of the immunoglobulin molecule when it is bound in an antibody-antigen complex. These are not found on T cells, or found only in minute amounts. T lymphocytes have the ability to bind sheep erythrocytes and form rosettes.

A variety of substances have been found to react at the surface of the various lymphoid populations and often lead to lymphocyte activation and proliferation. Some of these mitogens show selectivity for activating different cell populations; for example, Concanavalin A specifically activates T lymphocytes in the mouse, while bacterial lipopolysaccharide or

Table 1. Distinguishing Characteristics of T Lymphocytes,
B Lymphocytes, and Macrophages

Membrane Marker	T lymph.	B lymph.	Macroph.
IgG	−	+	−
Receptor for C_3(EAC rosettes)	−	+	+
Receptor for Ig or Ab-Ag complexes (Fc)	−	+	+
Thymus-specific antigens	+	−	−
Receptors for sheep red blood cells (E rosettes)	+	−	−
In vitro stimulation of DNA synthesis by mitogens			
PHA	+	−[a]	−
Con A	+	−	−
LPS	−	+	−
Anti-Ig	−	+	−
Specific binding to antigen-coated beads	−	+	−
MLC reactivity	+	−	−
GVH-inducing capacity	+	−	−
Adherence to surfaces (glass, plastic)	−[b]	−[c]	+
Phagocytic	−	−	+

[a]Some B lymphocytes may be recruited to divide secondarily due to factors elaborated by activated T lymphocytes. B cells may also be stimulated when the mitogen is attached to a solid support.
[b]Except for blast cells.
[c]Except for mature plasma cells or when immune complexes are attached to B cells.
GVIT is graft versus host; LPS is lipopolysaccharide (endotoxin); MLC is mixed lymphocyte culture.

endotoxin spontaneously activates B lymphocytes. Much more remains to be learned, but these markers have permitted cellular immunologists to be able to distinguish the components in the various responses, and to isolate, separate, and characterize each of the known cell populations. The point to be emphasized is that lymphocytes represent a highly heterogeneous population of cells, and experimental observations are only meaningful in terms of specific subpopulations that are responsible for them. Simplistic, unqualified generalizations, for example, on "lymphocyte" activation are really no longer tenable.

I. B Lymphocytes

Because the B cell product is so well known and characterized, a great deal more is known about the capabilities of this cell population. When animals are immunized even with a defined antigen, such as DNP-oval albumin, an almost continuous spectrum of antibodies are produced, which, while binding the DNP group, vary enormously with respect to affinity of binding and electrophoretic mobility.

The clonal selection theory, which is the theoretical cornerstone of modern cellular immunology, holds that the role of antigen is to select those cells that have an immunoglobulin on their surface to which that antigen can bind, and that such cells are selected by antigen to clone and proliferate. The implication is that a heterogeneous antibody response indicates a heterogeneity of different lymphoid cells specific for the antigen or hapten in question.

From extraordinary recent advances made by Askonas and Williamson (3), Krause (4), Eichmann (5), and Schlossman (6) using simpler and more defined antigens such as haptens coupled to synthetic polypeptides, e.g., DNP-(lys)9, or repeating sequence bacterial polysaccharides such as pneumococcal or streptococcal polysaccharides, it has recently been possible to confirm directly the validity of the clonal hypothesis. When antigens containing a limited number of determinants are used, it was observed that a more homogeneous electrophoretic pattern of antibodies was seen, in contrast to the broad smear seen with more complex antigens. When these antibodies were studied by isoelectric focusing, it was found in some cases that there were a very small number of bands characteristic of a homogeneous protein—the multiple banding is due not to primary sequence differences, but rather to slight changes in charge due to amidation (Fig. 2). When a small number of cells from animals immunized with the simple antigens are transferred to irradiated recipients and the antibody produced in recipient spleens is examined electrophoretically, one can see the same immunoglobulin pattern as is

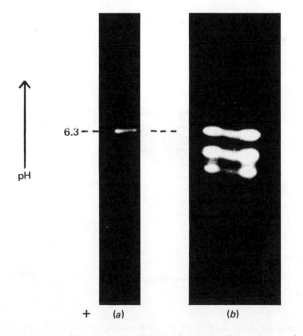

Fig. 2. Isoelectric banding of anti-DNP antibodies. (a). Early after passage of antibody-producing clone. (b). A later passage of the same clone [Williamson, A. R., and Askonas, B. A. *Nature* **238**:337 (1972)].

present in the donor. This demonstrates that antibody formation is clonal (Fig. 3).

The serial transfers can be carried on for several generations, the longest studied being the E9 clone observed by Askonas and Williamson to produce anti-DNP antibodies for almost nine serial transplant generations. If, in any passage of immune cells to the radiated recipient, the recipient is not challenged with the specific antigen, then the ability of the spleens to produce the homogeneous antibody disappears. Thus, antigen is required for selection and continued propagation of the clone, thereby confirming the continued need for selection by antigen.

Perhaps the most interesting observation in this system, when defined antigens are used, is that many individual mice of inbred strains produce antibody of the same homogeneous isoelectric focusing pattern to the same antigen. Thus, as many as 7 out of 11 individuals may produce the same antibodies, and hence possess the same clone specific to the same antigen. This bears importantly on the origin of antibody and immunologic diversity. If there were a very small number of genes controlling merely the various classes of chains, and diversity were produced by random mutation, it is very difficult to envision 7 out of 11 mice immunized with the same antigen undergoing simultaneous spontaneous mutations in the same genes to produce the same homogeneous antibody.

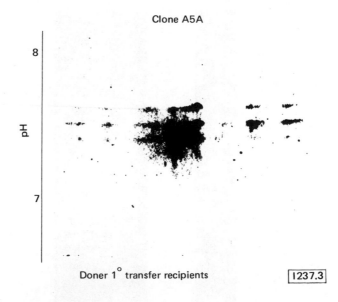

Fig. 3. Isoelectric focusing pattern of purified antibodies to streptococcal group A carbohydrate from a single mouse donor and from six irradiated mice immunized after reconstitution with hyperimmune spleen cells from the original donor [Eichmann, K. *Eur. J. Immunol.* **2**:301 (1972)].

Rather, the findings strongly argue that there are more than a very few germ-line genes, in this case a common germ-line gene carried in all mice of that strain responsible for the same homogeneous clone of antibody-forming cells to the same antigen. Thus, there are genes in the germ line not only for the constant regions that define the classes of antibody, but probably also for the variable regions that are responsible for antigen binding and specificity.

If proliferation is induced in antibody-containing B cells by contact with antigen, then we must explain what shuts the response off—what prevents every antigenic stimulus from leading to a multiple myeloma. The nature of the regulation of the immune response is of paramount importance and has not been possible to approach at the cellular level until very recently. Askonas and Williamson (7) have found that these antibody-forming clones have only limited potential for division (approximately 90 divisions), and after a maximum of about 300 days or 9 transfer generations, even in the presence of antigenic stimulation the clone dies. Put another way, as cells become increasingly differentiated, they appear to have a decreased potential for proliferation. On the basis of the observations that some immunoglobulin-secreting myelomas do not have detectable membrane-associated immunoglobulins, it is not implausible to regard the loss of surface receptors by T cells as a possible mechanism for preventing their infinite stimulation by antigen. Lymphoid cells without surface receptors should have no way of being stimulated even in the presence of antigen and would be expected to be terminal cells. In terms of the entire immune response, there are many other mechanisms by which regulation of the antibody response is affected: by antigen level, antibody feedback, and probably T cell suppression.

II. T Lymphocytes

The role and mechanism of action of T cells in cell cooperation and cell-mediated immunity in much less well understood than that of B cells. Clearly, one reason for this has been the general inability to identify the immunological receptor on T cells. While the subject remains highly controversial, it is the general finding that T cells have very little, if any, detectable surface immunoglobulin, and if the receptor is not an antibody, there are really no plausible hypotheses about its nature. It is clear, however, that T cells do possess immunological specificity, although recognition by T cells appears to require a larger antigenic determinant than an antibody molecule.

The known and suggested functions of T lymphocytes in vivo can be summarized as—

1. Enhancing Ab production to thymus-dependent antigens.

2. Mediating the IgM to IgG switch.

3. Initiating cell-mediated immune reactions, e.g., graft vs. host, allograft and tumor rejection.

4. Having adjuvant effects.

5. Mediating genetic control of Ir genes on immune responses.

6. Exerting suppressor effects on B lymphocytes and other T cells (tolerance? antigenic competition?).

And in vitro, various T cells act—

1. To proliferate in response to "carrier" determinants of antigens and to certain mitogens (e.g., Con A, PHA, MLC).

2. To produce mediators of cell-mediated immunity (e.g., MIF, ChF and cell-cooperation factors, adjuvant factor, and suppressor factors).

3. To mediate direct lymphocyte-target cell cytotoxicity.

4. Upon activation to possibly replicate viruses.

It is clear then that production of antibody to most protein and viral and bacterial antigen requires the participation of T cells, and the conversion from IgM to IgG antibody formation is apparently absolutely T cell dependent (1). T cells are involved in cell-mediated immunity and can produce some direct lymphocyte target cell cytotoxicity and soluble mediators of hypersensitivity, such as the migration inhibitory factor, chemotactic factor, lymphotoxin, etc. (8). These factors are produced in minute amounts and have not been well characterized biochemically; nevertheless, they have measurable and perhaps very important effects on other cells such as macrophages, somatic cells, and tumor cells. These factors are summarized in Table 2.

It need only be stated that the cell-mediated immune response arose 300 million years ago in evolution and must be doing something right. Two principal hypotheses have been adduced for its evolutionary selective value. The first holds that it is crucial for providing immunity against intracellular infection. Thus, while antibodies may be effective at preventing infection of cells or death by extracellular toxins and organisms, once inside cells these pathogens are refractory to the effect of antibodies. In that sense, perhaps all cell-mediated immune reactions may be looked at as rejection reactions. Paul Ehrlich (18) in 1909 suggested another evolutionary function for the cell-mediated immune response, namely, to provide for immune surveillance, that is, the continuous and systematic rejection of cells antigenically altered by neoplastic transformation.

It is currently emerging that T lymphocyte populations are not homogeneous and that several subpopulations of T cells may exist (9). From their behavior, there are long-lived and short-lived T cells and recirculating and nonrecirculating T cells. While the more common pro-

Table 2. Biological Activities of Products of Activated Lymphocytes

Product	Activity
Affecting macrophages	
Migration inhibitory factor (MIF)	Inhibits the migration of normal macrophages.
Macrophage aggregation factor (MAF)	Agglutinates macrophages in suspension.
Macrophage chemotactic factor (MCF)	Causes macrophages to migrate through micropore filter along gradient.
Macrophage resistance factor (postulated)	Possibly renders macrophages resistant to infection by certain bacteria.
Cytophilic antibodies	Confer on macrophages specific reactivity with antigen.
Affecting lymphocytes	
Blastogenic or mitogenic factor (BF or MF)	Induces blast cell transformation and TdR incorporation in normal Lymphocytes.
Potentiating factor (PF)	Augments or enhances ongoing transformation in MLC or antigen-stimulated cultures.
Cell-cooperation or helper factor	Produced by T cells, increases the number or rate of formation of Ab-producing cells in vitro.
Suppressor factor (postulated)	Inhibits activation of and/or antibody production by B cells.
Affecting granulocytes	
Inhibitory factor	Inhibits the migration of human buffy coat cells or peripheral blood leucocytes from capillary tubes or wells in agar plates.
Chemotactic factor	Causes granulocytes to migrate through micropore filter along gradient.
Affecting cultured cells	
Lymphotoxin (LT)	Cytotoxic for certain cultured cells, e.g., mouse L cells or HeLa cells.
Proliferation inhibitory factor (PIF) and cloning inhibitory factor (ClIF)	Inhibits proliferation of cultures cells without lysing them.
Interferon	Protects cells against virus infection.
Producing effects in vivo	
Skin reactive factor (SRF) (possibly a combination of several of the above activities)	Induces indurated skin reactions, histologically similar to delayed-type hypersensitivity reactions in normal guinea pig skin.
Macrophage disappearance factor	Injection interperitoneally causes macrophages to adhere to peritoneal wall.

perty of T cells that is studied in vitro is their ability to proliferate and incorporate thymidine, we have described earlier an antigen-sensitive lymphocyte that is nondividing (10). Such a cell, when activated, is capable of replicating viruses and producing mediators even in the presence of vinblastine or cytosine arabinoside.

One very interesting observation on the difference between proliferation and functional responses was made in studies on mixed lymphocytes interaction between histocompatible and histoincompatible mice. It has been shown by Festenstein (12) that Balb/c mice (H2^d) gave a mixed lymphocyte reaction with mitomycin-treated DBA cells (H2^d) of the same order of magnitude as CBA mice (H2^k), which differ by major histocompatibility locus. Howe (13) examined the function of these "activated" lymphocytes and found that only the H2 incompatible cells were effective in killing chromium-51-labeled DBA mastocytoma target cells. Dr. Kano in my laboratory, together with Dr. Howe, studied the ability of these various populations to replicate VSV virus. As seen in Table 3, while the degree of thymidine incorporation and presumably the number of dividing cells were comparable in these two mixed lymphocyte cultures, only one of the populations had immunological competence for cytotoxicity and viral production. The simplest interpretation, although by no means yet proven, is that there are at least two T cell populations, one of which proliferates, the other possibly being responsible for the cytotxicity and mediator production. In any case, these experiments emphasize the point that no longer can any single parameter, such as thymidine incorporation, be used to define immunological potential and function of T cells.

Table 3. Evidence of Distinct T Cell Populations

Mixed lymphocyte culture	^{14}C thymidine incorp	Cytotoxicity on ^{51}Cr-DBA	Virus plaque assay
Balb/c(H2^d) × DBA$_m$(H2^d)	+	−	−
CBA (H2^k) × DBA$_m$(H2^d)	+	+	+

III. Regulation

From the recent elegant studies on generation of immune responses to synthetic antigens in mice and guinea pigs, it has emerged that these reactions are under strict genetic control. The genes responsible have been termed immune response, or Ir, genes (14). Thus, strain 2 guinea pigs are capable of responding by antibody production and cell-mediated immunity to DNP-polylysine, whereas strain 13 animals are not.

The Ir 1 locus in the mouse, the best-studied immune-response gene, is located on the IXth linkage group, in the middle of the H2 major histocompatibility locus. It controls not simply the response to a single antigen, but the response of mice to a whole variety of synthetic antigens, some of which may have paradoxical chemical characteristics. The question is how can a single gene, for example, the PLL gene in guinea pigs, control the response to DNP-polylysine, a highly basic antigen, and limiting amounts of bovine serum albumin, the most acidic serum protein? By extremely elegant experiments, McDevitt and co-workers established that the regulatory effect of the Ir 1 gene is expressed exclusively by T lymphocytes, even though antibody in these situations is produced by B lymphocytes (15). Thus, by reconstituting responder T cells into nonresponder animals, responses were obtained.

In some very exciting recent work, Scholssman has studied the responses of guinea pigs to DNP-(lys) 14, a defined antigen that produces a monoclonal or homogeneous antibody response in strain 2 responder guinea pigs. When strain 13 nonresponder animals are injected with the same antigen in oil or in a weak adjuvant, they fail to respond. However, if they are given the antigen in complete Freund's adjuvant containing a human virulent mycobacterium as adjuvant, then the nonresponder pigs respond. And indeed, they respond by producing antibodies of the same electrophoretic mobility, i.e., the same clone as the responder animals. This would argue that the nonresponder animals have the same complement of B cell clones as responder animals and lack only specific T cells. And since the B cells obviously have the antibody receptor for DNP-(lys)14, it suggests that the T cell receptor, which is deficient in these animals, must be different from that antibody (16).

These results, together with other results in in vitro systems (17), suggest that adjuvants may augment antibody responses by causing sensitization to determinants in the adjuvant followed by antigenic stimulation of T cells by the antigens of the adjuvant, thereby releasing mediators which nonspecifically expand clone size of B cells responding to the antigen in question.

In addition to genetic control, we may again ask what other factors regulate the T cell response and prevent every T cell antigenic stimulation from resulting in lymphoma. Clearly, the availability of antigen, and possible excess of antigen, affect the T cell response. In addition, antibodies seem to exert a profound effect on cell-mediated immune responses, either preventing sensitization if they are developed first, or blocking the activity of already sensitized lymphocytes. It is this latter phenomenon, termed immunological enhancement, that has been postulated by the Hellstroms to be responsible for suppressing cell-mediated immunity against tumors (19). When lymphocytes in patients with tumors are studied in vitro for their ability to kill tumor cells from the patients or tumors of the same biological type they do so with a very high frequency.

However, when the cytotoxicity assay is done in the presence of the patient's serum, then most patients are found to have factors in their serum that block the cytotoxic reaction.

One further possible mechanism for regulation of T cell activity could be the fact that T cells in their resting state express poorly receptors for pharmacological agents. Perhaps following "activation" or during the process of lymphocyte differentiation into effector cells, receptors for such agents as insulin and histamine and perhaps cyclic AMP appear, which then allow them to be regulated by the pharmacological consequences of the reactions they mediate.

References

1. Katz, D., and Benacerraf, B. *Advan. Immunol.* **15**:1 (1972).
2. Unanue, E. *Advan. Immunol.* **15**:172 (1972).
3. Askonas, B. A., Williamson, A. R., and Wright, B. E. G. *Proc. Nat. Acad. Sci.* **67**:1398 (1970).
4. Eichmann, K., Lackland, H., Hood, L., and Krause, R. M. *J. Exp. Med.* **131**:207 (1970).
5. Eichmann, K. *Eur. J. Immunol.* **2**:301 (1972).
6. Schlossman, S. F., Ben-Efraim, S., Yaron, A., and Sober, H. A. *J. Exp. Med.* **123**:1083 (1966).
7. Williamson, A. R., and Askonas, B. A. *Nature* **238**:337 (1972).
8. Bloom, B. R. *Advan. Immunol.* **13**:102 (1971).
9. Cantor, H., and Asofsky, R. *J. Exp. Med.* **131**:235 (1970).
10. Bloom, B. R., Gaffney, J., and Jimenez, L. *J. Immunol.* **109**:1395 (1972).
11. Jimenez, L., Bloom, B. R., Blume, M. R., and Oettgen, H. F. *J. Exp. Med.* **133**:740 (1971).
12. Festenstein, H., Abbasi, H., Sachs, J. A., and Oliver, R. T. D. *Transpl. Proc.* **IV**:219 (1972).
13. Howe, M. (in press).
14. Benacerraf, B., and McDevitt, H. O. *Science* **175**:273 (1972).
15. Mitchell, G. F., Grumet, C. F., and McDevitt, H. O. *J. Exp. Med.* **135**:126 (1972).
16. Schlossman, S. (in press).
17. Maillard, J., and Bloom, B. R. *J. Exp. Med.* **136**:185 (1972).
18. Ehrlich, P. In *The Collected Papers of Paul Ehrlich*, F. Himmelweit, ed. vol. II, 550 (1957).
19. Hellstrom, I., and Hellstrom, K. E. *Cancer* **28**:1269 (1971).

COMPLEXITIES IN THE STUDY OF THE ROLE OF CYCLIC NUCLEOTIDES IN LYMPHOCYTE RESPONSES TO MITOGENS

CHARLES W. PARKER

Our interest in the possible role of cyclic nucleotides in immunological phenomena dates back to 1968, when we undertook studies on the effect of cyclic AMP and agents that raise intracellular cyclic AMP levels on DNA, RNA, and protein synthesis in PHA-stimulated and resting human lymphocytes. Our initial report indicating that sustained elevations of cyclic AMP inhibit lymphocyte transformation (9) was published shortly after the description of similar observations in tumor cells by Ryan and Heidrick (8). We subsequently conducted direct measurements of changes in cyclic AMP concentration in PHA-stimulated lymphocytes. We demonstrated that lymphocytes contain an adenylate cylase, which is hormonally responsive and stimulated by PHA and that PHA causes an early rise in intralymphocytic cyclic AMP concentrations—within 60 sec. (10–12). In longer-incubation experiments, cyclic AMP levels in PHA-stimulated cells eventually fell to below the level in control cells (at 6 and 24 hr), indicating variations in cyclic AMP metabolism in different parts of the cell cycle.

The early rise in cyclic AMP, occurring before other known metabolic changes in PHA-stimulated lymphocytes, suggested that cyclic AMP might be fulfilling the role of a secondary messenger for PHA. However, attempts to stimulate lymphocytes by adding graded amounts of cyclic AMP to the medium produced no more than two- to threefold increases in DNA synthesis. This was true even when lymphocytes were pulsed with cyclic AMP so as to mimic the time course of change in intracellular cyclic AMP concentrations produced by PHA itself (13). Moreover, using isoproterenol, PGE_1, or theophylline, we were unable to obtain any appreciable stimulation of radioactive thymidine incorporation. We considered a possible role for cyclic GMP, but were unable to show changes in cyclic GMP levels in PHA-stimulated cells, and addition

This research was supported by grants from the National Institutes of Health.

of cyclic GMP to the medium produced less stimulation than cyclic AMP itself.

Faced by an inability to fulfill the Sutherland postulates—for proving that cyclic AMP is acting as an intracellular messenger for an extracellular hormone (7), we were forced to conclude either that cyclic AMP is unimportant (which seemed illogical to us), that it is only part of the stimulus, or that PHA is acting on a unique adenylate cyclase and cyclic AMP pool not susceptible to other cyclic AMP stimulatory agents (12, 13). Almost two years later, the role of cyclic AMP is still not clear, but the possibility of a specialized cyclase now seems much more plausable than it did previously. Before discussing our most recent data, it is appropriate to consider very briefly how cyclic AMP affects other immunological systems and what some of the general problems in interpretation are.

The results of studies on the effects of cyclic nucleotides on other immunologic modalities have similar complexities. Increases in cyclic AMP inhibit antigen-mediated histamine and SRS-A release, lymphocyte cytotoxicity for target cells, antibody secretion, chemotaxis, lymphokine secretion, and lysozomal enzyme release (6). Yet there are examples in which graded amounts of cyclic AMP stimulate a response, as noted in the early studies of Braun and his colleagues (5), indicating a dualism in the action of cyclic AMP. Dualism might be explained in several different ways:

(1) It might be on the basis of a bell-shaped cyclic AMP dose-response curve in critical enzymatic reactions modulated by cyclic AMP.

(2) A biphasic action of cyclic AMP would not be unexpected in complex multistep reactions such as lymphocyte transformation, in which cyclic AMP might have opposing effects in different parts of the cell cycle (see below). It seems likely that some of the divergent effects of cyclic AMP on cell proliferation in different tissues depend on where the cells are in their cycle at the time the cyclic nucleotide is added. In this connection, it is interesting that PHA produces an early rise in lymphocyte cyclic AMP followed by a sustained fall, suggesting that different levels are appropriate at different stages of cell development.

(3) Paradoxical responses to cyclic AMP might also occur because of subcellular compartmentalization of cyclic AMP. If PHA had its own specialized adenylate cyclase in a localized area of the cell, stimulation by other agents that promote cyclic AMP accumulation could produce different, even antagonistic effects (see below). The concept of subcellular compartmentalization is supported by the recent studies of Davis, indicating metabolic channeling of carbamyl phosphate by separate carbamyl phosphate synthesizing enzymes in Neurospora (1).

(4) Another possible cause is cellular heterogeneity with different

cells programmed to respond in different ways to a cyclic nucleotide stimulus.

Regardless of the explanation, the important point is that many of the immunological responses currently under study are highly complex phenomena with special problems in interpretation that may not arise when tissue responses to rapidly acting hormones are studied. Before the role of cyclic nucleotides in these various immunological processes can be fully understood, it will be necessary to deal more directly with some of the inherent difficulties. The first problem is that of tissue heterogeneity and the complexities introduced by possible interactions among cells undergoing immunologic activation.

I.Tissue Heterogeneity and Cell-Cell Interaction

The problem of tissue heterogeneity is especially evident in studies of histamine release from mast cells in organs in which the mast cell represents only a small percentage of the total tissue present. Tissue cyclic AMP measurements in this situation do not necessarily reflect what is happening in the cell of interest. Moreover, there is the possibility of metabolic changes secondary to the release of mediators from neighboring cells. Heterogeneity also complicates the interpretation of studies with purified lympocytes, where mixtures of T and B lymphocytes are being used, and it is uncertain what percentage of cells is responding to the stimulatory agent. As discussed below, we now have evidence to indicate that when peripheral blood lymphocytes are passed over a nylon column so that the B cells are removed, the magnitude and shape of the cyclic AMP dose-response curve to PHA are markedly altered. Thus, the nature of the dose-response curve in the cells we are really interested in may not be evident until most of the non-responding cells have been removed. The fact that cells not undergoing mitogenesis also display changes in cyclic AMP levels on exposure to PHA does not prove that cyclic AMP is unimportant, since control may be exerted at a subsequent step in the activation sequence.

Despite the desirability of working with as well defined a cell population as possible, purification may remove cells that participate in the response and conceivably lead to misleading results. Most preparations of peripheral blood lymphocytes contain variable numbers of erythrocytes, monocytes, platelets, and polymorphonuclear leukocytes. The studies of Tarnvik (14) and Yachnin (16) indicate that non-lymphocytic cells are capable of acting as accessory cells which amplify the DNA synthetic response, the magnitude of the effect depending on the cell and the mitogen. It seems reasonable to assume that the action of accessory cells is to facilitate the formation of a matrix. Whether or not matrix formation is obligatory in lymphocyte transformation is uncertain.

Good to fair mitogenic responses can be obtained in the absence of accessory cells. However, it may be that once accessory cells have been removed, cell-cell interactions among the lymphocytes themselves can provide the necessary stimulus. If cell-cell interactions are important in lymphocyte stimulation, the method of preparing cells and the density at which they are used may be critical, particularly if an attempt is being made to demonstrate very early metabolic changes. In studies in organized tissues, there is no opportunity to manipulate cell density. With isolated single-cell suspensions of lymphoid cells, cell density can be varied over a wide range. As a rule, the metabolic effects of mitogens are studied in relatively dilute lymphocyte suspensions, yet in ordinary lymphocyte-transformation experiments, the cells settle and are in close juxtaposition throughout most of the culture period.

II. Possible Multiple (Simultaneous or Serial) Activation Events

Both for Concanavalin A (Con A) and PHA, there is evidence that the action of the mitogen must be sustained over a period of hours. Quite possibly, these mitogens must act at least two places in the cell cycle — very early, to initiate transformation, and later, when the cell is in mid or late G_1. The addition of a possible messenger for an early stimulatory event (such as cyclic AMP or cyclic GMP) would not necessarily produce the full transformation response, particularly if the delayed event involved an entirely different mechanism. An additional complexity is provided by the recent observations of Haber et al. (3), who have isolated fractions of PHA, which selectively stimulate the release or synthesis of lymphotoxin, interferon, or nucleic acid. The question is: Are the different PHA fractions stimulating different cells or different portions of the same cell? And if we are interested in studying lymphocyte replication, what early metabolic event should be chosen to prove that activation relevant to replication has taken place?

III. Subcellular Compartmentalization of Cyclic AMP

The Sutherland criteria for proving that cyclic AMP is mediating a metabolic change inside a cell break down if exogenous cyclic nucleotides are unable to penetrate the cell or if the cell is compartmentalized with respect to cyclic AMP. Recent immunofluorescence studies in our laboratory with human lymphocytes by Dr. J. Wedner in collaboration with Dr. F. Bloom of the National Institute of Mental Health strongly support the possibility of cyclic AMP compartmentalization (15). Using rabbit anticyclic AMP antibody and fluoresceinated goat antirabbit gamma globulin in a double fluorescent antibody reaction, we have been able to show that PGE_1, isoproterenol, and erythroagglutinating PHA give different cyclic AMP localization patterns in stimulated human lymphocytes. PGE_1 gives diffuse cytoplasmic fluorescence; isoproterenol gives diffuse nuclear and spotty cytoplasmic fluorescence; and PHA gives patchy cytoplasmic fluorescence. The specificity of the staining for cyclic AMP is indicated by cyclic AMP blocking experiments and a failure of normal rabbit gamma globulin to produce staining. Specificity for cyclic AMP is also suggested by studies in lymphocytes, salivary gland, brain, and liver, indicating correspondence between the intensity of immunofluorescent staining and the results of direct cyclic AMP measurements. We presume that what is being stained is protein-bound cyclic AMP and that the localization pattern is determined by the area of the cell in which adenylate cyclase activation is taking place. The validity of the nuclear isoproterenol localization pattern in human lymphocytes is supported by studies with isolated nuclei in which isoproterenol and epinephrine were shown to produce an increase in cyclic AMP, inhibitable with pro-pranolol, whereas PGE_1 and PHA were nonstimulatory (15). We interpret our data as indicating that isoproterenol, PGE_1, and PHA act on different adenylate cyclase molecules. If this is true, it is not surprising that PGE_1 and isoproterenol fail to have the same effect on DNA, and RNA synthesis as PHA and studies with these agents do not exclude an important role for cyclic AMP in activation by PHA. The cyclic AMP localization studies also provide a rational answer to the criticism that increases in cyclic AMP obtained with PHA are usually only two- to fourfold, which might be insufficient to produce major metabolic changes. From the fact that the accumulation of cyclic AMP in PHA-stimulated cells is sharply demarcated when PHA produces a twofold increase in cyclic AMP in the lymphocyte population as a whole, the local increase is probably more like 25- to 100-fold. Just how the local accumulation of cyclic AMP in a PHA-responsive area could produce a unique metabolic response is not altogether clear. Conceivably, PHA stimulation might activate a localized cyclic AMP dependent protein kinase with unique catalytic properties. The major role of cyclic AMP might be to aid in the translocation of calcium, which could, in turn, contribute to activation of the cyclic GMP system or provide energy for the transport of an informational polynucleotide (4) from the cell membrane to the interior of the cell. Obviously, until detailed metabolic studies can provide a plausible biochemical mechanism, the concept of local PHA action must be considered in the realm of hypothesis rather than established fact.

Other new data from this laboratory relate to what happens to mitogen effects on lymphocyte cyclic AMP concentrations when lymphocytes are fractionated. By passing Ficoll-Hypaque purified human peripheral blood lymphocytes over a long nylon column, we can largely or entirely remove cells with easily demonstrable surface immunoglobulin—presumably B cells (2). The procedure also removes the small numbers of monocytes and polymorphonuclear leukocytes that usually contaminate Ficoll-Hypaque preparations. When B cells are reduced from 20 to 5%, the shape of the dose-response curve to PHA and Con A is shifted, and the magnitude of the response is decreased (Table 1: Ficoll-Hypaque purified cells compared with Ficoll-Hypaque nylon cells). At intermediate- and high-mitogen concentrations, B cell depleted preparations now show a *fall* in cyclic AMP, whereas at lower concentrations, a rise in frequently demonstrable (not shown), particularly if measurements are made at 5 or 10 min. The two-cell preparations also display interesting differences in hormonal responses. The T cell–B cell mixture generally responds better to hormonal agents, but the converse is true at high concentrations of norepinephrine. Interestingly, preliminary immunofluorescence studies indicate that PHA can produce localized increases in cyclic AMP staining under conditions in which a fall in cyclic AMP is occurring in the cell population as a whole. We interpret these observations as indicating that PHA and Con A are able to either raise or lower cyclic AMP in T cells. Apparently, in unfractionated lymphocytes, the fall in cyclic AMP in the T cell segment is masked by the relatively small percentage of B lymphocytes that also are present.

The results when unpurified mixed leukocytes are studied are also of interest. As a rule, PHA-P and Con A produce a greater increase in cyclic AMP in mixed leukocytes than in more highly purified lymphocytes obtained by Ficoll-Hypaque density-gradient centrifugation or passage through a short nylon column. Thus, in the presence of accessory cells, the ability of the mitogens to promote cyclic AMP accumulation is mag-

Table 1. The Effect of Nylon Fiber Chromatography on Cyclic AMP Responses in Purified Human Lymphocytes

| | Cyclic AMP (pmoles per 10^7 cells)[a] | |
| | Ficoll-Hypaque cells | Ficoll-Hypaque nylon cells |
Agent		
Buffer	22	20
Isoproterenol 10 mM	123	78
Norepinephrine 10 mM	60	94
PGE_1 30 μM	180	84
PHA-P 1:10	25	9
PHA-P 1:50	29	10
PHA-P 1:250	34	12
Con A 80 μg/ml	52	16
Con A 16 μg/ml	42	19
Con A 3 μg/ml	26	16

[a]Mean of six or more experiments.
Note: One $\times$ 10^6 lymphocytes in 0.5 ml Gey's solution were incubated for 30 min at 37°. The cells were centrifuged, and after removal of the supernatant solution the cell pellets were frozen in liquid nitrogen. Cyclic AMP was determined by radioimmunoassay. Ficoll-Hypaque cells contained an average of 20% B cells as compared with 5% B cells in Ficoll-Hypaque nylon cells.

nified. Immunofluorescent studies indicate that the increase is taking place in lymphocytes, not the accessory cells themselves. While it is possible that the augmented response is a nonspecific effect of cell aggregation, the fact that accessory cells also potentiate mitogenesis should be kept in mind, and it may be that the greater increase in cyclic AMP is a manifestation (or cause) of more effective cell stimulation.

The final interpretation of these observations with respect to mitogenesis is not clear, but it is evident that PHA can produce either a rise or fall in cyclic AMP in lymphoid cells. It would be fascinating if it turned out that PHA can simultaneously stimulate and inhibit adenylate cyclase in different regions of the same cell. Matters would be simplified if one were to assume that in the activation of T cells, the fall in cyclic AMP is what is important, since that would explain the ability of agents that raise intracellular levels of cyclic AMP to inhibit lymphocyte tranformation and correlate with cyclic AMP effects on cell growth in established tissue culture lines. However, this formulation would ignore the very early localized accumulation of cyclic AMP in lymphocytes stimulated by PHA. My current view is that cyclic AMP does play a critical role in lymphocyte stimulation by mitogens and that both the rise and fall in cyclic AMP are important. Obviously, studies to date raise more questions than they answer, and much more work is needed.

Discussion of Dr. Parker's Paper

DR. AUSTEN: When you remove the B cells from your lymphocyte suspension with the nylon column and then stimulate the purified cells, you get an augmented effect with norepinephrine and a diminished response to PGE_1 and isoproterenol. What type of recognition mechanism is involved in the norepinephrine response?

DR. PARKER: Classically, norepinephrine is both an α and β stimulator. In some cells, it seems to be a better β stimulator than isoproterenol. Norepinephrine-induced increases in cyclic AMP in lymphocytes can be blocked by propranolol, indicating that a β receptor is involved, but the magnitude of the response may be influenced by concomitant α adrenergic stimulation, which would tend to lower cyclic AMP. We presume the pattern of isoproterenol and norepinephrine responsiveness in B and T cells involves differences in their β or α receptors or both. Actually, the response to norepinephrine in T cells is not greater throughout the entire dose range, but primarily at high norepinephrine concentrations.

DR. HADDEN: In the experiment with Ficoll-Hypaque and nylon-purified cells in which you saw a drop in cyclic AMP with PHA-P, is the dilution of PHA-P used mitogenic in those cells?

DR. PARKER: The dilutions used included concentrations that produce a good mitogenic response.

DR. ROBISON: Have you specifically studied the effect of changing the cell density on cyclic AMP per cell?

DR. PARKER: In short-term experiments, in the low-cell-density range (1 to 8 × 10^6 lymphocytes/ml), there is no consistent effect of cell density on cyclic AMP concentration in unstimulated cells. We have not systematically looked at the effect of higher cell concentrations, although limited data would suggest that there is no major change as the cell concentration undergoes an additional fourfold increase. The comparison in the presence of PHA is more complicated, because one must either change the PHA concentration or the quantity of PHA per cell. We are currently evaluating this question, and preliminary data indicate that the PHA dose-response curve is indeed influenced by cell concentration.

DR. WEISSMANN: I don't understand the conclusions that you have drawn from the fluorescent staining for cyclic AMP in air-dried cells. I gather that, in air-dried lymphocytes, the distribution of something that stains with immunofluorescence for cyclic AMP is redistributed, depending upon what you have done to the cell before you have air-dried it. Is that correct?

DR. PARKER: That is correct.

DR. WEISSMANN: What you are staining in the cell is, therefore, a large molecule that binds cyclic AMP discretely, and the redistribution is not of cyclic AMP necessarily, but of a binding site for this molecule in the air-dried cell. Is that right?

DR. PARKER: I would agree that the cyclic AMP being visualized by immunofluorescence must be presumed to be protein bound. However, I would doubt that the localization pattern is due to changes in the binding protein per se. The intensity and location of the fluorescence correlate quite well with the results of direct cyclic AMP measurements in whole lymphocytes and subcellular fractions. But even if you are correct and all of the subcellular localization results are due to differences in binding protein activity, the observation that different adenylate cyclase stimulating agents affect the cell in different ways would still be important, since it would imply an important functional difference in how these various agents work. Moreover, how could this occur if all of the agents act through the same adenylage cyclase molecules?

DR. CHANDRA: I would like to ask you about the Concanavalin A (Con A) effect. You showed a concentration-dependent increase of cyclic AMP by Con A. Can you tell us whether these concentrations of Con A influence mitogenesis?

DR. PARKER: We do not have extensive correlations between Con A stimulation of mitogenesis and changes in cyclic AMP concentrations. We have many such experiments with PHA.

DR. HIRSCHHORN: Your results with Con A are very interesting. There has been a recent report that passage through a nylon wool column removes a population of cells responsive to Con A, so that the column-purified cells, which had a decrease in cyclic AMP in response to Con A, would most likely not have responded to Con A with proliferation.

DR. PARKER: Your comment is well taken, although it is fair to point out that nylon purification also reduces the cyclic AMP response to high concentrations of PHA, whereas an increase in radioactive thymidine uptake with PHA still occurs.

DR. HIRSCHHORN: One thing worries me technically. What is the amount of contamination of the purified lymphocytes with platelets? I know it is fairly high for cell preparations prepared with Hypaque-Ficoll gradients even following three washes at 125 g × 10 min.

DR. PARKER: We routinely do two low-speed centrifugations (90 to 100 g x 7 min.) to remove platelets, and our preparations contain very few platelets. If the low-speed centrifugations are omitted, large numbers of platelets are present.

References

1. Davis, R. H. "Metabolite Distribution in Cells," *Science* **178**:835–840 (1972).
2. Eisen, S. A., Wedner, H. J., and Parker, C. W. "Isolation of Pure Human Peripheral Blood T Lymphocytes Using Nylon Wool Columns," *Immunological Communications* (in press).
3. Haber, J., Rosenau, W., and Goldberg, M. "Separate Factors in Phytohemagglutinin Induced Lymphotoxin, Interferon, and Nucleic Acid Synthesis, *Nature (New Biol.)* **238**:61–62 (1972).
4. Hall, M. R., Meinke, W., Goldstein, D. A., and Lerner, R. A. "Synthesis of Cytoplasmic Membrane-Associated DNA in Lymphocyte Nucleus," *Nature (New Biol.)* **234**:227–229 (1971).
5. Ishizuka, M., Gafini, M., and Braun, W. "Cyclic AMP Effects on Antibody Formation and Their Similarities to Hormone-Mediated Events," *Proc. Soc. Exp. Biol. Med.* **134**:963–967 (1970).
6. Parker, C. W. "Cyclic AMP." In *Inflammation: Mechanisms and Controls,* I. H. Lepow and P. A. Ward, eds. New York: Academic Press, pp. 239–259 (1972).
7. Robison, G. A., Butcher, R. W., and Sutherland, E. W. *Cyclic AMP.* New York: Academic Press, pp. 36–141 (1971).
8. Ryan, W. L., and Heidrick, M. L. "Inhibition of Cell Growth in Vitro by Adenosine-3′,5′-Monophosphate," *Science* **162**:1484–1485 (1968).
9. Smith, J. W., Steiner, A. L., and Parker, C. W. "Effects of Cyclic and Noncyclic Nucleotides on Stimulation by Phytohemagglutinin," *Fed. Proc.* **28**:566 (1969).
10. Smith, J. W., Steiner, A. L., Newberry, W. M., and Parker, C. W. "Cyclic Nucleotide Inhibition of Lymphocyte Transformation," *Clin. Res.* **17**:549 (1969).
11. Smith, J. W., Steiner, A. L., Newberry, W. M., and Parker , C. W. "Cyclic AMP in Human Lymphocytes. Alterations Following Phytohemagglutinin Stimulation," *Fed. Proc.* **29**:369 (1970).

12. Smith, J. W., Steiner, A. L., Newberry, W. M., and Parker, C. W. "Cyclic Adenosine-3',5'-Monophosphate in Human Lymphocytes. Alterations After PHA Stimulation," *J. Clin. Invest.* **50**:432–441 (1971).

13. Smith, J. W., Steiner, A. L., and Parker, C. W. "Human Lymphocyte Metabolism. Effects of Cyclic and Noncyclic Nucleotides on Stimulation by Phytohemagglutinin," *J. Clin. Invest.* **50**:441–448 (1971).

14. Tarnvik, A. "A Role for Red Cells in Phytohemagglutinin-Induced Lymphocyte Stimulation," *Acta Pathol. Microbiol. Scand. Section B.* **78**:733–740 (1970).

15. Wedner, H. J., Bloom, F. E., and Parker, C. W. "Compartmentalization of Cyclic AMP in Human Lymphocytes," *Clin. Res.* **20**:798 (1972).

16. Yachnin, S. "The Potentiation and Inhibition by Autologous Red Cells and Platelets of Human Lymphocyte Transformation Induced by Pokeweed Mitogen, Concanavalin A, Mercuric Chloride, Antigen and Mixed Leukocyte Culture," *Clin. and Exp. Immunol.* **11**:109–124 (1972).

THE EFFECT OF EXOGENOUS NUCLEOTIDES ON THE RESPONSE OF LYMPHOCYTES TO PHA, PWM, AND CON A

ROCHELLE HIRSCHHORN

I. Introduction

Since the discovery by Nowell in 1960 that phytohemagglutinin (PHA), an extract of the kidney bean, can stimulate peripheral blood leucocytes to enlarge and divide, and the demonstration in 1961 (Hastings et al., 1961; Carstairs, 1961) that the stimulated cells are lymphocytes, this cell culture system has been the basis for numerous experimental studies in immunology, cell biology, and genetics.

Lymphocyte stimulation can be achieved by a variety of stimulants. They include plant extracts, the principal ones being PHA (Nowell, 1960), pokeweed mitogen (PWM) (Farnes et al., 1964; Borjeson et al., 1966), Concanavalin A from the Jack bean (Con A) (Douglas et al., 1969; Powell and Leon, 1970), and those derived from numerous other less well-studied species (Douglas et al., 1969; Barker and Farnes, 1967); antilymphocyte sera (Grasbeck, et al., 1963; Holt et al., 1966); a streptococcal mitogen associated with but different from streptolysin S (SLSM) (Hirschhorn et al., 1964; Taranta et al., 1969); staphylococcal extracts, such as the exotoxin (Ling and Husband, 1964); Hg^{2+}(Schoepf et al., 1967) and Zn^{2+}(Ruhl et al., 1971); periodate (Novogrodsky and Katchalski, 1971); enzymes (Mazzei et al., 1966); antigen-antibody complexes (Bloch-Shtacher et al., 1968; Moller, 1969), and various mitogenic factors derived from cultured cells (Dumonde et al., 1969). These represent the majority of the currently known class of stimulants referred to as nonspecific, due to their ability to stimulate lymphocytes from any normal individual.

Another class of stimulants are called specific, because they act only on lymphocytes derived from individuals who demonstrate an immune response against the substance used. In most cases, this response is one of delayed hypersensitivity (Oppenheim, 1968).

This work was supported in part by NIH Grant #AI-10343.

Along with the antiallotypic antibodies, another type of stimulant not fitting the definition of specific or nonspecific is the allogeneic lymphocyte (Bach and Hirschhorn, 1964; Bain et al., 1964). Co-cultivation of lymphocytes, referred to as mixed lymphocyte culture (MLC), from genetically unrelated individuals results in cross-stimulation, referred to as the mixed lymphocyte response (MLR). Several alterations are common to the stimulation of lymphocytes by all of these agents. The cells enlarge; the nuclear material becomes euchromatic or less dense; the cytoplasm increases; and the cells divide. Synthesis of macromolecules increases markedly. With the nonspecific stimulants, an increase in RNA synthesis can be detected within 30 min and increases logarithmically. An increase in protein synthesis can be detected easily by 4 hr, and this continues to increase linearly for at least 48 hr DNA synthesis begins by 24 hr. after PHA exposure and reaches a peak between 48 and 72 hr. The peak of mitotic activity occurs about one day later. There is considerable asynchrony in DNA synthesis and mitosis as well as a lag prior to entry into the first S phase. For this reason, the cells have been considered to be in "G_0" (reviewed by Hirschhorn and Hirschhorn, 1973).

The experiments to be described all utilize the nonspecific mitogens as a model system to examine possible mechanisms of this stimulation. It may safely be assumed that activation by all of the mitogens must follow interaction of the agent with the lymphocyte cell membrane as a first step. Some of the early changes reported in activated lymphocytes are those relating to the cell membrane. Several studies have demonstrated changes in phospholipid metabolism early in the first hour after exposure to PHA. The major change after PHA appears to be increased incorporation of phosphate into phosphatidyl inositol (Fisher and Mueller, 1968). Later there is also an increased synthesis and accumulation of all the phospholipid fractions (Fisher and Mueller, 1969). A two- to threefold increase in the turnover of the fatty acid moiety of lecithin is seen within 1 hr of PHA stimulation and is initially restricted to plasma membrane (Resch et al., 1972). There is also increased incorporation of glucosamine into membrane glycoprotein (Hayden et al., 1970).

Increase of molecular transport across lymphocyte membranes by many different mechanisms has been demonstrated to occur within minutes after exposure to PHA. For example α-amino isobutyric acid (Mendelsohn et al., 1971; vandenBerg and Betel, 1971), 3-0 methyl glucose (Peters and Hausen, 1971), uridine (Peters et al., 1971), and other molecules (Robineaux et al., 1969) are taken up in greater amounts. Sodium and potassium ion exchanges increase, probably by means of increased activity of ouabain-sensitive ATPase (Quastel and Kaplan, 1968; Quastel and Kaplan, 1971). Calcium influx increases (Allwood et al., 1971; Whitney and Sutherland, 1972), probably mediated by its ATPase. Perhaps most important is the immediate increase in respiration-dependent pinocytosis, which we originally demonstrated with neutral red (Hirschhorn et al., 1968; Robineaux et at., 1969). The increased formation of pinocytotic vacuoles has been seen in electron micrographs (Inman and Cooper, 1963).

It therefore appears clear that there are numerous and important changes at the membrane very early in lymphocyte stimulation, which could potentially be responsible for the intracellular changes associated with activation.

In addition to the changes in the cell membrane, there appears to be a rearrangement of intracellular membranes such as those of lysosomes in that lysosomal membranes are more fragile and/or more permeable. Following homogenization of stimulated lymphocytes, an increased portion of lysosomal enzymes are found free in the supernatant fraction. Thus, within 2 to 4 hr. after stimulation, there appears to be a shift of lysosomal enzymes from the granular or lysosome-rich fraction to the free cell sap. Nonlysosomal enzymes do not show this alteration (Hirschhorn et al., 1968).

Among the earliest changes reported in stimulated lymphocytes are the acetylation of histones (Pogo et al., 1966) and the increased phosphorylation of nuclear proteins (Kleinsmith et al., 1966), including histones. These changes may be related to the observed increase of binding to the nucleus of acridine orange (Killander and Rigler, 1965) and actinomycin D (Darzynkiewicz et al., 1969).

Within 2 hr, there is an increased capacity of the DNA to act as template for RNA synthesis (Pogo, 1972; Hirschhorn et al., 1969). In addition, there is an increase in the activity of endogenous RNA polymerase I and II (Handmaker and Gratt, 1970; Pogo, 1972) in the nuclei of lymphocytes within a few hours after exposure to PHA.

In summary, our own work and that of others has demonstrated (1) alterations at the cell membrane, (2) alterations in translocation phenomena, possibly mediated by microtubules, (3) changes in transcription, (4) increased phosphorylation, and (5) lastly alteration in microtubular proteins with the formation and dissolution of the mitotic spindle.

All of these alterations in the stimulated lymphocyte are also seen in several other systems, where they are known to be mediated by changes in the levels of cyclic nucleotides. Thus, our own work and that of others, therefore, suggested that cyclic nucleotides might well modulate several parameters of lymphocyte stimulation. In addition, MacManus et al. (1971) have demonstrated that a population of thymic lymphocytes exists that responds to low doses of cyclic AMP by rapidly entering S phase or the DNA synthetic phase and followed by mitosis. These cells represent maximally 20% of thymic lymphocytes, and their relation to the circulating T-dependent, PHA-responsive lymphocyte is not clear.

We have therefore examined the effect of exogenous cyclic

nucleotide as well as agents presumed to affect the level of endogenous nucleotides upon stimulated and unstimulated cultured human peripheral blood lymphocytes.

When added 15 min prior to PHA, cyclic AMP, dcAMP, and theophylline each inhibited PHA-induced incorporation of ^{14}C-thymidine into DNA. Inhibition was evident when rates of DNA synthesis were determined both 24 to 48 hr and 48 to 64 hr after addition of PHA and the tested compound. Dibutyryl cAMP was the most effective compound inhibiting significantly between 10^{-4} to 10^{-5} M; theophylline was next, inhibiting between 10^{-3} to 10^{-4} M, and cyclic AMP was the least effective, requiring concentrations of 3.33×10^{-3} M for 50% inhibition.

To evaluate if these compounds were simply cytotoxic, cells were incubated with varying concentrations of cAMP as well as with no additive for 24 hr, washed and reincubated with PHA for another 24 hr ^{14}C-uridine was then added as a measure of RNA synthesis. It can be seen (Table 1) that there is no inhibition of uridine incorporation by preincubation with 3.3×10^{-3} M cAMP. This same concentration of cAMP is inhibitory for RNA synthesis when present simultaneously with PHA.

Cyclic GMP was even more effective than cAMP in inhibiting the increase in DNA synthesis induced by PHA, producing total inhibition at a concentration of 7×10^{-4} M.

Table 1. Effect of Preincubation with cAMP on Incorporation of ^{14}C-uridine By PHA-Stimulated Lymphocytes[a]

Treatment	cAMP	CPM ^{14}C-uridine
cAMP 0 to 24 hr	6.6×10^{-3} M	4636 ± 595
+ PHA 24 to 48 hr	3.3×10^{-3} M	4782 ± 487
PHA 24 to 48 hr	0	4874 ± 699
cAMP 24 to 48 hr	6.6×10^{-3} M	3824 ± 594
+ PHA 24 to 48 hr	3.3×10^{-3} M	4345 ± 357

[a]Lymphocytes were preincubated with cAMP, washed and incubated with PHA for 24 hr, and the incorporation of ^{14}C-uridine into RNA measured. Lymphocytes were also incubated simultaneously with cAMP and PHA to demonstrate inhibition of ^{14}C-uridine incorporation.

In order to evaluate the specificity of inhibition, adenosine, 5'-AMP, ATP, and ADP were also tested. Each of these compounds was effective in inhibiting PHA-induced increments in DNA synthesis at concentrations between 5×10^{-4} to 5×10^{-5} M. Not only were these compounds inhibitory, but ADP, ATP, and adenosine all inhibited protein synthesis to a greater extent than did cAMP (Table 2).

In contrast, in the absence of mitogens, exogenous cyclic AMP in our hands and that of others (Hirschhorn et al., 1970; Averner et al., 1972; Smith et al., 1971) has a biphasic effect. Although at high concentrations inhibition of macromolecular synthesis is seen, as with mitogens, lower

Table 2. Effect of Adenine Derivatives on Protein
Synthesis By Stimulated Lymphocytes[a]

Treatment	% of PHA response
PHA	100
PHA + cyclic AMP	81
PHA + ADP	70
PHA + ATP	35
PHA + adenosine	28
Control	24

[a]Peripheral blood lymphocytes were cultured for 28 hr with
PHA and the indicated compounds at concentrations of 3.3
× 10^{-3} M. ^{14}C-leucine was added during the last 4 hr of
culture, and TCA precipitable counts per minute were
determined. Results are expressed as a percentage of the
response seen with PHA alone.

concentrations of cyclic AMP causes a small degree of stimulation of
uridine and thymidine incorporation, two parameters used as measures of
lymphocyte transformation. Under the same conditions, we have been
unable to demonstrate a stimulation of uridine or thymidine incorporation by addition of cyclic GMP to cultures of unstimulated lymphocytes
in concentrations from 10^{-3} M to 10^{-11} M or by dibutyryl cyclic AMP.

The divergent response to cAMP seen in the presence and absence of
mitogens could reflect the existence of two populations of lymphocytes
in the peripheral blood—a large population, which responds to pHA and
whose response to the mitogen is inhibited by cAMP, and a much smaller,
minor population, which is stimulated by cAMP. The B-dependent
lymphocyte would fit this category. We therefore examined the effect of
cAMP upon stimulation by PWM and Con A as well as by PHA, since
there is evidence that these mitogens stimulate different, if overlapping,
populations. High doses of cAMP were found to inhibit PHA, PWM, and
Con A (Fig. 1) stimulation of lymphocytes to the same extent. This would
suggest that all cell populations responding to these different mitogens
are affected in the same manner. The response of unstimulated purified
B and T cell populations remains to be determined.

An alteration in the enzymes of nucleotide metabolism during
lymphocyte stimulation could also account for differential response of
stimulated versus unstimulated lymphocytes to exogenous nucleotides.
Elevation in the concentration of adenosine totally inhibited the increment in protein synthesis induced by PHA. Since it has been reported
that exogenous cAMP and noncyclic nucleotides are all converted to
adenosine, we have looked for alterations in one of the enzymes of the
degradative pathway for adenosine, adenosine deaminase. Two forms of
adenosine deaminase can be distinguished in circulating lymphocytes
using gel electrophoresis. One form is polymorphic and thus can be identified as corresponding to that seen in red blood cells (Spencer et al.,

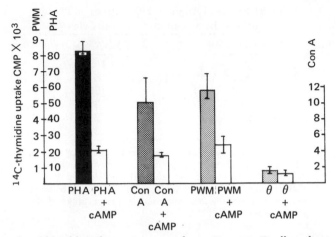

Fig. 1. Peripheral blood lymphocytes separated on a Hypaque-Ficoll gradient were incubated in tissue culture media with PHA (MR-68, 5 μg/ml), Concanavalin A (16 μg/ml), or pokeweed mitogen (0.01 ml/ml). Cyclic AMP at 2×10^{-3} M was added at 0 time, and the incorporation of ^{14}C-thymidine measured from 25 to 42 hr.   Different scales are indicated for the three mitogens.

1968). The other form is seen as a single band of activity traveling less anodally than the "red cell" form. On stimulation of lymphocytes by PHA, the lymphocyte adenosine deaminase becomes almost completely undetectable. This is in stirking contrast to what occurs for numerous other enzymes, such as 6PGD, G6PD, ICD, acid phosphatase, etc., all of which increase with stimulation. Another enzyme of adenine metabolism, adenylate kinase (converting ADP to AMP and ATP), is also increased in PHA-stimulated lymphocytes (Hirschhorn, 1972).

II. Discussion

High concentrations of exogenous cyclic nucleotides (10^{-3} M) and/or agents that presumably raise the level of cAMP inhibit the response of lymphocytes to mitogenic agents. This inhibition is most marked when DNA synthesis is the parameter used to measure lymphocyte stimulation. Additionally, other adenine-containing nucleotides and nucleosides are also inhibitory. This inhibition by *noncyclic* adenine-containing nucleotides and nucleosides and most potently by adenosine would suggest that the inhibition of macromolecular synthesis by cAMP is not specific for the cyclic compound (Robison et al., 1972). However, addition of agents that act to raise intracellular levels of cAMP also has been found to cause inhibition of lymphocyte transformation (Hirschhorn et al., 1970; Smith et al., 1971). Additionally, in some systems, exogenous noncyclic nucleotides have been found to cause elevation of the cyclic nucleotide (Sattin and Rall, 1970).

Therefore, it is still conceivable that *cyclic* nucleotides play a specific role in mitogen activation of lymphocytes and that very high levels of cAMP inhibit the second step in the chain of lymphocyte transformation following attachment of the mitogen to the receptor.

There is evidence for several other sites of action of cAMP in control of growth and division, which must be considered in interpreting the data. Firstly, cyclic AMP could primarily act at the level of the microtubules of the mitotic apparatus with failure of spindle formation and arrest just prior to mitosis. In support of this, cAMP has been found to inhibit the appearance of mitotic figures in synchronized lymphoblastoid lines when added at the beginning of the S or DNA synthetic phase, without at all affecting DNA synthesis (Millis et al., 1972). Secondly, cAMP could act upon the translocation of metabolites necessary for growth. Thirdly, cAMP could be acting in a trivial fashion such as diminishing the activity of thymidine kinase (the enzyme necessary for the measurement of DNA synthesis using the alternate pathway by which thymidine is incorporated) or transport of the nucleosides, as has been found in Chinese hamster cells (Hauschka et al., 1972).

The small degree of stimulation of macromolecular synthesis (particularly RNA) caused by addition of cAMP to nontransformed lymphocytes might suggest that cAMP is the second messenger for the stimulation of at least a population of nondividing lymphocytes to grow and divide. Thus, stimulation occurs at a critical intracellular concentration of cAMP and inhibition at a higher concentration. This is a tempting hypothesis, since it would provide a form of feedback control for the cellular proliferative response at delayed hypersensitivity. However, it is entirely possible that cyclic nucleotides act to stimulate an enzyme or enzymes involved in macromolecular synthesis in a discrete population of lymphocytes, rather than affecting mitogen action on T-dependent lymphocytes. For example, cAMP could be stimulating transcription by RNA polymerase as it does in prokaryotes (DeCrombrugghe et al., 1971).

In summary, the observed effects of exogenous cyclic and noncyclic nucleotides upon mitogen-stimulated and nonstimulated peripheral blood lymphocytes do not conclusively support the simple hypothesis that cAMP is a "second messenger" for the population of lymphocytes that respond to mitogens such as PHA. Changes in the levels of cyclic nucleotides do definitely modulate the rates of macromolecular synthesis in stimulated and nonstimulated lymphocytes by mechanisms that remain to be clarified.

Acknowledgment

My interest in cyclic AMP was first stimulated by Dr. Gerald Weissmann, with whom some of the work reported previously was done. I should also like to thank Professor Harry Harris of University College, London, in whose laboratory some of the enzyme studies were done, and Dr. D. A. Hopkinson and Dr. D. Swallow for their stimulating discussion and advice.

Discussion of Dr. Hirschhorn's Paper

DR. BLOOM: How long do you expect cyclic AMP to last?

DR. HIRSCHHORN: Not very long. I would guess that whatever it is doing, it does at the beginning. The reason for measuring it so late is that the system is highly artificial. You are adding a stimulus and expecting it to get into the cell at a particular concentration. At an early time, you expect that only 1 or 2% of the cells might be triggered. Indeed, it is possible that only 1 to 2% of the cells are sensitive. Measurements are made late with the expectation that the response will be amplified at that time.

DR. BOURNE: I wonder if you could comment on the relevance of adding exogenous nucleotides, as compared with stimulating enzyme production of endogenous nucleotides, for understanding lymphocyte responses. Lipolysis, for example, is inhibited in some circumstances by added cyclic AMP in a fashion that may have nothing whatever to do with cyclic AMP's role inside the cell.

DR. HIRSCHHORN: I don't know if you can classify them as artifacts. What you can say is that you don't know the relation between the amounts of exogenous cyclic nucleotides and their intracellular concentrations and that the addition of cyclic AMP may do more than increase the level of cyclic AMP in the cell. However, in several situations other than lipolysis such as in cell growth, the effects of adding exogenous nucleotides correlate well with the effects of agents altering endogenous cyclic nucleotide levels. Ideally, in all experiments, the actual intracellular concentrations of cyclic nucleotides should be known.

References

Allwood, G., Asherson, G. L., Davey, M. J., and Goodlord, P. J. *Immunol.* **21**:509 (1971).

Averner, M. J., Brock, M. L., and Jost, J.-P. *J. Biol. Chem.* **247**:413 (1972).

Bach, F. H., and Hirschhorn, K. *Sci.* **143**:813 (1964).

Bain, B., Vas, M. R., and Lowenstein, L. *Blood* **23**:108 (1964).

Barker, B. E., and Farnes, P. *Nature* **215**:659 (1967).

Bloch-Shtacher, N., Hirschhorn, K., and Uhr, J. W. *Clin. Exp. Immunol.* **3**:889 (1968).

Borjeson, J., Reisfeld, R., Chessin, L. N., Welsh, P. D., and Douglas, S. D. *J. Exp. Med.* **124**:859 (1966).

Carstains, K. *Lancet* ii:984 (1961).

Darzynkiewicz, Z., Bolund, L., and Ringertz, N. R. *Exp. Cell Res.* **56**:418 (1969).

DeCrombrugghe, B., Chen, B., Anderson, W. B., Gottesman, M. E., Perlman, R. L., and Paston, I. *J. Biol. Chem.* **246**:7343 (1971).

Douglas, S. D., Kamin, R., and Fudenberg, H. H. *J. Immunol.* **103**:1185 (1969).

Dumonde, D. C., Wolstencroft, R. A., Panayi, G. S., Matthew, M., Morley, J., and Howson, W. T. *Nature* **224**:38 (1969).

Farnes, P., Barker, B. E., Brownhill, L. E., and Fanger, H. *Lancet* ii:1100 (1964).

Fisher, D. B., and Mueller, G. C. *Proc. Nat. Acad. Sci.* **60**:1396 (1968).

Grasbeck, R., Nordman, C. T., and DeLaChapelle, A. *Lancet* ii: 385 (1963).

Handmaker, S. W., and Gratt, J. W. *Biochim. Biophys. Acta.* 199:95 (1970).

Hastings, J., Freedman, S., Rendon, O., Cooper, H. L., and Hirschhorn, K. *Nature* 192:1214 (1961).

Hauschka, P. V., Everhard, L. P., and Rubin, R. W. *Proc. Nat. Acad. Sci.* 69:3542 (1972).

Hayden, G. A., Crowley, G. M., and Jamieson, G. A. *J. Biol. Chem.* 245:5827 (1970).

Hirschhorn, K., Schriebman, R. R., Verbo, S., and Gruskin, R. H. *Proc. Nat. Acad. Sci.* 52:1151 (1964).

Hirschhorn, K., and Hirschhorn, R. In *Mechanisms of Cellular Immunity* (in press).

Hirschhorn, R. (unpublished observations, 1972).

Hirschhorn, R., Brittinger, G., Weissmann, G., and Hirschhorn, K. *J. Cell Biol.* 37:412 (1968).

Hirschhorn, R., Troll, W., Brittinger, G., and Weissmann, G. *Nature* 222:1247 (1969).

Hirschhorn, R., Grossman, J., and Weissmann, G. *Proc. Soc. Exp. Biol. Med.* 133:1361 (1970).

Holt, L. J., Ling, M. R., and Stanworth, D. S. *Immunochem.* 3:359 (1966).

Inman, D. R., and Cooper, E. H. *J. Cell Biol.* 19:441 (1963).

Killander, D., and Rigler, R. *Exp. Cell Res.* 39:701 (1965).

Kleinsmith, L. J., Allfrey, V. G., and Mirsky, A. E. *Science* 154:780 (1966).

Ling, N. R., and Husband, E. M. *Lancet* i:363 (1964).

MacManus, J. P., Whitfield, J. F., and Youdale, T. *J. Cell Physiol.* 77:103 (1971).

Mazzei, D., Novi, C., and Bazzi, C. *Lancet* ii:802 (1966).

Mendelsohn, J., Skinner, A. S., and Kornfeld, S. *J. Clin. Invest.* 50: 818 (1971).

Millis, A. J. T., Forrest, G., and Pious, D. A. *Biochem. Biophys. Res. Commun.* 49:1645 (1972).

Moller, G. *Clin. Exp. Immunol.* 4:65 (1969).

Novogrodsky, A., and Katchalski, E. *FEBS Letters* 12:297 (1971).

Nowell, P. C. *Cancer Res.* 20:462 (1960).

Oppenheim, J. J. *Fed. Proc.* 27:21 (1968).

Pogo, B. G. T. *J. Cell Biol.* 53:635 (1972).

Pogo, B. G. T., Allfrey, V. G., and Mirsky, A. E. *Proc. Nat. Acad. Sci.* 55:805 (1966).

Peters, J. H., and Hausen, P. *Eur. J. Biochem.* 19:509 (1971).

Powell, A. E., and Leon, M. A. *Exp. Cell Res.* 62:315 (1970).

Quastel, M. R., and Kaplan, J. G. *Nature* 219:200 (1968).

Quastel, M. R., and Kaplan, J. G. *Exp. Cell. Res.* 63:230 (1971).

Resch, K., Ferber, E., and Gelfand, E. W. Seventh Leucocyte Culture Conference (1972).

Robineaux, R., Buna, C., Anteunis, A., and Orme-Roselli, L. *Ann. Inst. Pasteur* 117:790 (1969).

Robison, G. A., Butcher, R. W., and Sutherland, E. W. In *Cyclic AMP.* New York: Academic Press, pp. 97–106 (1971).

Ruhl, H., Kirchner, H., and Bochert, G. *Proc. Soc. Exp. Biol. Med.* 137:1089 (1971).

Sattin, A., and Rall, T. W. *Mol. Pharmacol.* 6:13 (1970).

Schoepf, E., Schultz, K. H., and Gromm, M. *Naturwissenschaften* **5**:568 (1967).
Spencer, N., Hopkinson, D. A., and Harris, H. *Ann. Human Genet.* **32**:9 (1968).
Taranta, A., Cuppari, G., and Quagliata, F. *J. Exp. Med.* **129**:605 (1969).
vandenBerg, K. J., and Betel, I. *Exp. Cell Res.* **66**:257 (1971).
Whitney, R. B., and Sutherland, R. M. Seventh Leucocyte Culture Conference (1972).

CYCLIC AMP IN THE ACTIVATION OF HUMAN PERIPHERAL BLOOD LYMPHOCYTES IN IMMUNOLOGICALLY DEFICIENT PATIENTS AND IN HUMAN LYMPHOID CELL LINES

DAVID R. WEBB, JR., DANIEL P. STITES, JANICE PERLMAN, KATHRYN E. AUSTIN, and II. HUGH FUDENBERG

1. Activation of Peripheral Blood Lymphocytes by Plant Mitogens

A. Studies with PHA

It has been established that cAMP plays a pivotal role in the critical, early, biochemical events that initiate lymphocyte activation. However, it has remained unclear to what extent cAMP exerts its effects on subsequent biochemical events. On the one hand, work carried out by the late Werner Braun and his colleagues (Braun and Ishizuka, 1971; Ishizuka et al., 1971) has established that cAMP or compounds that alter intracellular cAMP levels in lymphocytes may enhance both in vivo and in vitro immune responses in mice. Several authors have reported increases in DNA synthesis (Hirschhorn et al., 1970), RNA synthesis (Duerner et al., 1972), accumulation of ATP (Cross and Ord, 1971), and phosphorylation of histones (Klein and Makman, 1972) in response to the exogenous administration of cAMP or its derivatives to lymphocyte cultures. On the other hand, reports of workers using phytomitogen-induced lymphocyte activation have consistently demonstrated suppression of macromolecular synthesis in the presence of exogenously administered cAMP or related compounds (Smith et al., 1971a). This occurs despite a demonstrated elevation of intracellular cAMP levels in lymphocytes following exposure to PHA (Smith et al., 1971b). Additionally, cAMP has been demonstrated to inhibit cytotoxic effects, for example, as well as blocking the release of effector substances in allergic responses (Bourne et al., 1972).

In an attempt to clarify these differences, we undertook a reinvestigation of the effects of phytomitogens and cAMP on lymphocyte activation with particular emphasis on DNA, RNA, and protein synthesis. This effort was necessary, as the ultimate goal of our investigations was to ascertain the possible role of cAMP in the activation of lymphocytes from immunologically deficient patients.

It was noted that virtually all previous reports in which PHA and cAMP had been studied involved the use of a single, presumably optimal, stimulating dose of mitogen. Goran Moller (1970) has suggested that the activation of lymphocytes is based on each cell reaching a certain threshold of membrane "hits" or signals (i.e., lymphocyte counting), and we therefore hypothesized that alterations in intracellular cAMP levels may be the mechanism by which the lymphocyte "counts" the number of membrane signals. From this we reasoned that a plausible explanation of the inhibition by cAMP of mitogen-induced lymphocyte transformation might be a hyperelevation of intracellular cAMP levels resulting in a shutting down of selected biochemical events, specifically DNA and RNA synthesis. According to this reasoning, if experiments were designed in which the concentrations of PHA were varied relative to the concentrations of exogenously added cAMP, then it should be possible to demonstrate cAMP-mediated enhancement of transformation at suboptimal doses of PHA, leading to increasing inhibition of transformation by cAMP at optimal and superoptimal doses of PHA.

Table 1 shows results of experiments in which human peripheral blood lymphocytes were exposed to varying doses of both PHA and cAMP for 72 hr, and the synthesis of DNA measured by incorporation of radioactive thymidine. Instead of stimulation or enhancement by cAMP occurring at suboptimal doses of PHA, inhibition was observed. At superoptimal doses of PHA, the addition of cyclic AMP results in enhancement of DNA synthesis. In Fig. 1, this dual effect may be seen more clearly as one dose of cyclic AMP 10^{-3} M as well as PHA alone are plotted. The high-dose recovery effect can clearly be seen. This type of experiment has now been performed over a dozen times always with the

Table 1. DNA Synthesis in Normal Human Peripheral Blood Lymphocytes 72 Hr After Exposure to PHA or PHA and cAMP[a]

cAMP (M)	PHA (μg/ml)							
	0	1	10	100	250	500	1000	2000
0	—	0.88	2.45[c]	32.29	33.81	18.88	8.14	2.38
10^{-3}	1.05	0.92[b]	0.57	0.70	1.70	1.71	1.68	1.07
10^{-4}	1.01	1.00	0.97	1.27	1.29	1.28	1.24	1.12
10^{-5}	1.49	0.96	1.39	0.83	1.13	0.98	1.16	1.06

DNA synthesis was measured by a 4-hr pulse of ^{3}H-thymidine from 68 to 72 hr. Each number represents the average of triplicate determinations.
[a]These values represent the ratio of mean counts per minute of experimental groups, divided by the appropriate control group.
[b]cAMP + PHA/PHA alone.
[c]PHA/no treatment.

same results, the only difference being a shift in the optimum dose of PHA over a range of 50 to 250 μg/ml. The experiment was repeated with measurement of RNA synthesis at 24 hr. Figure 2 shows the results of a typical experiment, which measured the effects of cAMP on PHA-induced RNA synthesis at 24 hr. Cyclic AMP again elicited high-dose recovery at superoptimal doses of PHA, whereas inhibition by cAMP was observed at suboptimal or optimal doses of PHA. Thus, it is possible to demonstrate a consistent enhancement of PHA-induced DNA or RNA synthesis by cyclic AMP only when large amounts (superoptimal doses) of PHA are used. There are several ways these data may be interpreted. For example, cyclic AMP could be involved in some sort of nutritive recovery effect, which has little to do with signal amplification, and to which only a certain portion of the lymphocyte population responds. This is unlikely, since at no time during the culture period could we detect any cytotoxicity as a result of exposure to high doses of PHA.

It is also possible that this effect is an artifact of the relatively impure PHA-M (Difco), which has been used for the majority of our experi-

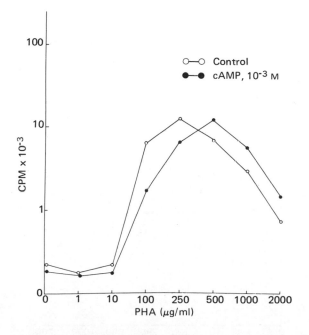

Fig. 1. The effects of PHA and cAMP on DNA synthesis in human peripheral blood lymphocytes measured at 72 hr. The lymphocytes were isolated by density centrifugation in Ficoell-Hypaque and cultured in RPMI1640 without serum. The culture system used was that described by Hartzman et al. (1972), in which the cells are grown in microtiter plates and then assayed using a multiple sample analyzer, which deposits the cells on cellulose filters. The cells are then analyzed for radioactivity by standard scintillation techniques. Each point represents the average of triplicate determinations.

ments. To test this, we obtained a highly purified preparation of PHA (kindly sent to us by Dr. Stanley Yachnin, University of Chicago), which has no hemagglutinating activity and low leukoagglutinating activity. Using this preparation, we observed the identical effects as were obtained with PHA-M (Webb, unpublished observation).

Another explanation of the results, and the one we favor, is that the population of cells enhanced by the addition of cAMP following exposure to high doses of PHA represents a discreet, unique population that responds in a positive manner (stimulation) to the continuous presence of cAMP, unlike the majority of lymphocytes in peripheral blood that respond negatively (inhibition) to the continuous presence of cyclic AMP. Thus, one may distinguish at least two populations on the basis of their response to PHA and exogenous cyclic AMP—one population being generally inhibited by cAMP and PHA, and the other being stimulated by cAMP and PHA.

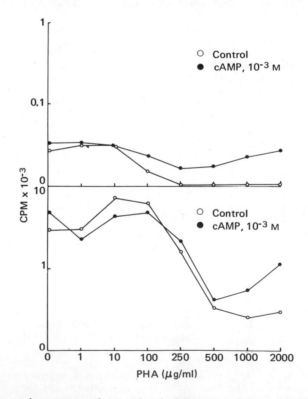

Fig. 2. RNA synthesis measured at 2 and 24 hr following exposure to PHA and cAMP. The top graph represents lymphocytes exposed to PHA alone at various doses (open circles) and PHA plus cAMP (closed circles) with RNA synthesis being measured 2 hr after the addition of the drugs. The lower graph shows the same experiment 24 hr after culture initiation. The culture system and assay are as in Fig. 1.

In terms of intracellular mechanisms, an explanation of these results is difficult. Measurement of intracellular cAMP levels at the different PHA doses was performed using the cyclic AMP binding protein assay of Gilman (1970) following exposure to PHA-M (Fig. 3). At a suboptimal PHA concentration (1 μg/ml), there is no increase in intracellular cAMP levels over the first 40 min following exposure to the mitogen; at the optimal concentration of PHA (100 μg/ml), the increase in cAMP levels is relatively small but significant; and if a superoptimal dose of PHA is used (1000 μg/ml), a very marked increase in intracellular cyclic AMP levels occurs. If a more purified preparation of PHA with lower leukoagglutinating activity is used (PHA-W, Burroughs-Wellcome), the optimal dose (10 μg/ml) is more stimulatory than the optimal dose of PHA-M, and superoptimal doses (50 μg/ml and 100 μg/ml) of PHA-W lead to very large increases in intracellular cAMP levels (see Fig. 4). If exogenous cyclic AMP is added at a concentration of 10^{-3} M to untreated or PHA-exposed lymphocytes, the intracellular concentration rises to greater than 100 pmoles cAMP per 10^6 cells and remains high over the entire early period following lymphocyte activation (1 hr). These data suggest that mechanistically as well, one can distinguish between populations of lymphocytes on the basis of their ability to respond to varying doses of PHA by the synthesis of intracellular cyclic AMP. It also supports the

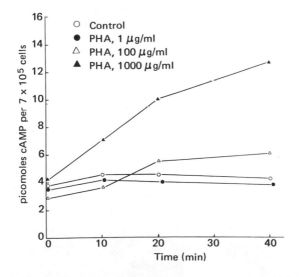

Fig. 3. Intracellular levels of cAMP following the addition of PHA-M. Cyclic AMP levels were measured using the technique of Gilman (1940). The lymphocytes were prepared from human peripheral blood by density-gradient centrifugation in Ficoll-Hypaque. The cells were maintained in Hanks BSS without serum supplement during the experiment at 37°C in 5% CO_2–95% air.

concept of the existence of a cell population that requires fairly large increases in intracellular cAMP levels in order for activation to occur.

Thus, it is possible that these cells responding in suboptimal to optimal doses of PHA produce relatively low amounts of cAMP compared to cells responding at superoptimal PHA concentrations, and the former cell population is inhibited by large increases in cAMP with regard to DNA and RNA synthesis. So it would appear that the cell population that responds to superoptimal doses of PHA and that is enhanced by the addition of cAMP may also be capable of responding to high doses of PHA by greatly increasing its intracellular concentration of cAMP as well.

While the idea of a discreet cell population is appealing, it must be pointed out that none of the experiments reported here clearly demonstates that such a discreet population exists. Indeed, at this point, we are unable to separate different populations on the basis of a PHA response resulting in DNA and RNA synthesis. Also, measurements of intracellular cAMP levels reflect the whole cell population as an average rather than discreet components. Thus, at this stage, the evidence for the absolute existence of a separate cell population that is enhanced by cAMP following exposure to PHA must remain circumstantial.

With these qualifications in mind, one may still ask the question if

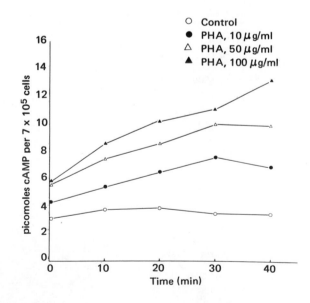

Fig. 4. Intracellular cAMP levels following the addition of PHA-W. Highly purified PHA (Burroughs-Wellcome, Lot K944) was used in this experiment, which was performed exactly as stated in Fig. 3.

discreet populations exist, what type of cell might they be (i.e., T or B cell) and how might they function? The two parts of the question are, of course, interrelated in that the identification of the cell type would, to a large extent, explain its function.

B. Activation of Lymphocytes from Immunologically Deficient Patients

With reference to the high-dose recovery effect, investigations were begun using peripheral blood lymphocytes from patients with immune deficiency. The two diseases studied to date are adult acquired hypogammaglobulinemia (AAH) and chronic lymphocytic leukemia (CLL). The rationale for studying the AAH is that there is some evidence of abnormal mitogen responses, which indicate a possible inactivation dysfunction (Douglas et al., 1968). CLL was chosen since these cells have greatly lowered mitogen responses, the principal circulating cell is predominantly a "B" type cell, and Johnson and Abell (1970) had already reported that cyclic nucleotides had some effect on these cells. Using CLL cells, we hoped to gain insight into the nature of the cell type involved in the high-dose enhancement we observed with normal lymphocytes.

While it is not the purpose of this report to dwell on the nature of the two above-mentioned diseases, a brief comment about the nature of each disease is in order. AAH is usually characterized as a lymphoid dysfunction of delayed onset characterized by lowered serum levels of all serum immunoglobulins as well as a predisposition toward repeated bacterial infections primarily involving the sinopulmonary system, variable absence of delayed hypersensitivity, and any number of associated disorders including lymphoproliferative malignancies (Barth et al., 1965). It had previously been reported by others working in our laboratory (Douglas et al., 1968) that these patients had lowered PHA responses at 72 hr after culture initiation. Using the same patients, the lymphocytes were isolated and tested using the PHA-cAMP checkerboard model. The results of one such experiment showing PHA treatment alone and exposure to cAMP (10^{-3} M) plus PHA are shown in Fig. 5. The results indicate that such patients, in our hands, give responses as measured by DNA and RNA synthesis which appear normal, and further, they possess the cells involved in the normal high-dose recovery elicited by exogenously administered cAMP. Four patients have been tested using this experimental design, and all have responded similarly. One exception is D.L. who, in addition to having AAH, also has a malignancy (metastatic melanoma). While his PHA dose-response curve fell at the low end of the normal range, we were unable to demonstate the high-dose enhancement effect. Thus, insofar as measurements of DNA and RNA synthesis in our laboratory are concerned, patients with AAH appear to respond normally to PHA in terms of DNA and RNA synthesis. In addition, these patients exhibit the cAMP-mediated high-dose enhancement of DNA and RNA synthesis. Concerning reports that suggest that AAH patients respond poorly at 72 hr after PHA addition, it may only be said that the culture techniques and assay system used are substantially different and may account for the dissimilarity of the results.

CLL is better characterized in terms of the principal cell type involved. This disease, which usually occurs in older people, is characterized by a marked increase in the number of small lymphocytes in peripheral blood, enlarged spleen and lymph nodes, and lymphocyte infiltration of the marrow (Silver, 1969). In addition, these patients have a greatly reduced response to mitogens such as PHA. The cell type involved has been characterized by Grey et al. (1971) as monoclonal, i.e., the cells from a single individual having a single Ig class (e.g., kappa IgM). This suggests that the principal cell type could be termed a B cell rather than a T cell. These patients were studied mainly because of the fact that the cell type involved is known, and thus the results should shed additional light on the high-dose recovery effect. The results of a typical experiment are shown in Fig. 6, in which the PHA-cAMP checkerboard experimental design was again used. The figure shows DNA synthesis at 72

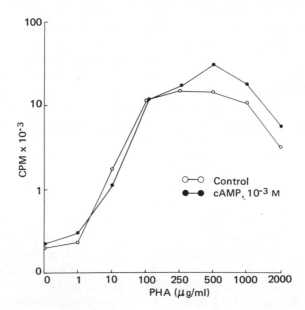

Fig. 5. DNA synthesis in lymphocytes isolated from a patient with adult acquired hypogammaglobulinemia 72 hr after the addition of PHA (open circles) or PHA plus cAMP (closed circles). The assay procedure and culture system were previously described in Fig. 1.

hr after treatment with PHA or with both cAMP (10^{-3}M) and PHA. The most striking finding is the total absence of the cAMP-mediated high-dose recovery, in addition to the generally reduced response to PHA. Preliminary results measuring 24-hr RNA synthesis, however, suggest that high-dose recovery does occur at least with RNA synthesis. We are, at present, unable to explain the difference observed between DNA and RNA synthesis. A possible explanation is that the cell population that responds to cAMP-mediated high-dose recovery in normals is incapable of completing the entire activation pathway in CLL.

C. Studies with Concanavalin A and Pokeweed Mitogen

In addition to the studies performed using PHA, parallel studies have been performed using Concanavalin A (Con A) and pokeweed mitogen (PWM) in experiments designed in the same way as those employing PHA. These experiments have been consistent in only one respect—their uniform failure to yield anything other than equivocal results. That is to say, unlike PHA in which a pattern of inhibition-stimulation can be detected, the pattern of effects of cAMP on Con A or PWM-induced DNA synthesis is random, as may be seen from Table 2. There is some

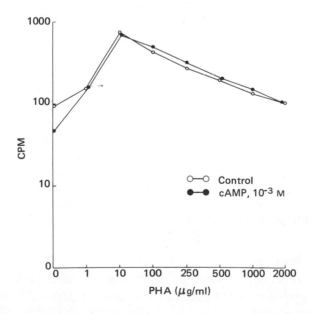

Fig. 6. DNA synthesis in lymphocytes isolated from a patient with chronic lymphocytic leukemia 72 hr after the addition of PHA (open circles) or PHA plus cAMP (closed circles).

Table 2. DNA Synthesis in Normal Human Peripheral Blood Lymphocytes
72 hr After Exposure to Concanavalin A
or Pokeweed Mitogen and Cyclic AMP[a]

cAMP	Con A (μ/ml)							
(M)	0	0.1	1	2.5	5	10	25	50
0	—	1.95[b]	13.93	4.23	1.33	0.63	0.34	0.25
10^{-4}	0.86[c]	0.68	0.94	0.85	0.96	1.15	0.98	1.15
10^{-5}	1.27	1.03	0.99	1.12	1.11	1.21	1.23	1.51
10^{-6}	0.87	0.89	0.94	0.83	0.78	1.14	0.81	1.27

	PWM(μg/ml)							
	0	0.001	0.1	10	25	50	100	250
0	—	1.20	2.67	6.77	7.92	7.96	6.39	6.96
10^{-3}	1.11	0.99	0.85	0.91	1.14	1.17	1.30	1.25
10^{-4}	1.47	1.03	1.26	1.01	1.03	0.98	0.99	1.03
10^{-5}	1.28	1.00	1.21	1.06	1.03	1.01	1.02	1.12

[a]These values represent ratios of mean CPM. The ratios are derived from triplicate determinations of ^{3}H-thymidine incorporation (in CPM in these cultures from 68 to 72 hr.
[b]Con A + cAMP/Con A.
[c]Con A/treated Groups.

tendency toward high-dose recovery of DNA synthesis by cAMP in PWM, but the data are not consistent from experiment to experiment.

One possible reason for this is illustrated in Fig. 7, which depicts intracellular cAMP levels following exposure to Con A or PWM. As can be seen, there is little, if any, stimulation of cAMP synthesis by PWM and none by Con A at doses that are optimal for DNA synthesis (where at least some effect may be seen with PHA).

Hadden et al. (1972) have shown that Con A and PHA may stimulate synthesis of cyclic GMP, and it may well be that the primary effect of Con A or PWM on cyclic nucleotide synthesis may be through changes in cyclic GMP levels. Indeed, although it is speculative at this point, it may well turn out that in activation of immunocompetent cells, modulation in the ratio of cyclic AMP to cyclic GMP is the critical early biochemical event that determines the subsequent biochemical pathways a cell will follow. Evidence presented by Strom et al. (1972) on the alteration of cytotoxic activity of lymphocytes by compounds effecting both cyclic AMP and GMP suggests that this may be the case.

D. Conclusions?

The question mark is intentional, as it should be obvious that the present results provide more questions than answers. The cAMP-mediated high-dose recovery effect observed in PHA-treated cultures still remains unresolved regarding the precise nature of the cell type involved and as to its possible biochemical significance in the process of lymphocyte activation.

It is clear that PHA can elevate intracellular cyclic AMP levels, but it is not clear why Con A and PWM do not. It is interesting to note that lymphocyte activation in AAH appears to be normal in terms of the magnitude of the response, the existence of the high-dose recovery effect, and the consistency of the effect on both RNA and DNA synthesis, although these patients are known to produce lower levels of immunoglobulins, suggesting that a defect exists somewhere along the activation pathway.

The response of the lymphocytes from patients with CLL is interesting in that no high-dose recovery may be elicited with DNA synthesis, while it may well occur at the level of RNA synthesis. This suggests that in some of these cells, there may be a defect in the early stages of the transformation process. Unfortunately, the results from these patients led to no insight into the nature of the cell type involved in the high-dose recovery effect.

With regard to cAMP effects in the absence of any phytomitogen, Table 3 is of interest. Several of the conference participants (MacManus and Whitfield, 1972; Rixon et al., 1970) have reported that cAMP or its derivatives such as dibutyryl cAMP have limited mitogenic capabilities.

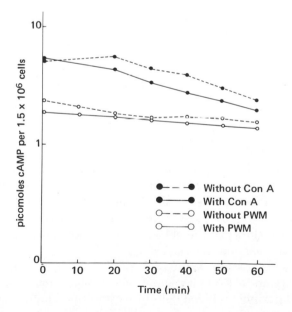

Fig. 7. Intracellular levels of cAMP in lymphocytes following exposure to either Concanavalin A (2.5 μg/ml) or pokeweed mitogen (25 μg/ml). The assay system used was the cAMP protein binding assay of Gilman.

In Table 3, we have compiled the results of a number of experiments involving measurement of DNA synthesis at 72 hr by pulse label with radiolabeled thymidine. The data are presented in terms of the number of times stimulation, suppression, or no effect was observed (as defined in the table) following exposure to different doses of cyclic AMP. From these data, there is a suggestion of a positive effect occurring at 10^{-3} M, no effect predominating at 10^{-4} M, and a tendency toward suppression at 10^{-5} M. Thus, one sees a spectrum of effects of cAMP on DNA synthesis supporting the concept of magnification of subsequent biochemical events resulting from modulations in cAMP levels. This is, of course, assuming that one is dealing with a uniform cell population, which is probably untrue.

It should be emphasized that in performing experiments involving cAMP-mediated events in lymphocytes, one must not lose sight of the fact that a variety of subpopulations may be present, and the end result may represent something like a rough average of effects rather than being indicative of what is occurring in the various cell populations.

Table 3. Effect of cAMP on DNA Synthesis[a]

cAMP (M)	Stimulation[b]	No effect[c]	Suppression[d]
10^{-3}	11	13	6
10^{-4}	2	15	7
10^{-5}	5	11	12

[a]Derived from experimental data in which DNA synthesis was measured three days after the addition of cyclic AMP by radioactive thymidine incorporation.
[b]Defined as any value greater than 1.2 times control values.
[c]Defined as any value between 0.90 to 1.2 times the control value.
[d]Defined as any value less than 0.90 times control values.

II. Lymphoblastoid Cell Lines

A. *General Characteristics*

Continuously replicating human lymphoid cell lines have been established in a number of different laboratories in recent years (Moore et al., 1966). The cells, of unidentified origin, from the peripheral blood, lymph nodes, or bone marrow are functionally and morphologically heterogenous. A majority of the cells retain a diploid chromosome number in culture (Huang and Moore, 1969).

The factors responsible for long-term growth of lymphoid cells in tissue culture are not entirely known. An increased, spontaneous in vitro proliferation of peripheral blood cells into established lines occurs in patients with a high percentage of circulating lymphocytes showing increased DNA synthesis (Glade et al., 1969). The low rate of spontaneous transformation of cells from normal subjects may be enhanced by the use of low doses of plant mitogens or by incubation in the presence of Epstein-Barr (EB) virus-containing extracts from established cultures. In fact, the presence of EB virus may be necessary for long-term proliferation in vitro, since virtually all lines examined contain either infectious virus or viral genome (Broder et al., 1970).

A partial characterization of the properties of these cells reveals the ability to synthesize several classes of immunoglobulins and free-light chains (Finegold et al., 1967), interferon production (Kasel et al., 1967), synthesis and release of several effector substances of cellular immunity such as lymphocytotoxin (Glade and Hirschhorn, 1970), migration inhibition factor (Smith et al., 1970), as well as synthesis of other substances.

The cultures grown in our laboratory have been derived from the peripheral blood of normal subjects (ES-1), and from patients with chronic granulomatous disease (CGD-3), mononucleosis (DW-2, DW-3), multiple myeloma (Mogler and MYG-1), AAH (Ag-16), and Waldenstrom's macroglobulinemia (Allen).

A variety of reports by many authors has verified that cyclic AMP or related drugs may influence established cell lines from a variety of sources (Burger et al., 1972). As part of an effort to understand more thoroughly the nature and capabilities of the lymphoid cell lines, experiments were undertaken in our laboratory to investigate the effect of cyclic AMP on DNA, RNA, and protein synthesis in the continuously replicating cell lines. In addition, measurements of adenyl cyclase activity and intracellular cAMP levels were carried out. It was hoped that since these cells were undergoing continuous division, they might be useful in studying cyclic AMP effects in already "activated" cells.

The initial experiments involved measurements of adenyl cyclase activity and cyclic AMP levels in the unsynchronized continuously replicating cell lines mentioned above as compared to normal, unstimulated human peripheral blood lymphocytes (Fig. 8). In general, the greatly reduced activity of the adenyl cyclase in the cell lines compared to the normal peripheral blood lymphocytes is also reflected in the lowered (roughly threefold) levels of intracellular cAMP. Thus, with the possible exception of the Allen line (macroglobulinemia), virtually none of the other lines had what could be called an active adenyl cyclase.

These data reflect what has been found by other investigators using different kinds of cell lines. It must be emphasized, however, that these results reflect activity in cells, the majority of which are in the S phase of the cell cycle, and it may well be that higher cAMP levels and greater adenyl cyclase activity might be detected at a different phase in the cell cycle. This is the case in at least one system, mouse fibroblasts, investigated by Burger et al. (1972).

Since the cells had lowered cAMP levels and adenyl cyclase activity, it was of interest to measure the effects of the addition of exogenous cAMP to these cultures after 24 hr. The data shown in Table 4 indicate the effects of such a regimen on protein, RNA, and DNA synthesis in several of the cell lines. The results are presented in terms of the ratio of experimental groups to untreated controls. The cells were exposed to 10^{-3} to 10^{-5} M cAMP in serum-free media for 24 hr. The general impression is that the various cell lines do respond differentially to cAMP, and differences in responsiveness occur depending on whether protein, RNA, or DNA synthesis is being measured.

The cell line ES-1 established from a normal subject exhibits a substantial increase in RNA synthesis, and to a lesser extent with DNA synthesis with no effect on protein synthesis. The two mononucleosis lines,

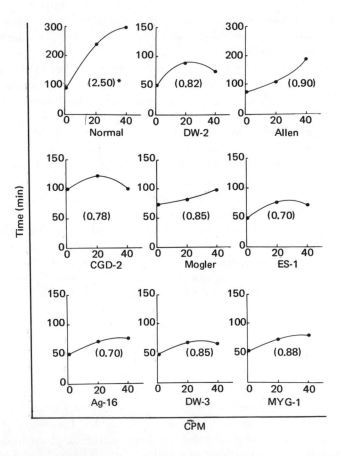

Fig. 8. Adenyl cyclase activity and cAMP levels in human lymphoblastoid cell lines as compared to isolated normal human peripheral blood lymphocytes. The graphs represent adenyl cyclase activity of crude cell membrane fractions prepared from various lymphoid cell lines. The assay system was that of Winchurch et al. (1972). The numbers in parentheses represent intracellular cAMP levels in picomoles per 1.5×10^6 cells as measured by the cAMP binding protein assay of Gilman (1970).

Table 4. Effects of Exogenous Cyclic AMP on Protein, RNA, and DNA Synthesis in Human Lymphocytic Cell Lines[a]

Product synthesized	Cell line							
	DW-2	Allen	CGD-3	MYG-1	Ag 16	DW-3	Mogler	ES-1
cAMP $10^{-3}M$								
Protein	1.45	0.83	1.07	1.12	1.46	1.15	0.88	1.03
RNA	1.15	0.87	1.20	1.52	1.19	1.34	0.98	5.98
DNA	1.46	1.06	1.26	1.90	1.48	1.08	0.69	1.72
cAMP $10^{-4}M$								
Protein	1.07	0.98	1.13	1.22	1.24	1.16	0.94	1.06
RNA	0.99	1.01	1.14	1.39	1.24	0.99	0.77	3.99
DNA	1.32	1.17	1.25	1.72	1.43	1.25	0.74	2.21
cAMP $10^{-5}M$								
Protein	1.17	0.88	1.12	1.39	2.50	1.18	1.11	1.14
RNA	0.92	0.73	1.22	1.30	1.27	1.07	0.96	1.24
DNA	1.09	0.77	1.37	1.58	1.54	1.25	0.80	1.34

[a]These values represent ratios of experimental groups exposed to cyclic AMP divided by controls receiving no treatment. Protein synthesis was measured using a radiolabeled amine acid mixture with a 4-hr pulse 20 to 24 hr after exposure to cyclic AMP. RNA and DNA synthesis were measured using labeled uridine and thymidine, respectively, with a 4-hr pulse 20 to 24 or 68 to 72 hr after exposure to cyclic AMP. All cultures were grown in microtiter plates and assayed using an automatic culture assay machine.

DW-2 and DW-3, show similar responses, which might be expected as they were derived from the same patient. The lines give some stimulation of protein synthesis (10^{-3} M, 10^{-5} M), little or no effect on RNA synthesis, and some stimulation of DNA synthesis (10^{-3} M, 10^{-4} M). The myeloma lines, MYG-1, Mogler, and Allen (macroglobulinemia) yield differing responses, with Allen and Mogler responding similarly and MYG-1 in a totally different manner. Cyclic AMP stimulates protein, RNA, and DNA synthesis in MYG-1. Mogler and Allen exhibit either inhibition or no effect of cAMP on protein, RNA, and DNA synthesis. In the cell line CGD-3, cAMP shows no effect on protein synthesis and enhancement of RNA and DNA synthesis. The remaining cell line, Ag-16, responds to cAMP with increased protein, RNA, and DNA synthesis similar to MYG-1.

These data reveal that even in activated cells, macromolecular synthesis can be affected by increases in cAMP levels. More importantly, not all cell lines respond in precisely the same way, which may reflect differences due to the diseases of the patients from which these lymphocytes were isolated, or may instead indicate the predominance of a particular circulating cell type in these patients. Of some interest to us is the possibility that the cell type that establishes itself in culture may correspond to one of the two populations, which may be seen in the PHA-cAMP experiments mentioned in Section I. Such considerations are speculative at this stage, however, and must remain so until further experiments can be performed.

III. Summary

In summary, we have shown that in normal human peripheral blood lymphocytes, at least two populations can be differentiated on the basis of their response to PHA and cyclic AMP. One population responding at suboptimal to optimal doses of PHA is either inhibited or unaffected by the addition of cAMP with respect to DNA and RNA synthesis. The other population responding at superoptimal concentrations of PHA is stimulated by cAMP addition. We term this effect high-dose recovery. While the significance of this latter population in terms of the process of lymphocyte activation remains unclear, it is important to note that subpopulations of lymphocytes may respond in varying ways to modulations in cAMP levels. Thus, it is clear that some cells respond to high doses of PHA by increased synthesis of cAMP, and further, these cells tolerate high intracellular concentrations of cAMP and, in fact, respond in a positive manner.

An examination of two immune deficiencies, hypogammaglobulinemia and chronic lymphocytic leukemia reveals normal response to PHA and cAMP in the former case (AAH) and a lack of high-dose recovery in the latter case (CLL). It would appear that in AAH the cAMP-activation system is intact, while in CLL either a cell type is missing (T cell?) or there is a fault in the activation process in these cells.

The lack of correlation between cAMP and other mitogens such as Con A and PWM will require further investigation. In this regard, the work of Hadden et al. (1972) offers a provocative suggestion concerning the role of cyclic GMP in lymphocyte activation. Indeed, the relative modulation of cyclic AMP and cyclic GMP levels may turn out to be the critical controlling factor in early biochemical events following lymphocyte stimulation.

The research presented concerning the human lymphoblastoid cell lines is provocative but largely preliminary and remains a fruitful area for further experimentation.

Discussion of Dr. Webb et al.'s Paper

Dr. WEISSMANN: Is it fair to say that judging from the three presentations we have heard, the role of cyclic AMP as a messenger in lymphocyte proliferation is no longer interesting?

Dr. PARKER: I think the answer to your question is no for several reasons:

(1) The early increase in cyclic AMP takes place before other metabolic changes can be shown.

(2) Judging by immunofluorescence studies, the pattern of stimulation by PHA is highly selective and clearly differs from what is seen with prostaglandins and catecholamines. Indeed, there is the paradox that cyclic AMP measurements in isolated cells may show a net *fall* at a time when local increases in cyclic AMP (or local AMP binding protein activity, if you prefer) are occurring. Failure to reproduce Sutherland's postulates cannot be considered as conclusive in this situation, since none of the other agents that raise cyclic AMP (PGE_1, isoproterenol, cyclic AMP, theophylline) can really simulate what PHA does.

(3) Even if we dismiss the early increase in cyclic AMP in PHA-stimulated cells as unimportant (which I am unwilling to do), the overall fall in cyclic AMP, which occurs early with isolated T cells over a broad PHA concentration range and after several hours with mixtures of T and B cells, is of considerable interest, since it correlates with what happens in malignant cell lines undergoing proliferation and with the observation that agents that raise intracellular cyclic AMP inhibit PHA-induced transformation. We have concentrated on the local increase in cyclic AMP, a very early event in stimulated cells, but the observation that cyclic AMP frequently falls in the cell population as a whole may have greater importance.

Dr. WEISSMANN: I would just suggest that neither you nor anyone else has shown where cyclase is localized. You are simply showing by immunofluorescence that a binding protein for cyclic nucleotide can be detected, by means of antibodies, in different locations in the air-dried cell. That has not been controlled by other cyclic nucleotides, for example.

I do not believe that any of the criteria required to prove that cyclic AMP is a messenger in lymphocyte activation has been fulfilled. Thus, we thought that the idea of cyclic AMP being the second messenger in lymphocyte activation was amusing and interesting four or five years ago, but now can no longer consider it of crucial importance. There is no critical evidence that cyclic AMP acts as a second messenger in lymphocyte activation.

DR. PARKER: On the contrary, our studies with isolated nuclei provide strong evidence that cyclic AMP is made in the nucleus, that the accumulation is differentially stimulated by catecholamines, and that a β adrenergic receptor is involved. And I don't think it is fortuitous that the same class of agents that stimulate cyclic AMP accumulation in the isolated nucleus causes a marked increase in cyclic AMP staining in the same region of the intact cell. Moreover, the staining is blocked by cyclic AMP and not by other nucleotides. On the other hand, I agree completely that at this point in time, there is no proof that cyclic AMP is a second messenger for PHA. But if you accept the concept of cyclic AMP compartmentalization, this possibility is very difficult to exclude.

DR. HADDEN: Dr. Parker, you have noted nodular localization of PHA-stimulated adenylate cyclase on the surface of lymphocytes, and I would like to address myself to the phenomenon of capping as it relates to this observation. Agglutinating agents act on the surface of the lymphocytes to cause the receptors to clump and, ultimately, to either be cast off or internalized. I would suggest that the clumping of adenylate cyclase activity that you see may be the result of receptor clumping seen prior to capping.

DR. PARKER: Dr. Wedner has not seen capping in short-incubation experiments (up to 30 min following exposure of cells to PHA). Most of the incubations have been at 0° or at room temperature, which would decrease the tendency to capping, and it is conceivable that capping might be demonstrable if altered incubation conditions were used. Perhaps Dr. Wedner would like to comment further.

DR. WEDNER: We have done experiments at 0°, room temperature, and at 37°, and the cyclic AMP staining pattern does not suggest capping. We have looked at times from 5 to 30 min and see no capping. We have also used a mushroom agglutinin which is not a mitogen, and it gives a completely different pattern, more like what is seen with PGE_1. Thus far, the pattern seen with PHA is unique, and we are now looking at other mitogens to see if they have similar patterns.

DR. BOURNE: Relative to Dr. Weissmann's comment, if the Sutherland criteria cannot be used, is there a set of criteria that will allow you to determine that this compartmentalization phenomenon is responsible for different responses to cyclic AMP?

DR. PARKER: The approaches are difficult. I think that it will be necessary to isolate the cell nuclei and develop better methods for measuring metabolic changes relevant to nuclear activation such as stimulation of DNA-dependent RNA synthesis. The effect of cyclic nucleotides alone and in combination with various macromolecules in the cytoplasm could then be examined. We also need to study the possibility that cyclic AMP might act locally at the level of the external cell membrane to stimulate the synthesis or release of an unidentified messenger for the nucleus. A possibility might be the low-molecular-weight plasma membrane DNA fraction described by Lerner and his colleagues in a human lymphoid cell line. Mutant cell culture lines such as the ones Tompkins and his colleagues have used in their studies of hepatic enzyme induction may also be helpful.

DR. JOHNSON: You did your work on isolated blood leukocytes. What happens if you make these measurements on thymocytes?

DR. PARKER: We have not looked at human thymocytes, and as far as I know, no one else has either.

DR. MACMANUS: While the situation in peripheral lymphocytes is unclear and we are tortured in trying to confirm Sutherland's postulates, the situation in thymocytes is quite clear. We can induce a new population of cells to synthesize DNA and continue into mitosis with agents that raise intracellular concentrations of cyclic AMP. We can also mimic the actions of these agents by treating the cells with cyclic AMP. Compounds that inhibit phosphodiesterase, such as caffeine, potentiate the cyclic AMP effects and also stimulate DNA synthesis and mitosis. I shall talk more about this tomorrow, but I want to say that the Sutherland postulates are confirmed in thymocytes. What these thymus cells have to do with thymus-derived peripheral lymphocytes is unclear at the moment.

DR. PLESCIA: I am wondering whether the question concerning the role of cyclic AMP as a second messenger in the activation of lymphocytes is not too broad. If we use the term "activation," we should be more specific—activation for what? I think the answer would then be quite different.

DR. WINCHURCH: Talking about cyclic AMP and lymphocytes, I think we ought to comment on the fact that cyclic AMP and agents that stimulate its accumulation have been shown to enhance certain immune responses. This is predominantly the work of Werner Braun, who showed that the polynucleotides act as adjuvants in immune responses and was able to link this activity to increases in cyclic AMP. He further showed that exogenous cyclic AMP caused increases in the number of antibody-producing cells, both in vivo and in vitro. I think Dr. Braun's data strongly implicate cyclic AMP as a second messenger in lymphocytes and more particularly in immune responses.

DR. AUSTEN: I wonder if Dr. Webb would comment further about the mechanism of cAMP-dependent high-dose reversal.

DR. WEBB: Only to say that we also have measured cyclic AMP levels over a 72-hr period using 1000 μg/ml PHA (high dose) and cAMP (10^{-3} M) with the naive thought that we could remove cyclic AMP once it has been added to the media. We found that these measurements could not be made, presumably because of the existence of cAMP receptors on the surface of the peripheral blood cells. Extremely high values for cyclic AMP were obtained, despite extensive cell washing. We assume we are measuring membrane-bound as well as intracellular cyclic AMP, although it is also possible that a portion of the extracellular cyclic AMP is internalized. What this has to do with the mechanism of the high-dose enhancement is uncertain. I can only say that certain cells will tolerate the continuous presence of a large amount of cyclic AMP and respond in a positive manner when PHA is also present. Whether or not these observations will lead to a better understanding of cAMP-mediated events and differences in lymphocyte subpopulations is not clear.

DR. HADDEN: Has the possibility been considered that surface-bound cyclic AMP might reduce PHA binding?

DR. WEBB: I think I would not, but we have not conducted direct binding measurements.

DR. PARKER: We have looked at whether cyclic AMP, catecholamines, and prostaglandins compete with radiolabeled PHA for lymphocyte binding sites, and they do not.

DR. ROBISON: Dr. Webb, did you do experiments similar to those of Dr. Hirschhorn?

DR. WEBB: We have not looked at all the possibilities. We have looked at theophylline using the checkerboard experimental model. Preliminary results suggest that if a proper concentration of theophylline is used, a high-dose effect can be demonstrated.

DR. HIRSCHHORN: I would like to suggest an alternative possibility, that at high doses of PHA, you are suppressing a PHA-responsive population, and that what you are seeing at high doses of PHA is the response of another population of cells that respond to elevations of cyclic AMP by proliferating.

DR. WEBB: That was the point I was trying to make, that cells responding to high doses of PHA and cAMP are a separate population of cells. More importantly, one may not, therefore, say that cyclic AMP itself is exclusively inhibitory in mitogen-induced DNA synthesis.

References

Barth, W. F., Asofsky, R., Liddy, T. J., Tanaka, Y., Rowe, D. S., and Fahey, J. L. "An Antibody Deficiency Syndrome," *Amer. J. Med.* **39**:319–334 (1965).
Bourne, H. R., Lichtenstein, L. M., and Melmon, K. L. "Pharmacologic Control of

Allergic Histamine Release in Vitro: Evidence for an Inhibitory Role of 3′,5′-Adenosine Monophosphate in Human Leukocytes," *J. Immunol.* **108**:695–705 (1972).

Braun, W., and Ishizuka, M. "Antibody Formation: Reduced Responses After Administration of Excessive Amounts of Nonspecific Stimulators," *Proc. Nat. Acad. Sci.* (U.S.A.). **68**:1114–1116 (1971).

Broder, S. W., Glade, P. R., and Hirschhorn, K. "Establishment of Long-Term Lines from Small Aliquots of Normal Lymphocytes," *Blood* **35**:539–542 (1970).

Burger, M. M., Bombik, B. M., Breckenridge, B., and Sheppard, J. "Growth Control and Cyclic Alterations in Cyclic AMP in the Cell Cycle, *Nature (New Biol.)* **239**:161–163 (1972).

Cross, M. E., and Ord, M. G. "Changes in Histone Phosphorylation and Associated Early Metabolic Events in Pig Lymphocyte Cultures Transformed By Phytohaemagglutinin or 6 N, 2′-0-Dibutyryl-Adenosine 3′, 5′-Cyclic Monophosphate," *Biochem. J.* (Britain) **124**:241–248 (1971).

Douglas, S., Kamin, R., and Fudenberg, H. H. "Human Lymphocyte Response to Phytomitogens in Vitro: Normal, Agammaglobulinemic and Paraproteinemic Individuals," *J. Immunol.* **103**:1185–1195 (1969).

Duerner, M. J., Brock, M. L., and Jost, J. "Stimulation of Ribonucleic Acid Synthesis in Horse Lymphocytes By Exogenous Cyclic Adenosine 3′, 5′-Monophosphate," *J. Biol. Chem.* **247**:413–417 (1972).

Finegold, I., Fahey, J. L., and Granger, H. "Synthesis of Immunoglobulins By Human Cell Lines in Tissue Culture," *J. Immunol.* **99**:839–848 (1967).

Gilman, A. "A Protein Binding Assay for 3′,5′-Cyclic Adenosine Monophosphate," *Proc. Nat. Acad. Sci.* (U.S.A.) **67**:305–312 (1970).

Glade, P. R., Hirshant, Y., Stites, D. P., and Chessin, L. N. "Infectious Mononucleosis: In Vitro Evidence for Limited Proliferation," *Blood* **33**:292–299 (1969).

Glade, P. R., and Hirschhorn, K. "Products of Lymphoid Cells in Continuous Culture," *Amer. J. Pathol.* **60**:483–493 (1970).

Grey, H. M., Rabellino, E., and Perofsky, B. "Immunoglobulins on the Surface of Lymphocytes. IV. Distribution in Hypogammaglobulinemia, Cellular Immune Deficiency, and Chronic Lymphatic Leukemia," *J. Clin. Invest.* **50**:2368–2375 (1971).

Hadden, J. W., Hadden, E. M., Haddox, M. K., and Goldberg, N. D. "Guanosine 3′, 5′-Cyclic Monophosphate: A Possible Intracellular Mediator of Mitogenic Influences in Lymphocytes," *Proc. Nat. Acad. Sci.* (U.S.A.) **69**:3024–3027 (1972).

Hartzman, R. J., Bach, M. D., Bach, F. H., Thurman, G. B., and Sell, K. "Precipitation of Radioactively Labeled Samples: A Semiautomatic Multiple-Sample Processor," *Cell. Immunol.* **4**:182–186 (1972).

Hirschhorn, R., Grossman, J., and Weissmann, G. "Effects of Cyclic 3′,5′-Adenosine Monophosphate and Theophylline on Lymphocyte Transformation," *Proc. Soc. Exp. Biol. Med.* **133**:1361–1365 (1970).

Huang, C. C., and Moore, G. E. "Chromosomes of 14 Hematopoietic Cell Lines Derived from Peripheral Blood of Persons With and Without Chromosome Anomalies," *J. Nat. Cancer Inst.* **43**:1121–1128 (1969).

Ishizuka, M., Braun, W., and Matsumoto, T. "Cyclic AMP and Immune

Responses. I. Influence of Poly A:U and cAMP on Antibody Format on Formation in Vitro," *J. Immunol.* **107**:1027–1035 (1971).

Johnson, L. D., and Abell, C. W. "The Effects of Isoproterenol and Cyclic Adenosine 3′,5′-Phosphate on Phytohemagglutinin-Stimulated DNA Synthesis in Lymphocytes Obtained from Patients with Chronic Lymphocytic Leukemia," *Cancer Res.* **30**:2718–2723 (1970).

Kasel, J., Hoose, A. T., Glade, P. R., and Chessin, L. N. "Interferon Production in Cell Lines Derived from Patients with Infectious Mononucleosis," *Proc. Soc. Exp. Biol. Med.* **128**:351–359 (1968).

Klein, M. I., and Makman, M. H. "Properties and Cellular Distribution of Cyclic AMP-Dependent Protein Kinase from Thymus," *J. Cell. Physiol.* **79**:407–412 (1972).

MacManus, J. P., and Whitfield, J. F. "Cyclic AMP Binding Sites on the Cell Surface of Thymic Lymphocytes," *Life Sci.* **11** (part II): 837–845 (1972).

Moller, G. "Induction of DNA Synthesis in Human Lymphocytes. Interaction Between Nonspecific Mitogens and Antigens," *Immunol.* (Britain) **19**:583–598 (1970).

Moore, G. E., Grace, J. T., Citron, P., Gerner, R., and Burns, A. "Leukocyte Cultures of Patients with Leukemia and Lymphomas," *N. Y. J. Med.* **66**:2757–2764 (1966).

Rixon, R. H., Whitfield, J. F., and MacManus, J. P. "Stimulation of Mitotic Activity in Rat Bone Marrow and Thymus By Exogenous Adenosine 3′,5′-Monophosphate (cyclic AMP)," *Exp. Cell Res.* **63**:110–115 (1970).

Silver, R. T. "The Treatment of Chronic Lymphocytic Leukemia," *Sem. Hematol.* **6**:344–355 (1969).

Smith, J. W., Steiner, A. L., and Parker, C. W. "Human Lymphocyte Metabolism. Effects of Cyclic and Noncyclic Nucleotides on Stimulation by Phytohemagglutinin," *J. Clin. Invest.* **50**:442–448 (1971a).

Smith, J. W., Steiner, A. L., Newberry, W. M., and Parker, C. W. "Cyclic Adenosine 3′,5′-Monophosphate in Human Lymphocytes. Alterations After Phytohemagglutinin Stimulation," *J. Clin. Invest.* **50**:432–441 (1971b).

Smith, R. T., Bauscher, J. A. C., and Adler, W. H. "Studies of an Inhibitor of DNA Synthesis and a Nonspecific Mitogen Elaborated By Human Lymphoblasts," *Amer. J. Pathol.* **60**:495–504 (1970).

Strom, T. B., Deissroth, A., Morganroth, J., Carpenter, C. B., and Merril, J. P. "Alteration of the Cytotoxic Action of Sensitized Lymphocytes By Cholinergic Agents and Activators of Adenylate Cyclase," *Proc. Nat. Acad. Sci.* (U.S.A.) **69**:2995–2999 (1972).

Webb, D. R., Stites, D. P., Perlman, J. D., Austin, K. E., and Fudenberg, H. H. (submitted for publication).

Winchurch, R., Ishizuka, M., Webb, D., and Braun, W. "Adenyl Cyclase Activity of Spleen Cells Exposed to Immuno-Enhancing Synthetic Oligo- and Polynucleotides," *J. Immunol.* **106**:1399–1401 (1971).

MITOGENIC ACTIVITY OF POLYNUCLEOTIDES ON THYMUS-INFLUENCED LYMPHOCYTES

ARTHUR G. JOHNSON and IHN H. HAN

I. Introduction

An adjuvant effect on antibody synthesis of isologous or heterologous nucleic acids has been known for some time (Jaroslow and Taliaferro, 1956; Merritt and Johnson, 1965). With the recent rapid acquisition of more precise knowledge as to how the immune system functions, the groundwork has been laid to define the cell types and the intracellular biochemical events affected by this adjuvant. In initiating investigation on this aspect, we have studied (Johnson et al., 1971) the characteristics of a nontoxic (Han et al., 1973), double-stranded homoribopolymer, polyadenylic acid complexed to polyuridylic acid (poly A:U). Similar efforts have emerged from Dr. Werner Braun's laboratory (Braun et al., 1971), with a concentration on experiments linking the poly A:U effect to activation of cyclic AMP in spleen cells (Ishizuka et al., 1971). In our laboratory, we have postponed measurements of cyclic AMP, electing first to define the precise cell type affected by this adjuvant.

The general characteristics that we have observed to date (Schmidtke and Johnson, 1971) include the capacity to increase antibody synthesis to a number of antigens by several different routes. In addition, cell-mediated immunity was enhanced (Friedman et al., 1969; Cone and Johnson, 1971). The adjuvant activity was apparent most readily when immunity was partially depressed, as for example by thymectomy (Cone and Johnson, 1971), or when the antigenicity of the material was low (Schmidtke and Johnson, 1971). The induction period was shortened generally by two to four days, and the extent of peak titer elevation was dependent on the antigen dose used in control animals, the lower the control, the larger fold increase by poly A:U. In our hands, the complexed form was required; neither poly A nor poly U alone incited statistically significant enhancement.

II. Cells Affected by Polynucleotides

The activity of many cells was found to be increased in the presence of poly A:U. Thus, the macrophage (Johnson and Johnson, 1971), T cell (Cone and Johnson, 1971), memory cell (Stout and Johnson, 1972), stem cell (McNeill, 1971), and even nonimmunocompetent cells as found in the salivary gland and pancreas (Han, personal communication) all showed increased activity under the influence of poly A:U. With our experimental approaches, the B cell appeared not to be affected, excepting as an expression as the end cell of an enhanced T cell population. Others (Campbell and Kind, 1971; Mozes et al., 1971), however, have reported evidence that poly A:U may affect this cell also. Thus, poly A:U appears to increase the capacity of many different cell types to respond with the function to which they are genetically committed.

The major effect we observed under our conditions was on the thymus-influenced T cell. This first was revealed as the capacity to restore immunocompetency in neonatally thymectomized mice. The data suggested that poly A:U was amplifying the residual T cells seeded to the peripheral lymphoid tissue prior to neonatal thymectomy. Evidence to support this hypothesis was gained using the experimental model wherein the X-irradiated mouse requires the adoptive transfer of defined numbers of both T and B cells to mount an immune response. To be effective, a minimum of 10^7T and B cells were necessary. However, on incubation with poly A:U in vitro for 1½ hr prior to transfer, 10- to 50-fold less T cells were found equal in this capacity. In comparable experiments, no effect of poly A:U was observed when 10^5, 10^6, or 10^7B cells were incubated with poly A:U. Similarly, if T cells were injected with poly A:U one day before B cells were injected alone, and vice versa, adjuvant action was stimulated only in the T cell population. In this instance, 10^7T cells incubated with poly A:U gave a response equal to 10^8T cells without A:U stimulus, again suggesting poly A:U might be increasing T cell numbers (Cone and Johnson, 1972).

III. Effect of Polynucleotides on DNA Synthesis in T Cells

Evidence that poly A:U served to increase antigen-mediated cell division first was obtained in studies of cell-mediated immunity (Friedman et al., 1969). Thus, when poly A:U was added to mixed human leukocyte cultures, uptake of tritiated thymidine was increased some fivefold. Similarly, when the adjuvant was added with PPD to human lymphocytes from tuberculin-positive donors, the uptake of tritiated thymidine by what presumably were T cells was increased approximately 12-fold. However, in control tests, cells exposed to poly A:U without antigen did not show any increase in cell division, suggesting antigen triggering of cells was mandatory to poly A:U action.

To test further whether or not poly A:U independent of antigen was mitogenic for T cells, thymuses were removed from four-week-old mice, the thymocytes freed by teasing the tissue and placed in culture in Eagle's medium with 10% fetal calf serum. Tritiated thymidine was added to separate cultures at either 0, 3, 6, 12, 18, 24, or 48 hr. Poly A:U alone without antigen did not appear to be mitogenic in that no uptake of ^{3}H over control cells was seen in cultures containing poly A:U, despite titration of the dose of adjuvant from 1×10^{-4} to 20 μg. Similar evidence was obtained with spleen cells. This conclusion also was reached by Jaroslow and Ortiz-Ortiz (1972) in their study of the effect of oligonucleotides on antibody synthesis in culture. On the other hand, when 5×10^{-5} μg poly A:U was added with antigen (sRBC) to 10^5 T cells, a 30% increase in uptake of ^{3}H was seen in multiple experiments when the label was added 2 hr after initiation of a culture period totaling 6 hr.

These experiments raised the question as to whether culture conditions as we employ them were conducive to division of the relatively unstable T cell. For example, poly A:U was much more effective in vivo than when added in vitro to spleen cell cultures. Figures 1 and 2 compare the activity of poly A:U on normal and primed spleen cells with that of a B cell mitogen, endotoxin. Both adjuvants exhibited a dose-response effect in that low doses enhanced the number of plaque-forming cells, while high doses inhibited the response. However, the enhancement observed in vitro routinely approximated only 150% of normal as compared to a much higher order of magnitude when injected in vivo.

Consequently, in vivo experiments, although more indirect, also were designed to test whether poly A:U per se was mitogenic. One approach utilized the drug vinblastine with the rationale that if lymphocytes in poly A:U treated animals were initiating division more rapidly than normal, a drug specifically interfering with dividing cells should inhibit the immune response earlier than animals given antigen alone. It was found (Cone and Johnson, 1972) that antibody synthesis in mice receiving antigen alone was not inhibited by a single injection of vinblastine given 6 hr after antigen injection. However, the drug was effective when given 12 hr after antigen, indicating division was not initiated until 12 hr after exposure to antigen. On the other hand, mice given 300 μg poly A:U at the time of antigen injection initiated cell division within 6 hr, as indicated by the fact that the immune response was 90% suppressed by a single injection of vinblastine given 6 hr after antigen plus poly A:U. Further verification that poly A:U increases the number of T lymphocytes in the spleen was obtained recently by Marchalonis et al. (1973). The number of theta-bearing anti-sRBC rosette-forming cells (RFC) in the spleen of mice receiving antigen alone was negligible, while injection of poly A:U with antigen increased this value to 50% of the total RFC.

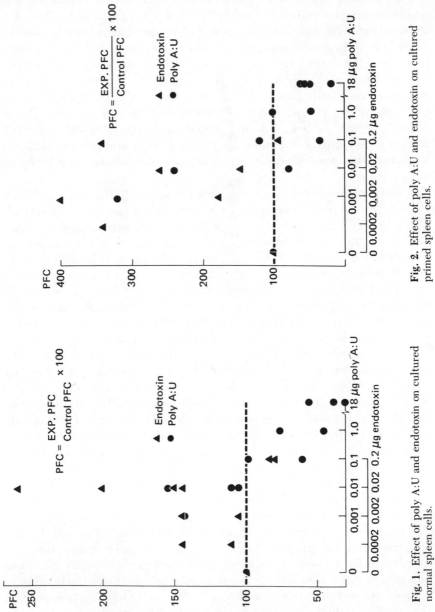

Fig. 1. Effect of poly A:U and endotoxin on cultured normal spleen cells.

Fig. 2. Effect of poly A:U and endotoxin on cultured primed spleen cells.

A second approach involved a comparison of the uptake of tritiated thymidine (0.5 microcuries intraperitoneally per gram body weight) by spleen cells in mice given poly A:U with or without antigen. Included in the experiment were mice treated additionally with 2.5 mg cortisone given two days before the antigen and/or poly A:U on the premise that the removal of steroid-sensitive B cells (Cohen and Claman, 1971) might permit better visualization of a mitogenic effect on T cells. The result of measurement of the counts per minute of ^{3}H in 50 μl of spleen homogenate from the various groups of mice is shown in Table 1. Under these conditions, a mitogenic effect of poly A:U per se was indicated in that this adjuvant alone stimulated almost twice the cpm of control mice injected with saline. This was increased to sixfold in corticosteroid-treated mice. Similar experiments utilizing thymus cells in place of spleen cells did not reveal a mitogenic effect of poly A:U. Thus, this adjuvant may exert its mitogenic effect mainly on peripheralized T cells.

Table 1. Influence of Poly A:U on Incorporation of ^{3}HT into Spleen Cells in Cortisone-Treated and Nontreated Balb/aj Mice

First injection[a]	Second injection[b]	Counts per minute
Saline	Saline	500
	sRBC	680
	Poly A:U	935
	sRBC + poly A:U	1100
Cortisone	Saline	240
	sRBC	825
	Poly A:U	1500
	sRBC + poly A:U	1560

[a]Given intraperitoneally four days before removal of spleen and assay.
[b]Given intraperitoneally two days before removal of spleen and assay.
Dosages: Sheep red blood cells (sRBC), 0.2 ml of a 10% solution; poly A:U, 300 μg poly A complexed with 300 μg poly U; cortisone, 2.5 mg; saline, 0.25 ml.

Discussion of Dr. Johnson's and Dr. Han's Paper

DR. SHEARER: Is the rosette assay the only assay you have used?

DR. JOHNSON: We measured plaque-forming cells (PFC) in early experiments and found a correlation between PFC and rosette-forming cells (RFC). The RFC measurement is much more sensitive than the PFC measurement, and we have preferred to use the former. Cone and Wilson (*Int. Arch. Allergy* **43**: 123–130, 1972) have shown poly A:U to increase both T and B cell rosette formation.

DR. SHEARER: What are the relative effects on immune responses to thymus-dependent antigens?

DR. JOHNSON: If antigen is incubated with spleen cells from nude mice (which have no T cells), antibody formation to T-dependent antigens does not occur. But, if as few as 10^3 T cells that have been stimulated with poly A:U were added, an immune response is obtainable. With the same number of unstimulated T cells, antibody formation does not take place.

DR. LICHTENSTEIN: Would you review what is known about lymphocyte cyclic AMP concentrations at various doses of poly A:U?

DR. JOHNSON: We have not measured cyclic AMP, but Dr. Winchurch has. As I recall, the increase in cyclic AMP was not large. I think the major link of poly A:U to cyclic AMP comes from the fact that theophylline produces a dramatic potentiation of poly A:U effects. This is a function of the dose of poly A:U and the dose of theophylline. Poly A:U also produces a variable but definite increase in lymphocyte adenylate cyclase activity. As I recall, the maximal increase is about two- to threefold.

DR. BOURNE: I have not been able to find any published measurements of cyclic AMP in cells exposed to poly A:U. I wonder if anybody here has done such measurements. We have measured cyclic AMP in mouse lymphocytes exposed to 300 μg of poly A:U per milliliter, and no changes were found. This was under circumstances in which the immunological effect of poly A:U was clear-cut.

DR. JOHNSON: We have not done any measurements of cyclic AMP as yet. Our philosophy has been to pinpoint the subpopulation of T cells that is responsive to poly A:U first, and then make measurements without noise from lymphocytes in the nonresponding population. That is where we hope our corticosteroid experiments are leading us.

DR. WEBB: Shortly before Dr. Braun's death, he and I had a discussion about the measurement of lymphocyte cyclic AMP levels following the addition of polynucleotides. He indicated that he did have data indicating an increase in cyclic AMP levels following the addition of poly A:U at concentrations that are immunoenhancing in vitro. The doses were quite a bit lower than the amounts used in vivo. The concentrations that we have used successfully in vitro in the Mishell-Dutton mouse spleen cell culture system are in the range of 10 to 100 ng/ml.

References

Braun, W., Ishizuka, M., Yajima, Y., Webb, D., and Winchurch, R. "Spectrum and Mode of Action of Poly A:U in the Stimulation of Immune Responses." *Biological Effects of Polynucleotides*, R. F. Beers and W. Braun, eds. New York: Springer-Verlag, pp. 139–156 (1971).

Campbell, P. A., and Kind, P. "Bone Marrow Derived Cells As Target Cells for Polynucleotide Adjuvants," *J. Immunol.* **107**: 1419–1423 (1971).

Cohen, J. J., and Claman, H. N. "Thymus-Marrow Immunocompetence. V. Hydrocortisone-Resistant Cells and Processes in the Hemolytic antibody response of Mice," *J. Exp. Med.* **133**:1026–1034 (1971).

Cone, R. E., and Johnson, A. G. "Regulation of the Immune System by Synthetic

Polynucleotides. III. Action on Antigen-Reactive cells of Thymic Orgin," *J. Exp. Med.* **133**:665–676 (1971).

Cone, R. E., and Johnson, A. G. "Regulation of the Immune System by Synthetic Polynucleotides. IV. Amplification of Proliferation of Thymus-Influenced Lymphocytes," *Cell. Immunol.* **3**:283–293 (1972).

Friedman, H. G., Johnson, A. G., and Pan, P. "Stimulatory Effect of Polynucleotides on Short-Term Leucocyte Cultures," *Proc. Soc. Exp. Biol. Med.* **132**:916–918 (1969).

Han, I. H., Johnson, A. G., Cook, J., and Han, S. S. "Comparative Biological Activity of Endotoxin and Synthetic Polyribonucleotides," *J. Infec. Dis.* (in press).

Ishizuka, M., Braun, W., and Matsumoto, T. "Cyclic AMP and Immune Responses. I. Influence of Poly A:U and cAMP on Antibody Formation in Vitro, *J. Immunol.* **107**:1027–1035 (1971).

Jaroslow, B. N., and Taliaferro, W. H. "Restoration of Hemolysin-Forming Capacity in X-Irradiated Rabbits by Tissue and Yeast Preparations," *J. Infec. Dis.* **98**:75–81 (1956).

Jaroslow, B. N., and Ortiz-Ortiz, L. "Influence of Poly A-Poly U on Early Events in the Immune Response in Vitro," *Cell. Immunol.* **3**:123–132 (1972).

Johnson, A. G. "Stimulation of the Immune System by Homopolyribonucleotides." *Biological Effects of Polynucleotides*, R. F. Beers and W. Braun, eds. New York: Springer-Verlag, pp. 157–177 (1971).

Johnson, H. G., and Johnson, A. G. "Regulation of the Immune System by Synthetic Polynucleotides. II. Action on Peritoneal Exudate Cells," *J. Exp. Med.* **133**:649–664 (1971).

Marchalonis, J. J., Cone, R. E., and Rolley, R. T. "Amplification of Thymus-Influenced Lymphocytes By Poly A:U: Inhibition of Antigen Binding By Antiserum to Immunoglobulin Light Chain," *J. Immunol.* **110**:561–556 (1973).

McNeill, T. A. "The Effect of Synthetic Double-Stranded Polyribonucleotides on Haemopoietic Colony Forming Cells in Vitro," *Immunol.* **21**:741–750 (1971).

Merritt, K., and Johnson, A. G. "Studies on the Adjuvant Action of Bacterial Endotoxins on Antibody Formation. VI. Enhancement of Antibody Formation by Nucleic Acids," *J. Immunol.* **94**:416–422 (1965).

Mozes, E., Shearer, G. M., Sela, M., and Braun, W. "Conversion with Polynucleotides of a Genetically Controlled Low Immune Response to a High Response in Mice Immunized with a Synthetic Polypeptide Antigen." *Biological Effects of Polynucleotides*, R. F. Beers and W. Braun, eds. New York: Springer-Verlag, pp. 197–214 (1971).

Schmidtke, J. R., and Johnson, A. G. "Regulation of the Immune System by Synthetic Polynucleotides. I. Characteristics of Adjuvant Action on Antibody Synthesis," *J. Immunol.* **106**:1191–1200 (1971).

Stout, R. D., and Johnson, A. G. "Regulation of the Immune System By Synthetic Polynucleotides. V. Effect on Cell-Associated Immunoglobulin Receptors and Immunological Memory," *J. Exp. Med.* **135**:45–67 (1972).

EFFECTS OF ADENYL CYCLASE-STIMULATING FACTORS OF VIBRIO CHOLERAE ON ANTIBODY FORMATION

RICHARD A. WINCHURCH

I. Introduction

Immune adjuvants are those substances that are capable of enhancing an immune response. A wide variety of adjuvants has been and is currently in use, ranging in composition from simple inorganic materials (Warren et al., 1969) through complex macromolecules such as the polynucleotides (Braun and Nakano, 1967) and including the heterologous chemical structures of myobacteria and bacteria (Freund et al., 1948; Munoz, 1964). The mechanisms by which adjuvants exert their effects on the cells of the immune system are at best poorly understood. Furthermore, it is unlikely, and indeed unnecessary, that all adjuvant materials possess a common mode of action. The immune response is a complex reaction involving interactions between cells and immunogens, cells and soluble mediators (Mackler, 1971) and among various cell types themselves (Miller and Mitchell, 1968). Accompanying these interactions are biochemical events within lymphoid cells, some of which are related to the energy processes required to drive the cellular metabolism, and some of which are probably related to the reception and transmission of the antigenic signal itself. Adjuvants may act by influencing any or all of these interactions and chemical processes. By studying the effects of adjuvants possessing similar properties, we hope to learn something about their mode of action and thus obtain a more complete knowledge of the cellular interactions and biochemical events of the immune response.

The studies reported here deal with the effects and properties of one adjuvant substance, HS, a somatic antigen fraction isolated from *Vibrio cholerae*, Ogawa 41. This material is, in itself, a potent immunogen and has proved to be an interesting tool in investigations into the mode of action of certain adjuvants. In a manner reminiscent of bacterial endotoxin, it stimulated increases in antibody-forming cells to both sheep and horse erythrocytes. In in vitro studies, the cholera fraction was mitogenic for mouse spleen cells in a manner similar to phytohemagglutinin. Finally, it was found to stimulate increased activity in the membrane-bound adenyl cyclase of mouse spleen cell cultures. It is generally believed that adenyl cyclase activity plays a role in the intracellular metabolism of a varity of cell types (Breckenridge, 1970) and may influence the transcription and translation of DNA (Dokas and Kleinsmith, 1971). Since adenyl cyclase is known to be a membrane-associated enzyme and is regulated by pharmacologic receptors on the cell surfase (Robison et al., 1971), studies were undertaken to show whether or not the cholera fraction exerted its influence on lymphocytes through these receptors. The data indicate that at least one effect of the cholera fraction HS can be blocked by appropriate adrenergic blocking agents.

These findings are discussed, and a hypothesis is described, which might explain the mode of action of this type of adjuvant. In addition, possible relationships between adjuvant activity and the stimulation of immunocompetant cells by antigen alone are described.

II. Materials and Methods

A. Agar Plaque Assay

Mice were injected intravenously with 1 to 2×10^8 sheep or horse red blood cells (sRBC or hRBC) in a volume of 0.1 ml with or without the *Vibrio cholerae* fraction. Two days later, the spleens were removed, teased in Eagle's minimum essential medium (MEM), and the number of direct plaque-forming cells enumerated by a previously described modification of the method of Jerne and Norden (1963).

B. Determination of DNA Synthesis in Cell Cultures

Spleens from animals presensitized to the cholera fraction or normal spleens were aseptically removed and gently teased in MEM. After removal of large clumps, the cell suspension was adjusted to a concentration of 2.5 to 3×10^6 cells/ml in a complete medium consisting of MEM, 10% calf serum, 2 mM 1-glutamine, and containing 100 units penicillin and 100 μg streptomycin per milliliter. Two milliliter aliquots were seeded in 12×75 mm plastic culture tubes and incubated at 37°C in a humid atmosphere of 6% CO_2 in air. Sixteen hours before termination of the cultures, 1 μ Ci tritiated thymidine (New England Nuclear Corp., Boston, Mass.) specific activity 2.0 Ci/mmole in 0.1 ml was added to each culture. At the time of assay, the cultures were centrifuged, the supernatant medium drawn off, and the incorporation of radioactivity into acid precipitable material assessed according to a previously published procedure (Winchurch and Actor, 1972).

C. Adenyl Cyclase Assay

Cell cultures were prepared and maintained as described above except that the cell concentration was adjusted to 6 to 10×10^6 cells/ml, and 5 ml aliquots were seeded in 17×100 plastic culture tubes. At the termination of the incubation period, the cultures were centrifuged and all media removed. The cell pellet was then quick-frozen in alcohol-dry ice and stored at $-60°C$ overnight. A crude enzyme fraction was prepared by breaking the cells in a glass homogenizer with a motor-driven teflon pestle and centrifuging the homogenate at $20,000 \times g$ for 20 min. The supernatant was discarded, and the enzyme-containing pellet was resuspended in 0.6 ml of a reaction mixture consisting of 40 mM tris-HC1, pH 7.6, 3.3 mM $MgSo_4$, 5 mM theophylline, 10 mM Phosphoenol pyruvate, 5 mM Clelands reagent, 9 units pyruvate kinase, and 2 mM ^{3}H-ATP (New England Nuclear Corp.) specific activity 8.3 mCi/mmole. The enzyme reaction was carried out at $31°C$ for 20 min. The cyclic AMP generated in the reaction was assayed by the method of Krishna (Krishna et al., 1968). Cellular protein in each assay was digested with 0.1 N NaOH and quantitated by the method of Lowry (Lowry et al., 1951).

D. Sources of Reagents

Phytohemagglutinin (PHA-M) was obtained from Difco Laboratories, Detroit, Mich. It was reconstituted with 5 ml sterile water and used at a concentration of 0.01 ml/ml of culture. The preparation of the *Vibrio cholerae* fraction, HS, has been described elsewhere (Jensen et al., 1972). Essentially, *V. cholerae,* Ogawa 41, was grown overnight on solid agar at $37°C$. The cells were broken using glass beads and debris removed by centrifugation at $27,000 \times g$ for 60 min. The supernatant was put onto a DEAE cellulose column and eluted with 10 mM tris buffer, pH 7.2, containing 10 mM $MgCl_2$ and 100 mM NH_4Cl. The first peak eluted contained the highly immunogenic cholera fraction used in the present studies.

Chemical reagents were obtained from Sigma Chemical Company, St. Louis, Mo. Propranolol was obtained from Ayerst Laboratories Inc., New York, N.Y.

III. Results

A. The Adjuvant Properties of the Vibrio Cholerea Fraction

Table 1 summarizes the effects of the *Vibrio cholerae* fraction on antibody responses in mice. In this and in similar experiments, the antigen, either sheep or horse red blood cells, was injected intravenously via the tail vein, while the adjuvant was administered intraperitoneally. HS caused a fourfold increase in the number of antibody-forming cells to sRBC, and 1½-fold increase when hRBC was the immunizing antigen. The similarity between the cholera fraction and bacterial endotoxin

Table 1. The Effect of Vibrio cholerae Fraction HS on the Two-Day Hemolysin Response to Sheep Red Blood Cells (sRBC) and Horse Red Blood Cells (hRBC) in CFW Mice

	Mean PFC ± SE per 10^8 spleen cells assayed	
Treatment group[a]	Against sRBC	Against hRBC
25 μg HS, ip	50.5 ± 16	44.2 ± 18
25 μg HS, ip + 10^8 sRBC, iv	819.8 ± 135	127.2 ± 27
10^8 sRBC, iv	176.8 ± 53	33.4 ± 19
25 μg HS, ip + 10^8 hRBC, iv	52.0 ± 18	349.0 ± 45
10^8 hRBC, iv	27.6 ± 15	169.0 ± 26
Saline control	23.4 ± 13	17.6 ± 5

[a]Five mice per group.

becomes evident when the nonspecific responses against a heterologous antigen were measured. The injection of sRBC along with cholera fraction caused an increase of nearly threefold in the response to hRBC, an antigen which cross-reacts poorly, if at all, with sRBC.

B. *The Effect of the Cholera Fraction on DNA Synthesis*

An increase in the uptake and incorporation of tritiated thymidine was observed when both normal and presensitized mouse spleen cells were cultured in the presence of the HS. While this finding was expected in the case where the splenic lymphocytes were presensitized to the cholera material, it was not expected in the case of normal lymphocytes,

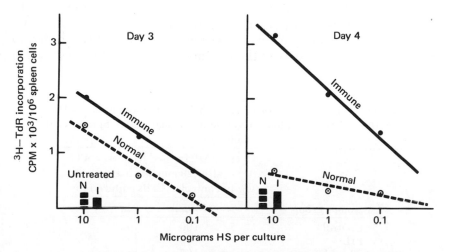

Fig. 1. The effect of HS on tritiated thymidine incorporation in normal and presensitized mouse spleen cell cultures.

Table 2. The Effect of Vibrio Cholerae Fraction HS on the Incorporation of Tritiated Thymidine in Normal Mouse Spleen Cell Cultures

Exp. No.	Strain	Treatment group[a]	^{3}H-TdR incorporation (mean CPM)	
			Day 2	Day 3
1	Balb/c	Control	838	2,329
		PHA−M	10,021	6,130
		20 µg HS	3,653	7,651
2	A/J	Control	1,324	806
		PHA	9,272	7,114
		20 µg HS	9,388	4,443

[a]Four cultures per group.

particularly since we could demonstrate no background of antibody production against *V. cholerae* in any of the mouse strains studied.

The data in Fig. 1 and Table 2 summarize these mitogenic effects of the cholera fraction and compare these with the effects of PHA.

C. The Effects of PHA and the Cholera Fraction on Adenyl Cyclase Activity

The effects of the HS on the adenyl cyclase activity of mouse lymphocytes were investigated by incubating cultures with the cholera material for varying periods of time and subsequently measuring the adenyl cyclase activity in a cell-free lysate obtained from the cultures. The data in Table 3 show the results of incubation for 15 min and 1 hr. Both PHA and HS have no effect on adenyl cyclase after 15 min., but after 1 hr each of these substances increases adenyl cyclase activity about 1½-fold. Subsequent experiments with the cholera fraction indicated that incubation for periods of up to 1 hr frequently produced no elevation of adenyl cyclase, but longer periods of up to 2 or 4 hr led to marked increases in activity of the enzyme. Typical results are shown in Fig. 2.

Table 3. The Influence of PHA and Vibrio Cholerae Fraction HS on the Activity of Adenyl Cyclase in Normal Balb/c Spleen Cell Cultures

Treatment group	Cyclic AMP formed from ^{3}H−ATP after 20 min incubation (% of controls)[a] Cells cultured for	
	15 min	1 hr
Control	100	100
PHA	95	147
20 µg HS	107	154

[a]Duplicate determinations.

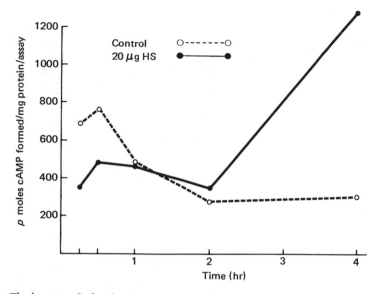

Fig. 2. The kinetics of adenyl cyclase stimulation by *Vibrio cholerae* fraction HS in normal Balb/c spleen cell cultures.

D. The Effect of Adrenergic Blockers on Cholera Fraction-Induced Thymidine Incorporation

Lymphocytes were treated in vitro with mitogenic concentrations of HS in the presence or absence of propranolol, a potent blocker of beta adrenergic activity. At concentrations of 10^{-5} M, propranolol caused a modest but significant decrease in the ability of the cholera material to stimulate DNA synthesis in lymphocytes ($P = 0.05$). Typical effects of propranolol are summarized in Table 4. The effect of propranolol appeared to be due to its pharmacologic properties and not to any toxic effects. (It should be mentioned that phenoxybenzamine, an alpha blocker, also decreased thymidine incorporation, but in this case toxic effects were evident and were presumed to be the reason for the decrease.)

Table 4. Inhibition by a Beta Adrenergic Blocker of HS-Induced Tritiated Thymidine Incorporation in Balb/c Spleen Cell Cultures

Treatment group	^{3}H-TdR incorporation measured on day 3[a]
Control	277 ± 49
20 μg HS	1051 ± 115
20 μg HS + 10^{-5} M propranolol	761 ± 73
10^{-5} M propranolol	254 ± 46

[a]Mean CPM $\pm$ SD.

IV. Discussion

The interaction of an immunogen with an immunocompetant cell leads to alterations in the metabolism and biosynthetic processes of that cell. Translation of RNA, synthesis of protein, and replication of DNA are believed to be followed by transformation of the cell and ultimately mitosis resulting in proliferation of the cell in clonal fashion. These events are evident in vitro (Hirshhorn et al., 1963); Chalmers et al., 1967) and are thought to be part of a sequence essential to the immune response. Many agents that interfere with these events are immunosuppressive. On the other hand, adjuvants are agents that enhance immune responses, and some adjuvants may act by influencing any or all of these cellular events in a positive manner. While the synthetic activities of lymphocytes are essential to immune responses, they are, nonetheless, secondary events that follow the interaction of the immunocompetant cell with the stimulating antigen. The question to which we must address ourselves regards the nature of the interaction of a cell with antigen and whether or not adjuvants might exert their effects by influencing this interaction.

There is evidence that immune events may be directed by cAMP-dependent mechanisms. Ishizuka et al. have shown that the activities of immunoenhancing polyribonucleotides, notably poly adenylate: poly uridylate complexes (poly A:U), are further enhanced by simultaneous administration of aminophylline, a potent inhibitor of the phosphodiesterase which degrades 3', 5'-cyclic AMP to 5'-AMP (Ishizuka et al., 1970). In addition, these investigators have demonstrated that addition of exogeneous cyclic AMP or its butryl derivatives to antigen-stimulated mouse spleen cell cultures results in a significant increase in the number of antibody-forming cells. Similar effects have been obtained in in vivo experiments (Braun et al., 1971a). Winchurch et al. showed that this activity of the polynucleotides may be related to their stimulatory effects on the adenyl cyclase of spleen cell cultures (Winchurch et al., 1971). Further evidence for the hypothesis that cyclic AMP plays a role in the activity of lymphocytes is found in the work of Parker and his colleagues. Using human peripheral blood leucocyte cultures, these workers showed that the mitogen, phytohemagglutinin, increased the activity of adenyl cyclase and elevated intracellular cyclic AMP levels (Parker et al., 1971; Smith et al., 1971). Whitfield and MacManus and co-workers demonstrated that the effects of several hormones on thymocytes are mediated through the action of cyclic AMP (MacManus et al., 1971; Franks et al., 1971).

Additional evidence that cyclic AMP influences nucleic acid metabolism has been obtained by Gallo and Whang-Peng (1970) as well as by Braun and Winchurch (Braun and Winchurch, unpublished data). Although contrary reports have appeared in the literature (Hirshhorn et al., 1970), these workers showed that addition of cyclic AMP or its dibutyryl derivatives to PHA-stimulated cultures resulted in significant increases in the uptake and incorporation of tritiated thymidine.

While these studies do not prove conclusively that the induction or generation of an immune response is necessarily mediated through cAMP-dependent mechanisms, they do present data compatible with such a hypothesis.

Our investigations using the *Vibrio cholerae* fraction provide evidence in further support of the involvement of cyclic AMP in the induction of immune responses and may point toward a possible mechanism for this activity. The cholera fraction, in a manner similar to that of the polynucleotides and bacterial endotoxin, stimulates increases in antibody-forming cells. The similarity between the cholera fraction and endotoxin is especially evident from the nonspecific increases in antibody-forming cells to heterologous horse red blood cells in animals injected with sheep red blood cells along with the cholera fraction. It might be pointed out that Braun has shown that epinephrine, an alpha and beta adrenergic agonist, and chlorpromazine, a compound known to block alpha acitvity and potentiate stimualtion of vascular beta receptors (Jarvick, 1970), both stimulate similar nonspecific increases in the numbers of antibody-forming cells (Braun et al., 1970).

Our studies also indicate that the cholera fraction is mitogenic for normal mouse splenic lymphocytes. Earlier studies have shown that poly A and its oligomers can exert an influence on PHA-induced DNA synthesis (Braun et al., 1971b). While differences exist between the conditions under which poly A and the cholera fraction affect DNA synthesis (poly A seems to require prior stimulation by PHA to exert its effects), the fact remains that both adjuvant substances affect this phase of lymphocyte metabolism.

Finally, our data show that the cholera fraction stimulates increases in the adenyl cyclase activity of mouse spleen cell cultures. Again, earlier studies have indicated that poly A has a similar effect. We believe that it is not inconsistent to assume that some of the biologic activities due to the cholera fraction, poly A, and perhaps even PHA are similar and that these activities follow the sequence: (1) stimulation of membrane-bound receptors functionally linked to adenyl cyclase, (2) increases in adenyl cyclase activity and in intracellular cyclic AMP levels, (3) de-repression of the lymphocyte genome through the activation of histone kinases by cyclic AMP and subsequently the phosphorylation of nuclear histones, and (4) translation of RNA, synthesis of protein, and replication of DNA followed by proliferation of the cell. Since both poly A and the cholera fraction have been shown to influence both early and late events in this sequence, it is logical to conclude that their effect is probably on one of the initial steps in the sequence, e.g., the stimulation of the adenyl cyclase receptors. If this is the case, then all of the subsequent demonstrable effects including adjuvant activity may be due to this initial stimulation.

Evidence for the interaction of cyclic AMP and nuclear histones has been presented. Cross and Ord have demonstrated that in PHA-stimulated pig leucocyte cultures, adenyl cyclase stimulation may be linked to the phosphorylation of certain nuclear histones (Cross and Ord, 1971). Langan has presented extensive data on this aspect of cell metabolism and has proposed a model to explain how intracellular increases in cyclic AMP levels lead to the activation of histone kinases which, in turn, phosphorylate lysine-rich nuclear histones (Langan, 1970). According to Langan, this results in unlocking of the genome for subsequent transcription and translation. If Langan's hypothesis is correct and if certain adjuvants act by influencing an initial step in the above-mentioned sequence, e.g., stimulation of adenyl cyclase receptors, then their mode of action is easily explained.

The question that remains, however, is the following: Which receptors on the lymphocyte surface are responsible for activating adenyl cyclase and turning on lymphocyte activity? Two alternative explanations seem to be indicated. First, activation of adenyl cyclase is a consequence of the interaction of antibodylike receptors on the lymphocyte surface with antigenic determinants of the immunogen. It is known that conformational changes occur in the invariant portion of the membrane-bound immunoglobulin chains after interaction of the combining sites with determinants on the immunogen (Bretscher and Cohn, 1968). By some unknown mechanism, these changes might activate the membrane-bound adenyl cyclase, increase intracellular cyclic AMP levels, and thus initiate the sequence perhaps in a manner similar to that of PHA. In this case, certain adjuvants might act by facilitating the interaction of antigens with the lymphocyte antibodylike receptors. Alternatively, conformational changes might occur as a result of the interaction of antigen, T cells, and B cells. Adjuvants might facilitate the cell-to-cell interaction and thus effect the same net result.

A second possibility is that activation of adenyl cyclase occurs through the interaction of receptors (X receptors) distinct from the antibodylike receptors with portions of the immunogen ("X" determinants) structurally unrelated to antigenic determinants. In this case, the function of antigen-antibody receptor interaction would be simply to provide the specificity of the response, while the activation of lymphocytes would occur through the second nonspecific interaction. This second interaction between the X determinant and X receptors would be more likely to occur with those lymphocytes that have been brought into close proximity with the immunogen through the specific event and would drive these cells into synthetic activity and proliferation. In this hypothetical scheme as diagrammed in Fig. 3, adjuvants might serve to augment the nonspecific "hits" by providing additional X determinants

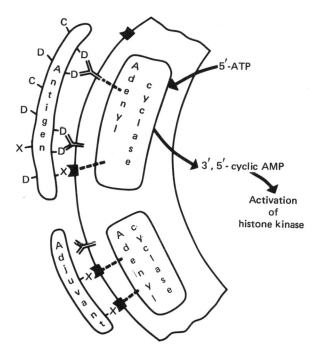

Fig. 3. Diagrammatic representation of a section of lymphocyte plasma membrane showing the interaction of specific antigenic determinants (C or D) with antibodylike receptors and the interaction of hypothetical X determinants with receptors for X.

which, after interacting with additional X receptors, would increase the efficiency of the initial signal due to the immunogen. The specific event (involving interaction of antibody receptors with antigenic determinants C or D), while concerned mainly with bringing together the appropriate lymphocyte and immunogen, could also play a minor role in the activation of adenyl cyclase in the manner described above.

This hypothesis would explain the nonspecific effects of endotoxinlike materials. Random interactions of these adjuvants with the lymphocyte population would activate some cells in the absence of specific antigen and result in increases in the number of antibody-forming cells to heterologous antigens. The effects due to epinephrine and chlorpromazine might be due to similar nonspecific activation of lymphocytes.

While the identities of the hypothetical X determinant and the receptor for X are not known, certain possibilities are indicated. Epinephrine is a typical beta adrenergic agonist, and chlorpromazine has been shown to block alpha adrenergic activity and potentiate stimulation of beta receptors. The X determinant may be structurally similar to adrenergic agonists and may, in fact, be similar to a beta agonist. It is not impossible that immunogens and certain adjuvants carry exposed surface structures, which resemble adrenergic agonists in both molecular form and charge. While there is no direct evidence that immunogens or adjuvants bear adrenergiclike structures, the presence of adrenergic receptors on lymphocytes has been established. Smith et al. have shown that propranolol, a beta adrenergic blocker, inhibits PHA-induced increases in cyclic AMP in human lymphocyte cultures (Smith et al., 1971). In these experiments, phentolamine, an alpha blocker, was without effect. Isoproterenol, a potent beta agonist, increased lymphocyte cyclic AMP in a manner that could be selectively blocked by beta blockers. Bourne and Melmon have presented similar data on the beta adrenergic responses of lymphocytes (Bourne and Melmon, 1971). Hadden et al. have shown that both alpha and beta agonists can, under appropriate conditions, enhance PHA-induced thymidine incorporation (Hadden et al., 1971). Bitensky et al. have presented data showing that *E. coli* endotoxin stimulated increases in epinephrine-responsive adenyl cyclase in both mouse spleen and liver cells (Bitensky et al., 1971). The enhanced activity due to endotoxin was measured in vitro and was evident even in tissues from adrenalectomized animals, indicating a direct and sustained activity of endotoxin on the cells themselves. These data point toward a possible synergism between endotoxinlike materials and adrenergic stimulators.

Our data concerning the blocking effects of propranolol on cholera fraction–induced DNA synthesis are in agreement with the concept that lymphocytes respond to adrenergic stimulation. Furthermore, these data also indicate that agents other than catacholamines can provoke an adrenergiclike response in lymphocytes. While it is evident that this response is, in many respects, similar to that due to antigenic stimulation, there is no direct evidence indicating that antigenic stimulation is a direct consequence of adrenergiclike stimulation. Further evidence is needed. Nonetheless, the concept that certain adjuvants and immunogens act on lymphocytes through cAMP-dependent steps and that adrenergic mechanisms govern these events may provide a useful tool for future investigations.

V. Summary

The biologic effects of a somatic antigen fraction prepared from *V. cholerae* were examined. In addition to being a potent immune adjuvant, the cholerae fraction stimulated increases in the uptake and incorporation of tritiated thymidine in mouse splenic lymphocytes.

This mitogenic activity was partially blocked by propranolol, an agent that blocks beta adrenergic activity. In addition, the cholerae material stimulated adenyl cyclase activity in mouse spleen cell cultures. These findings along with recent reports in the literature are discussed, and a hypothesis is presented to explain the activity of certain types of immune adjuvants and to relate this activity to adrenergic stimulation of immunocompetant cells.

Discussion of Dr. Winchurch's Paper

DR. MORSE: I would like to ask whether HS has any specificity for either T or B cells and whether you would say something about the nature of HS.

DR. WINCHURCH: We have not studied HS effects on isolated B and T cells. HS is a somatic antigen fraction composed predominantly of polysaccharide and lipid. It might be described as an endotoxinlike material.

DR. MORSE: Why don't you call it endotoxin or *Vibrio cholerae* endotoxin?

DR. WINCHURCH: That is a good question. Many of its properties are like endotoxin. My terminology is only a matter of convenience.

DR. LICHTENSTEIN: How can you be sure it isn't cholera toxin? The properties are much the same.

DR. WINCHURCH: There are a number of criteria that indicate that it is not classic cholera enterotoxin. We did the usual toxin tests and found no toxin activity. Just this morning I learned from Dr. Nelson, who has also worked with the material, that he has detected no toxin activity.

DR. KALINER: I have two questions. First, what is the data showing that antigen activates adenylate cyclase. Secondly, are there other activators of adenylate cyclase that act as adjuvants? Theophylline produces effects on cyclic AMP by phosphodiesterase inhibition rather than by stimulating adenylate cyclase. And there is some evidence that it produces effects on cells outside of the cyclic AMP system.

DR. WINCHURCH: The only two adjuvants that I know of, which have been studied with regard to adenyl cyclase activity, are HS and the polynucleotides. I don't say that all adjuvants operate via this mechanism. For all I know, these are the only two that do. On the other hand, any material that can stimulate lymphoid cells to proliferate and mount an immune response can be classified as an adjuvant, including substances that stimulate lymphocyte metabolism. As far as antigen stimulation of adenylate cyclase is concerned, I have heard of one report indicating that antigen is able to stimulate adenylate cyclase activity, but I have not seen it in print.

DR. BLOOM: Did I understand correctly that you were saying that the action of adjuvant can readily be explained by saying that the cells are full of X?

DR. WINCHURCH: I like the way you put that.

DR. BLOOM: The comment was rhetorical.

DR. WINCHURCH: "X" is a hypothetical structural feature, which, we think, may be common to antigens and some adjuvants. One needs to postulate some way an antigen can turn on a lymphoid cell. As I see it, there are two alternatives. First, the interaction of the antigenic determinants with immunoglobulin receptors activates the cell, presumably through cyclic AMP. Alternatively, most protein or polysaccharide antigens might contain small portions of their structure (X), which fortuitously interact with hormonal receptors on the lymphocyte membrane and activate adenylate cyclase. One could further postulate that adjuvants may also contain molecules resembling "X" that provide an enhanced signal to the cell.

References

Bitensky, M. W., Gorman, R. E. and Thomas, L. "Selective Stimulation of Epinephrine-Responsive Adenyl Cyclase in Mice by Endotoxin," *Proc. Soc. Exp. Biol. Med.* **138**:773–775 (1971).

Bourne, H. R., and Melmon, K. L. "Adenyl Cyclase in Human Leucocytes. Evidence for Activation by Separate Beta Adrenergic and Prostaglandin Receptors," *J. Pharmacol. Exp. Ther.* **178**:1–7 (1971).

Braun, W., and Nakano, M. "Antibody Formation: Stimulation by Polyadenylic and Polycytidylic Acids," *Science* **157**:819–821 (1967).

Braun, W., Yajima, Y., Jimenez, L., and Winchurch, R. In *Developmental Aspects of Antibody Formation and Structure,* J. Sterzl, ed. New York: Academic Press (1970).

Braun, W., Ishizuka, M., Winchurch, R., and Webb, D. "On the Role of Cyclic AMP in Immune Responses," *Ann. N.Y. Acad. Sci.* **185**:417–422 (1971a).

———. "Cells and Signals in Immunological Nonresponsiveness," *Ann. N.Y. Acad. Sci.* **181**:289–298 (1971b).

Breckenridge, B. "Cyclic AMP and Drug Action," *Ann. Rev. Pharmacol.* **10**:19–34 (1970).

Bretscher, P. A., and Cohn, M. "Minimal Model for the Mechanism of Antibody Induction and Paralysis by Antigen," *Nature* **220**:444–448 (1968).

Chalmers, D. G., Cooper, E. G., Evans, C., and Topping, N. "Quantitation of the Response of Lymphocytes in Culture to Specific and Nonspecific Stimulation," *Int. Arch. Allergy* **32**:117–130 (1967).

Cross, M., and Ord, M. "Changes in Histone Phosphorylation and Associated Early Metabolis Events in Pig Lymphocyte Cultures Transformed by Phytohemagglutinin or 6-N, 2'-o-Dibutyryl-Adenosine 3',5' -Cyclic Monophosphate," *Biochem. J.* **124**:241–248 (1971).

Dokas, L., and Kleinsmith, L. "Adenosine 3', 5' -Monophosphate Increases Capacity for RNA Synthesis in Rat Liver Nuclei," *Science* **172**:1237–1238 (1971).

Franks, D. J., MacManus, J. P., and Whitfield, J. F. "The Effect of Prostaglandins on Cyclic AMP Production and Cell Proliferation in Thymus Lymphocytes," *Biochem. Biophys. Res. Commun.* **44**:1177–1183 (1971).

Freund, J., Thomson, K. J., Hough, H. B., Sommer, H., and Pisani, T. "Antibody Formation and Sensitization with the Aid of Adjuvants," *J. Immunol.* **60**:383–398 (1948).

Gallo, R. C., and Whang-Peng, J. "Enhanced Transformation of Human Immunocompetent Cells by Dibutyryl Adenosine 3',5' -Monophosphate," *J. Nat. Cancer Inst.* **47**:91–94 (1971).

Hadden, J. W., Hadden E. M., and Good, R. A. "Adrenergic Mechanisms in Human Lymphocyte Metabolism," *Biochim. Biophys. Acta* **237**:339–347 (1971).

Hirschhorn, K., Buck, F., Kolodny, R., Firshein, I., and Hashem, N. "Immune Response and Mitosis of Human Peripheral Blood Lymphocytes in Vitro," *Science* **142**:1182–1187 (1963).

Hirschhorn, R., Grossman, J., and Weissmann, G. "Effect of Cyclic 3',5'-Adenosine Monophosphate and Theophylline on Lymphocyte Transformation," *Proc. Soc. Exp. Biol. Med.* **133**:1361–1365 (1970).

Ishizuka, M., Gafni, M., and Braun, W. "Cyclic AMP Effects on Antibody Formation and Their Similarities to Hormone-Mediated Events," *Proc. Soc. Exp. Biol. Med.* **134**:963–967 (1970).

Jarvick, M. E. In *The Pharmacological Basis of Therapeutics* L. S. Goodman and A. Gilman, eds. New York: MacMillan (1970).

Jensen, R., Gregory, B., Naylor, J., and Actor, P. Isolation of Protective Somatic Antigens from *Vibrio cholerae* (Ogawa) Ribosomal Preparations," *Infec. Immunol.* **6**:156–161 (1972).

Jerne, N. K., and Norden, A. A. "Plaque Formation in Agar by Single Antibody-Producing Cells," *Science* **140**:405 (1963).

Krishna, G., Weiss, B., and Brodie, B. J. "A Simple, Sensitive Method for the Assay of Adenyl Cyclase," *J. Pharmacol. Exp. Ther.* **163**:379–385 (1968).

Langan, T. A. In *Role of Cyclic AMP in Cell Function*, P. Greengard and E. Costa, eds. *Advances in Biochemical Psychopharmacology*, vol. 3, pp. 307–323. New York: Raven Press (1970).

Lowry, O. H., Rosenbrough, N. J., Farr, L., and Randall, R. J. "Protein Measurement with the Folin Phenol Reagent, *J. Biol. Chem.* **193**:265–275 (1951).

Mackler, B. "Role of Soluble Lymphocyte Mediators in Malignant-Tumor Destruction," *Lancet* ii:297–301 (1971).

MacManus, J. P., Whitfield, J. F., and Yondale, T. "Stimulation by Epinephrine of Adenyl Cyclase Activity, Cyclic AMP Formation, DNA Synthesis and Cell Proliferation in Populations of Rat Thymic Lymphocytes," *J. Cell Physiol.* **77**:103–116 (1971).

Miller, J. F. A. P., and Mitchell,G. F. "Cell-to-Cell Interaction in the Immune Response. Hemolysin-Forming Cells in Neonatally Thymectomized Mice Reconstituted with Thymus or Thoracic Duct Lymphocytes," *J. Exp. Med.* **128**:801–820 (1968).

Munoz, J. "Effect of Bacteria and Bacterial Products on Antibody Response." *Advances in Immunology* vol. 4, pp. 397–440, F. J. Dixon and J. H. Humphrey, eds. New York: Academic Press (1964).

Parker, C. W., Smith, J. W., and Steiner, A. L. "Early Effects of Phytohemagglutinin (PHA) on Lymphocyte Cyclic AMP Levels," *Int. Arch. Allergy* **41**:40–46 (1971).

Robison, G. A., Butcher, R. W., and Sutherland, E. W. "Adenyl Cyclase As an Adrenergic Receptor, *Ann. N.Y. Acad. Sci.* **139**:703–723 (1967).

—— *Cyclic AMP*. New York: Academic Press (1971).

Smith, J. W., Steiner, A. L., Newberry, Jr., W. M., and Parker, C. W. "Cyclic
    Adenosine 3',5'-Monophosphate in Human Lymphocytes. Alterations after
    Phytohemagglutinin Stimulation," *J. Clin. Invest.* **50**:432–441 (1971).
Warren, J., Kende, M., and Takani, K. "The Adjuvant Effect of Powered Perric
    Oxide: Enhancement of Response to Mycoplasma Pneumoniae and
    Respiratory Syncytial Virus Vaccines," *J. Immunol.* **102**:1300–1308 (1969).
Winchurch, R., Ishizuka, M., Webb, D., and Braun, W. "Adenyl Cyclase Activity
    of Spleen Cells Exposed to Immunoenhancing Synthetic Oilgo- and
    Polynucleotides," *J. Immunol.* **106**:1399–1400 (1971).
Winchurch, R., and Actor, P. "The Effects of an Immunoenhancing Bacterial
    Product on the Adenyl Cyclase Activity of Mouse Spleen Cells," *J. Im-
    munol.* **108**:1305–1311 (1972).

PHARMACOLOGIC REGULATION OF ANTIBODY RELEASE IN VITRO: EFFECTS OF VASOACTIVE AMINES AND CYCLIC AMP

H. R. Bourne, K. L. Melmon, Y. Weinstein, and G. M. Shearer

I. Experimental Background

Vasoactive hormones, employing intracellular cyclic AMP as a "second messenger," can regulate the expression of both immediate and delayed (cell-mediated) immune responses in vitro. Histamine, betaadrenergic catecholamines, and certain prostaglandins inhibit both the IgE-mediated release of histamine and other mediators from basophils and mast cells (Lichtenstein and Margolis, 1968; Koopman et al., 1970; Ishizaka et al., 1971; Bourne et al., 1971b, 1972) and the ability of sensitized lymphocytes to kill allogeneic cells (Henney and Lichtenstein, 1971; Henney et al., 1972). We have recently found that the same vasoactive hormones can modulate the expression of a third type of immunologic response: the release of antibody from lymphocytes. In the hemolytic plaque model of the humoral immune response (Jerne, 1963), as in the models of immediate and delayed hypersensitivity, the effect of these hormones is inhibitory. Furthermore, each of the hormones appears to inhibit plaque formation by stimulating the accumulation of cyclic AMP in lymphocytes.

This chapter will first outline the pharmacologic approach we have used to investigate the regulatory roles of vasoactive hormones and cyclic AMP on histamine release from human basophils and the cytolytic activity of sensitized mouse lymphocytes. (These experiments are described in greater detail in the chapters of this book by L. M. Lichtenstein and C. S. Henney, respectively.) In the next section we apply the

This work was supported by U. S. Public Health Service Grants HL-09964 and GM-16496 and agreement 06-035. The experiments were performed at the Weizmann Institute of Science during the tenure of a National Institutes of Health Special Fellowship 1 FO, 3 HL 50, 244-01 (KLM), and an Established Investigatorship of the American Heart Association (HRB).

same pharmacologic approach to the model of hemolytic plaque formation, and finally, we attempt to place these results in a broader perspective. Although the conclusions to be drawn from effects of drugs on
model systems in vitro must be qualified, our results raise interesting
questions relating to control of immunologic responses in vivo. In addition, our experiments imply the presence of receptors for vasoactive
hormones on immunocompetent cells. This implication forms the basis of
an entirely different experimental approach, aimed at detecting hormone receptors on leukocytes by binding cells to insolubilized hormones,
and using the binding phenomenon as a means of separating cells with
different immunologic functions. This approach will be discussed in the
following two chapters by K. L. Melmon and G. Shearer.

A. *Cyclic AMP in Immediate and Delayed Hypersensitivity*

The now familiar "second messenger" concept of cyclic AMP has led
to a rather formal set of criteria for implicating the cyclic nucleotide as
an intracellular mediator of the effect of a hormone on cell function
(Sutherland et al., 1968), which can be summarized briefly as follows:

(1) Cyclic AMP itself, or a suitable analogue, should reproduce the
hormone's effect on cell function.
(2) Theophylline or other inhibitors of intracellular degradation of
cyclic AMP by phosphodiesterase should reproduce the hormone's effect
and act synergistically with submaximal concentrations of the hormone
itself.
(3) Adenyl cyclase should be present in broken preparations of the
tissue and should be stimulated by the hormone.
(4) When the cell or tissue is exposed to the hormone, hormone congeners, or pharmacologic antagonists, the cell's cyclic AMP content
should rise or fall appropriately with the altered cell function thought to
be regulated by cyclic AMP.

In 1968, Lichtenstein and Margolis reported that catecholamines and
methylxanthines inhibited the antigen-induced release of histamine from
human leukocytes in vitro and hypothesized that the inhibition was
mediated by intracellular cyclic AMP. Later, Henney and Lichtenstein
found that the same drugs inhibited the cytolysis of allogeneic target
cells by sensitized mouse splenic lymphocytes (1971). Subsequently, in
collaboration with these workers, we established the inhibitory role of
cyclic AMP in both systems by applying the four formal criteria listed
above. Dibutyryl cyclic AMP inhibited both histamine release and
cytolytic activity; theophylline, an inhibitor of phosphodiesterase in
leukocytes and other tissues, produced inhibitory effects consistent with
the hypothesis; adenyl cyclase in broken leukocytes was responsive to

the appropriate hormones (Bourne and Melmon, 1971; Bourne et al., 1971b, 1972; Henney et al., 1972).

Any of these lines of evidence, taken alone, would constitute relatively weak evidence for the inhibitory role of cyclic AMP (Bourne, 1972). Consequently, we carefully compared the changes in leukocyte cyclic AMP content with the functional effects of a variety of hormones, hormone congeners, and their pharmacologic antagonists (i.e., we applied the fourth criterion listed above). Catecholamines, histamine, and certain prostaglandins stimulated cyclic AMP accumulation and inhibited both histamine release and lymphocyte cytolytic activity. A number of other hormones, which stimulate adenyl cyclase in other tissues, had no effect on leukocytes; these included glucagon, adrenocorticotrophic hormone, and serotonin (Bourne, 1972). All three types of hormones that affected leukocyte function were well known for their effects on the cardiovascular system, predominantly as vasodilators, but non had been thought of as potential regulators of immunologic responses. For this reason, their effects will be reviewed individually.

1. Catecholamines

The effects of catecholamines on leukocyte cyclic AMP, histamine release, and lymphocyte cytolytic activity fit into a pattern that implies a beta adrenergic receptor in human basophils and mouse lymphocytes. The order of potency of adrenergic agonists was similar to that described by Ahlquist (1948) for beta adrenergic receptors in other tissues. Isoproterenol and epinephrine were about equally potent, producing stimulation of cyclic AMP accumulation and inhibition of leukocyte function at concentrations below 1×10^{-7} M. Norepinephrine was considerably less potent, and phenylephrine, a predominantly alpha adrenergic agonist, had negligible effects. Propranolol, a beta adrenergic blocking agent, consistently prevented the effects of catecholamines on both cyclic AMP and leukocyte function, while phentolamine, an alpha adrenergic blocking agent, was inactive (Bourne and Melmon, 1971; Bourne et al., 1972).

In retrospect, these results were not surprising, since the stimulatory effects of catecholamines on adenyl cyclase in other tissues have nearly always been consistent with the presence of a beta adrenergic receptor (Sutherland et al., 1968). In a few tissues, alpha adrenergic agents have been found to inhibit cyclic AMP production. In fact, the IgE-mediated release of histamine and slow-reacting substance of anaphylaxis (SRS-A) from human lung is said to be augmented by alpha adrenergic stimulation, which opposes the beta adrenergic inhibition of histamine release in that system (see the chapter by Kaliner and Austen). In leukocytes, however, this appears not to be the case. We have attempted, without success, to inhibit cyclic AMP accumulation in leukocytes with alpha adrenergic stimulation (unpublished data), and Lichtenstein has found no effect of alpha adrenergic agents on histamine release (personal communication).

2. Histamine

Because histamine stimulated adenyl cyclase in brain and myocardium (Klein and Levey, 1971), and because we were investigating the effects of drugs on release of the amine, it appeared reasonable to look for an effect of histamine on leukocyte cyclic AMP. Exogenous histamine did prove to stimulate accumulation of cyclic AMP in leukocytes and did inhibit release of endogenous histamine (Bourne et al., 1971b, 1972). Later, we found that histamine had similar effects on cyclic AMP and cytolytic activity of mouse lymphocytes (Henney et al., 1972).

Pharmacologically, the leukocyte "receptor" for histamine has not been as well characterized as the catecholamine receptor. We found that the two receptors were presumably not identical, since propranolol failed to block any of the effects of histamine. Diphenhydramine, a histamine antagonist, did prevent the effect of histamine on leukocyte cyclic AMP, but only when present at high concentrations (1×10^{-3} M) (Bourne et al., 1972). More recently, a new histamine antagonist, burimamide, has been shown to prevent histamine's effect on gastric mucosa, a tissue where the older antagonists were not effective (Black, 1972). In another chapter of this book, Lichtenstein reports that burimamide also blocks the effect of histamine on cyclic AMP accumulation in leukocytes and the amine's inhibition of leukocyte function. We have confirmed the blocking effect of burimamide in human leukocytes (unpublished data).

3. Prostaglandins

Because prostaglandins affect the activity of adenyl cyclase in many tissues, we tested them in both human leukocytes and mouse lymphocytes. The E class prostaglandins (E_1 and E_2) stimulated adenyl cyclase in broken cells and cyclic AMP accumulation in intact cells. They also inhibited both histamine release and lymphocyte cytolytic activity (Bourne et al., 1971a, 1972; Henney et al., 1972). Effects on cyclic AMP accumulation and leukocyte function were detectable at low concentrations of the prostaglandins (1×10^{-9} M or slightly higher) and were not prevented by propranolol, indicating that the catecholamine and prostaglandin receptors were probably separate and distinct (Bourne and Melmon, 1971).

The relatively minor chemical difference between E and F prostaglandins makes them pharmacologically quite different in many other systems. For example, in many vascular beds, the E prostaglandins cause vasodilation, whereas the F's cause vasoconstriction. Similarly, $PGF_{1\alpha}$ and $PGF_{2\alpha}$ had little or no effect on cyclic AMP formation or function of leukocytes in either of the systems tested. More recently, careful comparison of the dose-response curves of seven different prostaglandins showed excellent correlation between effects on cyclic AMP and either histamine release from human cells or cytolytic activity by murine spleen cells (Lichtenstein et al., 1973a).

B. *Cholera Enterotoxin as a Pharmacological Tool*

The classical pharmacologic approach, using hormone congeners and antagonists, presented an obvious difficulty. Both human peripheral blood leukocytes and murine splenic leukocytes were heterogeneous. We had measured cyclic AMP in a mixed population of cells, whereas only a small proportion of these cells (either the basophils or specifically sensitized lymphocytes) was responsible for the effects inhibited by hormones. Furthermore, the use of theophylline and dibutyryl cyclic AMP to support the conclusion that intracellular cyclic AMP inhibited leukocyte function depended on the assumption that both drugs were producing their effects through the cyclic AMP system (presumably by inhibiting phosphodiesterase and mimicking the action of the endogenous nucleotide, respectively). These considerations led us to test the effect of cholera enterotoxin, the diarrheogenic bacterial protein of *Vibrio cholerae,* on leukocytes. The toxin probably causes cholera by stimulating production of cyclic AMP in gut mucosa (Pierce et al., 1971). Several characteristics of the toxin make it a potentially ideal agent for testing hypotheses relating cyclic AMP to cell function.

(1) At very low concentrations, the toxin produces a characteristically delayed increase in adenyl cyclase activity and accumulation of cyclic AMP in every mammalian cell so far tested.

(2) No effect of the toxin on any cell function has yet been shown to be separate from its stimulation of cyclic AMP accumulation.

(3) Specific inhibitors of the toxin's effects are available (a canine antiserum and a cholera toxoid) (Pierce et al., 1971).

Cholera toxin did inhibit IgE-mediated histamine release from human basophils, as well as cytolytic activity of sensitized lymphocytes. In both systems, the toxin was quite potent, with maximal inhibitory effect at 10 ng/ml, or about 1×10^{-10} M (molecular weight = approximately 90,000) (Pierce et al., 1971). In contrast to all the other drugs tested, the toxin's effects on both cyclic AMP and leukocyte function were delayed. They were detectable at 30 min and increased progressively thereafter. Finally, in both systems, the toxin's effects were abolished by toxin antiserum and cholera toxoid at concentrations that also prevented the toxin's effect on cyclic AMP (Lichtenstein et al., 1973b). Thus, experiments with the toxin strongly confirmed our hypotheses relating cyclic AMP to function of basophils and lymphocytes.

The usefulness of such a pharmacologic tool is increased if it can be used to disprove hypotheses, as well as confirm them. On the basis of experiments using prostaglandins, theophylline, and dibutyryl cyclic AMP, two regulatory roles had been attributed to cyclic AMP in a third subpopulation of leukocytes, the neutrophils. Weissmann and co-workers (1971) had proposed that neutrophil cyclic AMP prevented release of lysosomal hydrolases following phagocytosis, and we ourselves had proposed that cyclic AMP inhibited the ability of neutrophils to kill *Candida albicans* (Bourne et al., 1971a). We used cholera enterotoxin to test both hypotheses and found both wanting (Bourne et al., 1973).

Although cholera enterotoxin, like PGE_1, stimulated accumulation of cyclic AMP in human neutrophils, it had no effect whatever on postphagocytic release of lysosomal hydrolases, even at concentrations many fold higher than necessary for the effect on cyclic AMP. The toxin did consistently inhibit neutrophil candidacidal activity, but was much less effective than either theophylline or dibutyryl cyclic AMP. We concluded from these experiments that intracellular cyclic AMP probably had little or no effect on neutrophil hydrolase release, and that any effect of the endogenous nucleotide on microbicidal activity was small. More importantly, we concluded that the formal criteria (outlined above) proposed for determining the role of cyclic AMP in mediating a hormonal effect must be applied with caution, since the effects of methylxanthines and dibutyryl cyclic AMP may be misleading.

II. Effects of Drugs on Hemolytic Plaque Formation

We have recently investigated the possibility that vasoactive hormones, acting through intracellular cyclic AMP, modulate a third expression of the immune response: the release of specific antibody from lymphocytes of immunized animals. For these experiments, we chose an established in vitro model of antibody release, the formation of hemolytic plaques by splenic leukocytes from mice immunized with sheep red cells (Jerne et al., 1963).

Although this work will be the subject of a more detailed communication (Melmon et al., 1973a), the methods may be briefly reviewed. Spleen cells were obtained from three mouse strains six days after intraperitoneal injection of 3×10^8 sheep red blood cells (sRBC), and then washed and suspended in Eagle's minimal essential medium (pH 7.4). The cells were exposed to drugs for 15 min at 37°C before samples were removed for measurement of cyclic AMP (Gilman, 1970) or hemolytic plaque assays (Jerne et al., 1963). The plates were incubated at 37°C for 1 hr, and for an additional hour in the presence of guinea pig complement. Direct plaque-forming cells (PFC) were counted after the 2-hr incubation. Triplicate plates were made from each drug incubation, and each was coded and counted separately by two investigators. Each value reported here is the mean of at least two drug incubations and is expressed as percent inhibition of plaque formation, compared with control incubations (no drug).

In preliminary experiments, aliquots of the diluted incubates containing both cells and drugs were added to plain agar plates. During experiments with large numbers of samples, however, it was noted that a delay in reading PFC resulted in progressive loss of inhibition by the drugs, particularly when they were present at low concentrations during the preincubation. Diluting the incubate with the agar solution at the time of plating appeared to have lowered the concentration of some drugs (e.g., dibutyryl cyclic AMP) so that they were ineffective. For this reason, all the data reported here are from experiments in which drugs were added to both the top and bottom layer of agarose at concentrations precisely matching the corresponding drug incubation in the original incubates.

Increasing concentrations of adrenergic agents produced dose-related inhibition of plaque formation by spleen cells from Balb/BL mice (Fig. 1). The relative potencies of these agents were characteristic of effects produced via a beta adrenergic receptor. Isoproterenol and epinephrine were most potent, followed by norepinephrine; phenylephrine was almost inactive. The adrenergic drugs showed a similar rank order in stimulating cyclic AMP accumulation (results not shown). Propranolol, a beta adrenergic blocking agent, produced some

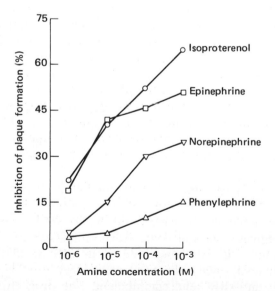

Fig. 1. Inhibition by adrenergic amines of 19S (direct) plaque formation of Balb/BL spleen leukocytes. Isoproterenol (O———O), epinephrine (□———□), norepinephrine (▽———▽), or phenylephrine (△———△) were added at the indicated concentrations to duplicate leukocyte samples, and plaque formation was assessed as described in the text.

inhibition of plaque formation when used alone at high concentrations. At a lower concentration (1×10^{-5} M), however, propranolol partially blocked the effects of isoproterenol (1×10^{-4} M) on leukocyte cyclic AMP and inhibition of plaque formation, reducing both by about 50% (Table 1; cyclic AMP data not shown).

Table 1. Pharmacologic Inhibition of Plaque Formation[a]

Drug	Concentration (M)	Inhibition (%)	
		Exp. 1	Exp. 2
Histamine	1×10^{-5}	41	40
Isoproterenol	1×10^{-5}	43	42
	1×10^{-4}	47	—
Phenylephrine	1×10^{-4}	16	—
PGE$_2$	1×10^{-5}	—	30
Dibutyryl cyclic AMP	1×10^{-4}	50	—
Theophylline	1×10^{-2}	—	17
+ Histamine	1×10^{-5}	—	74
+ Isoproterenol	1×10^{-5}	—	69
+ PGE$_2$	1×10^{-5}	—	64
Propranolol	1×10^{-5}	14	—
+ Isoproterenol	1×10^{-4}	28	—

[a]Experiments performed with spleen cells from Balb/BL mice immunized with sRBC. For details, see text.

The effects of other agents on plaque formation by Balb/BL cells are shown in Table 1. In addition to the catecholamines, both histamine and the E prostaglandins inhibited plaque formation. Theophylline had little effect when used alone, but appeared to potentiate the effects of the other agents. Dibutyryl cyclic AMP, a lipid-soluble analogue of the endogenous nucleotide, also inhibited plaque formation.

When three prostaglandins were compared in this system, the results were similar to those described for histamine release: PGE$_1$ and PGE$_2$ were effective in elevating cyclic AMP and inhibiting plaque formation, while PGF$_{2\alpha}$ was less effective, except at high concentrations (data not shown).

All of these experiments were consistent with the postulate that antibody release (direct plaque formation) was inhibited by vasoactive hormones, which acted by increasing the cyclic AMP content of mouse spleen cells. Several sorts of control experiments were necessary. We showed that the drugs were not acting on the sRBC or inhibiting the action of complement, since none of the agents prevented lysis of sRBC exposed to goat anti-sRBC and complement. The drug effects were not limited to Balb/BL mice, since the same results were found with C57 BL and Balb/c spleen cells (to be reported in detail in separate communication). In addition, preliminary experiments indicate that the same drugs also inhibit formation of indirect (7S) plaques.

An important experiment implicating cyclic AMP as a "second messenger" in this system was that with cholera enterotoxin (Fig. 2). The toxin failed to affect either leukocyte cyclic AMP or plaque formation when preincubated with leukocytes for 10 min as expected (see Section I.A). A 90-min preincubation with the toxin, however, produced both a substantial increase in cyclic AMP and inhibition of plaque formation. Both these effects of the toxin were prevented by antitoxin.

The evidence that cyclic AMP mediates the hormonal inhibition of antibody release appears conclusive:

(1) The order of potency of adrenergic agonists and prostaglandins in inhibiting plaque formation was the same as that for stimulating accumulation of cyclic AMP, and propranolol produced similar results in both systems.

(2) Dibutyryl cyclic AMP inhibited plaque formation, and theophylline potentiated the effects of hormones.

(3) Cholera enterotoxin caused a delayed increase in leukocyte cyclic AMP and a delayed inhibition of plaque formation, both of which were blocked by a specific antitoxin.

III. Discussion

The biologic significance of these results is not so certain, since in vitro experiments do not tell us that the same events actually occur in the

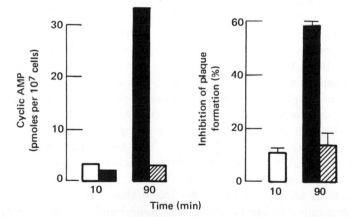

Fig. 2. Effects of cholera enterotoxin on cyclic AMP accumulation (left) and inhibiton of 19S (direct) plaque formation (right) in Balb/BL spleen cells. Leukocytes were incubated with no drug (open bars), cholera enterotoxin, 10 ng/ml (black bars), or with cholera enterotoxin, 10 ng/ml, plus antitoxin added simultaneously, 1:1000 dilution of canine antiserum, (crosshatched bars). Cyclic AMP data represent the mean of at least two determinations, differing by not more than 8%. Plaque inhibiton data represent the mean, ± SD per cent inhibition, of five experiments with each drug incubation performed in duplicate. As indicated, cholera enterotoxin was added to leukocytes 10 or 90 min before plating or measurement of cyclic AMP.

whole animal. Nevertheless, the parallel results with the same vasoactive hormones, working through intracellular cyclic AMP, as inhibitors of the expression of three immune responses—IgE-mediated release of histamine and other mediators of immediate hypersensitivity, the specific cytolytic activity of lymphocytes, and release of antibody from lymphocytes—deserve further comment. We shall consider the possible implications of these experiments under two headings: (1) The presence in lymphocytes of pharmacologic receptors for vasoactive hormones ("first messengers") and (2) the consistent inhibitory action of cyclic AMP (the "second messenger") in the same cells.

A. Receptors for Vasoactive Hormones

By the criteria of classical pharmacology, at least two subtypes of lymphocytes contain receptors for histamine, beta adrenergic catecholamines, and prostaglandins. On the one hand, these hormones inhibit the immunologically specific cytolysis of allogeneic target cells, which is thought to be a function of the T lymphocyte. Similarly, the mitogenic response of lymphocytes to phytohemagglutinin, also thought to be a T cell response, is inhibited by catecholamines and prostaglandins (Boume et al., 1971c; Smith et al., 1971a, 1971b) (the effect of histamine has not yet been tested). The release of humoral antibody (measured by formation of hemolytic plaques), however, is considered a function of a second major subtype of lymphocyte, the B cell. Our experiments show that B cell function is also inhibited by vasoactive hormones.

Granted that both classes of lymphocytes contain hormone receptors, we still lack firm clues as to the biologic meaning of these receptors. An important first step to understanding would be to determine whether hormone receptors are present on all lymphocytes or confined to certain subpopulations. We have recently shown that insolubilized histamine and catecholamines attached to Sepharose beads will bind human leukocytes and splenic lymphocytes from the mouse strains used in our experiments with plaque formation (Melmon et al., 1972; Shearer et al., 1972). As outlined in the next chapter (Melmon et al.), it appears likely that the binding has pharmacologic specificity, i.e., that the hormone-coated beads are bound to leukocyte membrane receptors for the two types of amine. Using the binding phenomenon to separate splenic leukocytes on the basis of their amine receptors, we found that a substantial number of leukocytes failed to bind to either hormone-bead conjugate. About half of the plaque-forming cells, however, did bind to the hormone-coated beads. This suggests that the ability to produce specific antibody develops in parallel with the amine receptors and is consistent with the ability of the free amines to inhibit plaque formation (Melmon et al., 1973b; see Melmon's chapter for detailed discussion).

The binding studies also suggested that the subpopulation of mouse splenic leukocytes with amine receptors includes a cell type capable of controlling the humoral antibody response of precursor cells (Shearer et al., 1972; see discussion by Shearer et al, in this book). We do not know whether this "controller cell" is a T or B lymphocyte or similar to the cells described by Gershon and others (1972) that are able to modulate the immune response of other lymphocytes.

B. Why Does Cyclic AMP Inhibit Immunologic Responses?

Intracellular cyclic AMP appears to inhibit a variety of immunologic responses, including release of inflammatory mediators like histamine, mitosis of lymphocytes in response to antigen or phytohemagglutinin, cytolytic activity of lymphocytes, and, as shown here, release of humoral antibody. These inhibitory effects contrast sharply with the generally stimulatory role of cyclic AMP in fat cells, liver, myocardium, pancreatic islets, and many other tissues (Sutherland et al., 1968). The biologic meaning of cyclic AMP's consistently inhibitory effects in leukocytes can be considered from two points of view, the one concentrating on biochemical regulation of cells, the other focusing on interactions between cells that may serve to regulate immunologic and inflammatory responses.

One obvious parallel with the direction of cyclic AMP's effects on leukocyte function lies in the numerous recent demonstrations that cyclic AMP inhibits growth of many cells in tissue culture (discussed by several other authors of this book). The biochemical mechanism(s) by which cyclic AMP slows cell growth are no clearer than the mechanisms by which the nucleotide inhibits leukocyte functions. It is possible, however, that both kinds of effect reflect a similar mode of action of intracellular cyclic AMP.

In this context, the demonstration by Daniel et al. (1973) that a steroid-sensitive line of lymphoma cells is also killed by dibutyryl cyclic AMP appears highly suggestive. These workers used the lymphocidal effect of dibutyryl cyclic AMP to select a variant lymphoma cell line, which proved insensitive to dibutyryl cyclic AMP. The resistant cells were shown to lack the cytoplasmic cyclic AMP binding protein and cAMP-dependent protein kinase, both of which were present in the parental cell line.

More recent experiments have shown that PGE_1, which stimulates accumulation of cyclic AMP in both cell lines, slows the growth of the parental (dibutyryl cAMP-sensitive) cells, but not that of the resistant cells. In addition, exogenous dibutyryl cyclic AMP inhibited the incorporation of radioactive leucine and uridine into acid-precipitable macromolecules of sensitive, but not resistant, cells (Daniel, Bourne, and Tomkins, in preparation). These results suggest that the effects of both exogenous and endogenous cyclic AMP on growth of lymphoma cells are dependent on the presence of a cytoplasmic protein capable of binding cyclic AMP. Absence of the intracellular "receptor" for cyclic AMP in the resistant cells makes the nucleotide biologically inactive.

Since the cAMP-dependent protein kinase has been suggested, but not proved, to mediate virtually all of the effects of cyclic AMP in mammalian cells (Kuo and Greengard, 1969), a variant cell population deficient in the protein kinase might be used as a negative control to determine how, in fact, cyclic AMP acts to slow cell growth. In addition, to the extent that growth and biochemistry of mouse lymphoma cells can serve as a model for functions of normal lymphocytes, such an experimental approach may provide useful clues to the action of cyclic AMP in modulating the immune response.

The inhibitory effects of cyclic AMP on immunologic model systems are suggestive from an additional point of view. We have suggested elsewhere (Bourne, 1972) that the nucleotide may mediate a variety of servomechanisms (negative feedback pathways), which control inflammatory and immune responses in vivo. The ability of exogenous histamine to inhibit the allergic release of basophil histamine (Bourne et al., 1971b, 1972) exemplifies one possible way in which a mediator of inflammation could also serve to limit the extent and intensity of an inflammatory response. Present evidence that histamine, catecholamines, and prostaglandins mediate and inhibit inflammatory responses ranges from conclusive to fragmentary (Spector and Willoughby, 1965; McKinney and Lish, 1966). The impressive interweaving of inflammatory and immune responses in both host defenses and causation of many diseases suggests, however, that servomechanisms must exist. Both endogenous vasoactive hormones and cyclic AMP amy eventually prove to play important roles in controlling or coordinating phenomena as diverse as inflammation, allergy, and humoral and cell-mediated immunity.

Discussion of Dr. Bourne et al.'s Paper

DR. HENNEY: I would like to offer additional data suggesting that adenylate cyclase stimulation of plasma cells inhibits antibody secretion. If mice are injected with sheep erythrocytes (sRBC) intraperitoneally, then spleen cell suspensions taken from these mice 11 days later are able to lyse the sRBC in vitro. Two cells types are required for such cytolysis, an antibody secreting plasma cell and a normal B lymphocyte effector cell. The lytic activity of these spleen cell suspensions is inhibited by PGE_1, isoproterenol, and cholera enterotoxin. The plasma cell requirement can be negated by the addition of its product, anti-sRBC antibody. Cytolysis still occurs, but lysis is not affected by any of the adenylate cyclase–stimulating agents. It is thus clear that PGE_1, isoproterenol, and cholera enterotoxin all inhibit secretion from the plasma cell. I now have a question for Dr. Bourne. Are both γ-G and γ-M antibody-secreting cells equally suppressed by the adenylate cyclase–stimulating agents?

DR. BOURNE: We have only looked at direct plaque-forming cells.

DR. LICHTENSTEIN: Did you examine the differential susceptibility of cells to depression of plaque formation as a function of the stage of the immune response? In other words, at different times following immunization, are there differences in cell susceptibility?

DR. BOURNE: Most of these studies were done at six days, and we don't know.

DR. PERPER: Have you treated cells with the various drugs and transferred them into X-irradiated recipients to see if the cells function in vivo?

DR. BOURNE: Yes, we have, and they do.

References

Ahlquist, R. P. "A Study of Adrenotropic Receptors," *Amer. J. Physiol.* **153**.586–600 (1948).

Black, J. W., Duncan, W. A. M., Durant, C. J., Ganellin, C. R., and Parsons, E. M. "Definition and Antagonism of Histamine H_2-Receptors," *Nature* **236**:385 (1972).

Bourne, H. R. "Leukocyte Cyclic AMP: Pharmacological Regulation and Possible Physiological Implications." In *Prostaglandins in Cellular Biology*, P. Ramwell and B. Pharriss, eds. New York: Plenum Press, pp. 111–150 (1972).

Bourne, H. R., Epstein, L. B., and Melmon, K. L. "Lymphocyte Cyclic Adenosine Monophosphate (AMP) Synthesis and Inhibition of Phytohemagglutinin-Induced Transformation," *J. Clin. Invest.* **50**:10a (abstract) (1971c).

Bourne, H. R., Lehrer, R. I., Cline, M. J., and Melmon, K. L. "Cyclic 3',5'-Adenosine Monophosphate in the Human Leukocyte: Synthesis, Degradation, and Effects on Neutrophil Candidacidal Activity, *J. Clin. Invest.* **50**:920–929 (1971a).

Bourne, H. R., Lehrer, R. I., Lichtenstein, L. M., Weissmann, G., and Zurier, R. "Effects of Cholera Enterotoxin on Adenosine 3',5'-Monophosphate and Neutrophil Function: Comparison with Other Compounds Which Stimulate Leukocyte Adenyl Cyclase," *J. Clin. Invest* (in press—1973).

Bourne, H. R., Lichtenstein, L. M., and Melmon, K. L. "Pharmacologic Control of Histamine Release in Vitro: Evidence for an Inhibitory Role of 3',5'-Adenosine Monophosphate in Human Leukocytes," *J. Immunol.* **108**:695–705 (1972).

Bourne, H. R., and Melmon, K. L. "Adenyl Cyclase in Human Leukocytes: Evidence for Activation By Separate Beta Adrenergic and Prostaglandin Receptors," *J. Pharmacol. Exp. Ther.* **178**:1–7 (1971).

Bourne, H. R., Melmon, K. L., and Lichtenstein, L. M. "Histamine Augments Leukocyte Adenosine 3',5'-Monophosphate and Blocks Antigenic Histamine Release," *Science* (Washington) **173**:743–745 (1971b).

Daniel, V., Litwack, G. and Tomkins, G. "Cyclic AMP–Induced Cytolysis of Cultured Lymphoma Cells and the Isolation of Resistant Variants," *Proc. Nat. Acad. Sci.* (U.S.A.) (in press—1973).

Gershon, R. K., Cohen, P., Hendin, R., and Liebhaber, S. A., "Suppressor T Cells," *J. Immunol.* **108**:586 (1972).

Gilman, A. G. "A Protein Binding Assay for Adenosine 3',5'-Cyclic Monophosphate," *Proc. Nat. Acad. Sci.* **67**:305 (1970).

Henney, C. S., Bourne, H. R., and Lichtenstein, L. M. "The Role of Cyclic 3',5'-Adenosine Monophosphate in the Specific Cytolytic Activity of Lymphocytes," *J. Immunol.* **108**:1526 (1972).

Henney, C. S., and Lichtenstein, L. M. "The Role of Cyclic AMP in the Cytolytic Activity of Lymphocytes," *J. Immunol.* **107**:610 (1971).

Ishizaka, T., Ishizaka, K., Orange, R. P., and Austen, K. F. "The Pharmacologic Inhibiton of the Antigen-Induced Release of Histamine and Slow Reacting Substance of Anaphylaxis (SRS-A) from Monkey Lung Tissues Mediated by Human IgE," *J. Immunol.* **106**:1267 (1971).

Jerne, N. K., Nordin, A. A., and Henry, C., "The Agar Plaque Technique for Recognizing Antibody-Producing Cells." In *Cell Bound Antibodies,* B. Amos and H. Koprowski, eds. Philadelphia, Pa.: Wistar Institute Press, p. 109 (1963).

Koopman, W. J., Orange, R. P., and Austen, K. F. "Immunochemical and Biologic Properties of Rat IgE: Ill. Modulation of the IgE-Mediated Release of Slow-Reacting Substance of Anaphylaxis by Agents Influencing the Level of Cyclic 3',5'-Adenosine Monophosphate," *J. Immunol.* **105**:1096 (1970).

Kuo, J. F., and Greengard P. "Cyclic Nucleotide-Dependent Protein Kinases. IV. Widespread Occurrence of Adenosine 3',5'-Monophosphate-Dependent Protein Kinase in Various Tissues and Phyla of the Animal Kingdom," *Proc. Nat. Acad. Sci.* (U.S.A.) **64**:1349 (1969).

Lichtenstein, L. M., Bourne, H. R., and Henney, C. S. "Effect of Cholera Toxin on in Vitro Models of Immediate and Delayed Hypersensitivity," *J. Clin. Invest.* (in press—1973a).

Lichtenstein, L.M., Gillespie, E., Henney, C. and Bourne, H. R. to "Effects of a Series of Prostaglandins on in Vitro Models of the Allergic Response and Cellular Immunity." *Prostaglandins* (in press—1973b).

Lichtenstein, L. M., and Margolis, S. "Histamine Release in Vitro: Inhibiton by Catecholamines and Methylxanthines," *Science* **161**:902 (1968).

McKinney, G. R., and Lish, P. M. "Interaction of Beta Adrenergic Blockade and Certain Vasodilators in Dextran-Induced Rat Paw Edema," *Proc. Soc. Exp. Biol. Med.* **121**:494 (1966).

Melmon, K. L., Bourne, H. R., Weinstein, Y., and Sela, M. "Receptors for Histamine Can Be Detected on the Surface of Selected Leukocytes," *Science* **177**:707 (1972).

Melmon, K. L., Bourne, H. R., Weinstein, Y., Shearer, G. M., and Bauminger, S. "Release of Antibody from Leukocytes in Vitro. Control By Vasoactive Hormones and Cyclic AMP" (submitted for publication—1973a).

Melmon, K. L., Weinstein, Y., Shearer, G. M., Bourne, H. R., and Bauminger, S. "Separation of Specific Antibody-Forming Cells By Their Receptors to Endogenous Hormones" (submitted for publication—1973b).

Pierce, N. F., Greenough, III, W. B. and Carpenter, C. C. J. "*Vibrio cholerae* Enterotoxin and Its Mode of Action," *Bacteriol. Rev.* **35**:1 (1971).

Shearer, G. M., Melmon, K. L., Weinstein, Y., and Sela, M. "Regulation of Antibody Response By Cells Expressing Histamine Receptors," *J. Exp. Med.* **136**:1302 (1972).

Smith, J. W. Steiner, A. L., Newberry, Jr., W. M., and Parker, C. W. "Cyclic Adenosine 3′,5′-Monophosphate in Human Lymphocytes: Alterations After Phytohemagglutinin Stimulation," *J. Clin. Invest.* **50**:432 (1971a).

Smith, J. W., Steiner, A. L., and Parker, C. W. "Human Lymphocyte Metabolism: Effects of Cyclic and Noncyclic Nucleotides on Stimulation by Phytohemagglutinin,: *J. Clin. Invest.* **50**:442 (1971b).

Spector, W. G., and Willoughby, D. A. "Chemical Mediators" In *The Inflammatory Process,* B. Zweifach, L. Grant, and R. McClusky, eds. New York: Academic Press, pp. 427–448 (1965).

Sutherland, E. W., Robison, G. A., and Butcher, R. W. "Some Aspects of the Biological Role of Adenosine 3′,5′-Monophosphate (Cyclic AMP)," *Circulation* **37**:279 (1968).

Weissmann, G., Dukor, P., and Zurier, R. B. "Effect of Cyclic AMP on Release of Lysomal Enzymes from Phagocytes," *Nature (New Biol.)* **231**:131 (1971).

LEUKOCYTE SEPARATION ON THE BASIS OF THEIR RECEPTORS FOR BIOGENIC AMINES AND PROSTAGLANDINS: RELATION OF THE RECEPTOR TO ANTIBODY FORMATION

K. L. Melmon, Y. Weinstein,
G. M. Shearer, and H. R. Bourne

I. Introduction

In the previous chapter (Bourne et al., 1973), the effects of biogenic amines and the E series prostaglandins on various immunologic processes were discussed. All stimulators of cyclic AMP accumulation in leukocytes appeared to inhibit at least three separate immunologic events. Histamine, beta catecholamines, and prostaglandin E_1 and E_2 inhibited IgE-induced release of histamine and other mediators of the inflammatory process from basophils and mast cells. The same drugs also inhibited lysis of allogeneic target cells (Henney et al., 1972) by sensitized lymphocytes, and they inhibited release of antibody from the splenic lymphocytes of mice immunized with sheep erythrocytes (Melmon et al., 1973a).

Separate receptors for histamine, beta catecholamines, and prostaglandins on human and probably also on murine leukocytes were demonstrated by a series of pharmacologic experiments also summarized in the last chapter (Bourne et al., 1973). That the location of some of the receptors for the hormones is the surface of the cell membrane has been suggested by studies in which the small endogenous hormones were linked to large carriers that could not enter the cell. When the complexes

This work was supported by U.S. Public Health Service grants HL 09964 and GM 16496 and agreement 06-035.

were incubated with dispersed leukocytes (Weinstein et al., 1973; Melmon et al., 1973b) or placed on isolated myocardium (Venter et al., 1972), pharmacologic effects of the hormone resulted. The effects were not exhaustively tested, but at least in part mimicked the action of the free hormone.

Lymphoid cell populations have been separated into various functional categories by empirical procedures, which take advantage of some physical property of the cell, e.g., sedimentation in albumin gradients or velocity sedimentation (see the next chapter—Shearer et al., 1973). More recently, the cells have been separated by insolubilizing antigens or haptens and attracting those cells with specific receptors for the chemicals or a portion of the chemicals to which they can respond physiologically (Wizzell and Anderson, 1969; Singhal and Wizzell, 1970; Wofsy et al., 1971). We reasoned that systematic insolubilization of amines and prostaglandin might be possible by linking them via their amino groups (amines) or their carboxyl groups (prostaglandin) to a carrier protein's carboxyl group or free amine, respectively (Melmon et al., 1972). The carrier could then be linked to either insoluble cross-linked agarose (Sepharose), nylon mesh or glass particles. If receptors to the hormones were on the cell surface, then the insolubilized substances might be used to attract some of the cells to the beads. Thus, separation of lymphocytes might be accomplished on the basis of a physiologic receptor for an agonist—a substance that was already known to influence the immunologic function of the cell.

Chromatography of human peripheral leukocytes and murine splenic leukocytes on columns of insolubilized histamine, norepinephrine (whose amino function is protected), isoproterenol, or prostaglandin E_1 or E_2 has been accomplished. We shall present evidence that such separation is related to the hormone and not its carrier or Sepharose, and that the binding of the cells is probably via a cell surface receptor for the hormone. This chapter will

(1) discuss the various techniques that we have used to insolubilize the hormones;

(2) present evidence that use of the hormone-carrier-Sepharose preparations results in separation of cells largely on the basis of the presence of receptors for the biogenic hormone;

(3) give one illustration of how cell separation can be used to study and confirm the function in immune processes of the cells with receptors; and

(4) suggest some approaches that can be used to study the development of receptors on immunocompetent cells, the possible roles of the biogenic amines and prostaglandins in the immune process, and the interrelationship between mediators of inflammation, cyclic AMP and modulation of immune processes.

II. Insolubilized Hormones

A. Preparation

Histamine (H), isoproterenol (I), norepinephrine (NE), and prostaglandins E_1 (PGE$_1$), E_2 (PGE$_2$), and $F_{2\alpha}$(PGF$_{2\alpha}$) were the first hormones we used. They were linked separately to either rabbit serum albumin (RSA), bovine serum albumin (BSA), or a copolymer (P) of DL-alanine with L-tyrosine prepared by random copolymerization of N-carboxy-α-amino acid anhydrides according to the method of Sela et al. (1967). The copolymer had a low molecular weight (ag 1840), a random structure, a molar ratio of DL-alanine (3) to L-tyrosine (1), and two reactive terminals, one for binding the amine and the other for binding the conjugate to Sepharose (Weinstein et al., 1973).

Essentially two methods were explored in depth for linking the hormone conjugates to one of the carriers: First, The amine was dissolved and incubated with carrier and carbodiimide-HCI (ECDI). This procedure conjugated the amines via an amide bond to the carbodiimide-activated carboxyl groups of the carrier. Prostaglandins managed in the same way were probably linked via their carboxyl groups to free amino functions of the carrier. This binding process drastically reduces the basicity of the amino group of the hormone and therefore, as we shall show, the binding capacity of the hormone conjugate. The hormone-carrier complexes were then linked to agarose activated by CNBr by the method of Porath et al. (1967). The second method employs conjugates made with glutaraldehyde (GA). Using this method, the carrier was first linked to activated Sepharose, and the resultant conjugate reacted with dissolved amine in the presence of glutaraldehyde. This process preserves the basic function of the hormone, since the imino bonds formed by the interaction of the aldehyde with the amine groups are completely reduced to alkyl amines by sodium borohydride. The change brought about by the second mechanism was the conversion of primary amines to secondary amines. In the case of norepinephrine, the amine may have been chemically converted from norepinephrine to a substance more resembling epinephrine (Fig. 1). Both preservation of the basic function of the amine and conversion to another chemical hormone substance may have contributed to the binding and pharmacologic properties of the insolubilized drug. Details of the methods of preparation and handling of the preparations with cell suspensions are recorded elsewhere (Melmon et al., 1972; Weinstein et al., 1973). A list of preparations, their chemical composition, and their relative ability to bind human leukocytes are found in Table 1.

A third method of insolubilizing hormones has only recently been explored. A diazotization process described by Venter et al. (1972) allows linkage of the hormone via its ring structure at a carbon adjacent to an

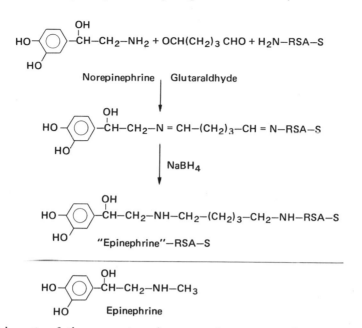

Fig. 1. Schematic of the conversion of norepinephrine to a molecule resembling epinephrine during the conjugation of the amine with either protein or polymer via glutaraldehyde. The end result is N-alkylated norepinephrine.

hydroxyl group. This procedure avoids substitution on the amino function of the hormone and may be ideally suited to a study of both structure-action relationships of the binding and pharmacologic properties of the insolubilized catechols.

B. Cell Binding and Pharmacologic Properties of the Insolubilized Hormones

1. Studies Using Human Peripheral Leukocytes

Human cells were prepared from sedimenting heparinized venous blood from normal volunteers in 3% Dextran 250. The cells were washed and used in two types of experiments. The first was designed to assess the characteristics and specificity of binding of leukocytes to the Sepharose preparations; the second was designed to yield a rough estimate of the distribution of receptors on various leukocytes.

a. Binding of Leukocytes to Sepharose. Different batches of Sepharose beads (0.05 ml of a 25% suspension of agarose in phosphate buffer) were incubated with 5 to 10 × 10⁶ leukocytes for 15 min with gentle shaking at 37°C. After incubation, the beads were examined under the microscope. One hundred consecutive beads were examined from each specimen and scored as binding if they held 20 cells or more per bead, and as negative if fewer cells were bound. Two additional types of experiments were conducted. In one, the cells were incubated with each

Melmon et al.

Table 1. Chemical Composition and Cell Binding Activity
of Sepharose Preparations

Preparation[a]	Nanomoles[b]		Ratio drug per carrier	% beads binding > 20 cells
	Drug	Carrier		
Histamine-R-S	135	4.4	3.05	87 ± 7
Histamine-R-S(GA)	460	8.7	51	71 ± 10
Histamine-P-S	320	162	1.98	23 ± 12
Isoproterenol-R-S	65.1	2.2	30	28 ± 7
Norepinephrine-R-S	154	2.2	70	3 ± 2
Norepinephrine-R-S(GA)	287	8.7	33	86 ± 6
Norepinephrine-P-S(GA)	52	162	.33	0
Prostaglandin E_2-R-S	6.06	2.2	2.7	3 ± 5
Prostaglandin E_2-P-S	40	162	0.25	0

[a]R = rabbit serum albumin; P = copolymer of DL-alanine and L-tyrosine; S = Sepharose.
[b]Represents nanomoles in 0.3 ml of 25% suspension (w/v) of the Sepharose preparation in
  PBS. This amount was used in experiments to characterize binding and to determine the
  pharmacologic properties of the preparation.

of several drugs at various concentrations for 10 min at 37°C before the
Sepharose-drug conjugate was added, and then incubated for an additional 15 min. The binding of the beads was assessed. In the second type
of experiment, the cells were incubated with a Sepharose-drug conjugate, and the free drugs were added after 15 min of incubation. The mixture was incubated for an additional 15 min, and the binding was
assessed. These latter experiments tested the ability of the drugs to displace cells after they had been bound to Sepharose.

    b. Preparation of columns. 24×10^6 leukocytes were incubated with
various Sepharoses (0.6 ml of a 25% suspension of agarose) for 15 min at
37°C. The suspension was then poured into a plastic column with a small
sponge rubber tip. The columns were washed until the unbound cells
were separated from the column. These unbound cells were then incubated with an additional Sepharose (0.3 ml Sepharose with 1 ml incubate) for 15 min at 37°C and later poured into a second column. In some
experiments, the bound cells were eluted from the columns by adding 1
ml portions of various drugs, and gently shaking and resuspending the
column contents with a Pasteur pipette.

C. Interactions of Sepharose Preparations with the Leukocytes

1. Binding

    By visual assessment, none of the control Sepharoses bound cells.
Neither Sepharose alone, activated Sepharose, nor Sepharose to which
RSA, BSA, or the copolymer was attached appeared to bind cells. When
either histamine or norepinephrine was linked directly to activated
Sepharose, they were no more capable of binding cells than the controls
(Melmon et al., 1972; Weinstein et al., 1973). When histamine was con-

jugated to RSA and attached to Sepharose by either the ECDI or glutaraldehyde (GA) method, the majority of beads were coated with leukocytes, but not platelets or erythrocytes. The binding of leukocytes seemed proportional to their incidence in the incubation mixture. The majority of binding took place within 12 min and was virtually complete in 15 min, whether the incubation was conducted at 37° or 4°C. The same time course of binding was observed for the other drug-Sepharose conjugates.

Iso-R-S did not bind as well [28 ± 7% (SD) beads with greater than 20 cells per bead] as H-R-S (87 ± 7%), H-R-S (GA) (71 ± 10%), or N-R-S (GA) (86 ± 6%) (Tables 1 and 2). Only 23% of H-P-S beads bound cells. All preparations of drugs linked to polymer and Sepharose bound significantly fewer cells than the corresponding drug-RSA-Sepharose preparation. Of interest was the remarkable difference between the binding affected by N-R-S, which never exceeded 5% of the beads binding more than 20 cells per bead, and the binding of cells by N-R-S (GA), which was equivalent to H-R-S. Prostaglandin E_2 attached to RSA and Sepharose did not appear to attract cells during the incubation. The amount of each type of Sepharose and the approximate molarity of protein and drug (histamine, PGE_2, norepinephrine, or isoproterenol) in the incubation mixtures are recorded in Table 1. The differences in binding between these Sepharose-drug-protein conjugates and the polymer-drug conjugates could not be attributed to differences in opportunity of the cells to make contact with the agonist. In fact, the number of moles of norepinephrine in the ECDI conjugates was more than twice that in the GA conjugates, yet only the latter [N-R-S (GA)] bound cells. However, the approximate molarity of the PGE_2-RSA-Sepharose preparation was lower than that of the other drug conjugates. This may have been responsible in part for the lack of apparent binding of cells to the prostaglandin E_2 (Table 1).

The scoring of beads as positive or negative for binding is arbitrary and could be somewhat misleading. Several instances of binding of some cells were consistently seen with isoproterenol and PGE_2 conjugates to RSA and Sepharose. However, because the number of cells was less than 20 per bead, they were scored as negative.

Further characteristics of binding were studied using those Sepharose-drug conjugates that readily bound large numbers of cells, i.e., H-R-S, H-R-S (GA), and N-R-S (GA) (Table 1). Binding of cells was pH dependent. Binding was maximal at about pH 7, and sharply decreased at higher pH values for both H-R-S and N-R-S (GA). Binding of H-R-S differed from H-R-S (GA) in that the latter was able to continue binding cells as the pH rose above 8. N-R-S did not bind cells over the whole range of pH values tested.

Drug-conjugate-Sepharose preparations using carriers of BSA or RSA had essentially similar properties. H-R-S had 87 ± 5% of beads with

Table 2. Characteristics of Binding of Human Leukocytes by Drug-RSA-Sepharose

Drug in preincubation[a]	Concn. (M)	Percent of beads binding > 20 cells[b]		
		H-R-S	H-R-S(GA)	N-R-S(GA)
Control		87 ± 7	71 ± 10	86 ± 6
Antihistamines				
Diphenhydramine	(10^{-3})	28 ± 22[c]	5	52 ± 6
	(10^{-4})	33 ± 7[c]	30	85 ± 7
Artazoline	(10^{-3})	21 ± 13[d]	12	79 ± 5
	(10^{-4})	57	37	—
Tripelennamine	(10^{-3})	31 ± 7[d]	—	85 ± 10
	(10^{-4})	38	56	—
Pyrilamine	(10^{-3})	24 ± 4[d]	2 ± 7[c]	69 ± 7
	(10^{-4})	44	30 ± 10	87
Adrenergic Blocking Agents				
Propranolol	(10^{-3})	84 ± 7	76 ± 11	11 ± 8[c]
	(10^{-4})	—	—	37 ± 6[c]
Phentolamine	(10^{-3})	45	—	53 ± 38
	(10^{-4})	95 ± 1	95 ± 4	71 ± 10
Agonists				
Histamine	(10^{-3})	64 ± 29[d]	40	86 ± 5
	(10^{-5})	91 ± 2	—	—
Epinephrine	(10^{-3})	84 ± 13	70 ± 26	60 ± 27
D,L-isoproterenol	(10^{-3})	86 ± 16	83	56 ± 19[d]
D-isoproterenol	(10^{-3})	87	50	39 ± 6[c]
L-isoproterenol	(10^{-3})	81	70	23 ± 12[c]
Norepinephrine	(10^{-3})	82 ± 21	88	84 ± 6
Phenylephrine	(10^{-3})	87 ± 6	82	75 ± 12
Prostaglandin E$_2$	(10^{-3})	85 ± 12	72 ± 6	90 ± 7

[a] 5×10^6 cells were preincubated for 10 min at 37°C with drug as indicated in 0.45 ml minimal essential media, pH 7.4. 0.05 ml of 25% v/v mixture of drug-Sepharose were then added; the cells were then incubated for 10 min with gentle shaking.
[b] Data are expressed as mean or mean + SD. Standard deviation was calculated only when 7 or more points were obtained. Means without SD are from 3 points. Each point represents the mean of triplicates. Variance between duplicate samples is < 6%. Statistical comparisons were only determined between controls and groups with more than 7 points.
[c] P value < 0.001 compared to control (paired test).
[d] P value < 0.01 compared to control (paired t test).

more than 20 cells per bead; 89 ± 9% of H-B-S beads had more than 20 cells. Similar curves of pH effects on binding, and prevention of binding or displacement of bound cells by drugs, were obtained with either protein carrier. Clumping of cells did not occur under these circumstances and, therefore, could not have contributed to the results of the binding experiment.

2. Specificity of Cell Binding

Experiments that sought to determine whether binding of cells by H-R-S, H-R-S (GA), and N-R-S (GA) was specific are summarized in Table 2. Binding of cells by H-R-S was significantly inhibited by preincubation of the cells with histamine, each of four antihistamines, and histamine-RSA alone. The inhibition was dose dependent and was not duplicated by comparable concentrations of catecholamines, prostaglandins, or competitive antagonists of catecholamines. The general pattern of prevention of binding seen in H-R-S was also seen when using H-R-S (GA) or H-P-S. Once bound, cells could be displaced from the H-R-S by high concentrations of diphenhydramine or histamine-RSA (0.25 mg/ml incubate). The displacement by H-R was not as complete as when the chemical was preincubated with cells before adding H-R-S. Prevention of binding by H-R was caused by clumping of the leukocytes by H-R.

The binding of leukocytes to the Sepharose-drug conjugate constructed by linking norepinephrine to RSA via glutaraldehyde conformed with that of a beta adrenergic agonist, and was clearly separable from the binding behavior of the histamine conjugate (Table 2). The binding of cells to N-R-S (GA) was not prevented by comparable concentrations of histamine or histamine antagonists, but was significantly inhibited by either D- or L-isoproterenol and the beta adrenergic blocking agent, proparanolol. Epinephrine produced slight inhibition of binding. Phentolamine, an alpha adrenergic blocking agent, interfered with binding in some experiments, but not consistently. Norepinephrine and phenylephrine, predominantly alpha-adrenergic agonists, did not inhibit binding of cells by N-R-S (GA).

These results are parallel to our previous experiments measuring the effects of histamine and adrenergic amines on leukocyte cyclic AMP accumulation (Bourne et al., 1973), in which specific amine antagonists clearly separated the effects of histamine and catecholamines, presumably by preventing interaction with separate receptors. In those experiments also, the rank order of potency of adrenergic agonists— isoproterenol and epinephrine > norepinephrine >> phenylephrine (=0)—was characteristic of actions on beta rather than alpha adrenergic receptors.

D. Effects of Insolubilized Drugs on Leukocyte Cyclic AMP

All the Sepharose preparations were tested for their ability to increase the cyclic AMP content of leukocytes. All failed to do so, except the N-P-S (GA), which increased the cyclic AMP in four of six experiments (Table 3) using human leukocytes and in a similar number of experiments using primed or unprimed (Balb/BL) murine spleen leukocytes. To be certain that minor amounts of catecholamines were not being dissociated from the Sepharose, both the supernatant of N-P-S (GA) and the supernatant from incubations of the same preparation with leukocytes were tested, and did not alter cyclic AMP content of the cells. Furthermore, the stimulation by N-P-S (GA) was almost completely inhibited by propranolol (10^{-4} M), minimally inhibited by phentolamine,

Table 3. Leukocyte Cyclic AMP After Exposure to Sepharose Preparations[a]

Drug or sepharose preparation	Experiment number					
	1	2	3	4	5	6
Control	27 ± 5[b]	40	40 ± 3[b]	25 ± 5[b]	29 ± 5[b]	34 ± 4[b]
Histamine 1 x 10^{-4} M	94	114	—	—	—	—
Isoproterenol 1 x 10^{-4} M	90	135	119	96	—	—
Epinephrine 1 x 10^{-4} M	—	—	—	—	86 ± 9	95 ± 5
Prostaglandin E_1 1 x 10^{-5} M	267	430	251	—	—	—
R-S	32	40	39	—	—	—
P-S	22	30	26 ± 5	20 ± 6	—	—
H-R-S	25	25	—	—	—	—
H-R-S (GA)	30	—	—	—	—	—
H-P-S	23	—	—	—	—	—
N-R-S	23	39	—	—	—	—
N-R-S (GA)	—	—	—	—	28 ± 3	30 ± 6
N-P-S (GA)	36	59	28 ± 1	12 ± 1	49 ± 6[d]	49 ± 7[d]
N-P-S (GA) + propranolol 10^{-4}	—	—	—	—	24 ± 7	30 ± 6
Supernatant of N-P-S (GA)[c]	—	—	38 ± 0	17	27 ± 5	32 ± 5
Iso-R-S	28	—	—	—	—	35 ± 2
PGE_2-R-S	36	32	38	—	35 ± 7	40 ± 6
PGE_2-P-S	—	—	—	—	86 ± 3[d]	83 ± 8[d]

[a]Each tube held drug at the stated molarity or 0.3 ml 35% (w/v) suspension of the indicated Sepharose preparation in 1 ml PSB containing 1 x 10^7 leukocytes. Theophylline, 1 x 10^{-2} M, was present in all tubes. Cyclic AMP was measured after 15 min incubation at 37°C and is expressed as picomoles per 1 x 10^7 cells.

[b]Mean ± SD of four determinations. Values shown without standard deviation are the mean of duplicate determinations, differing by not more than 6%.

[c]In Experiments 3 and 5, cells were incubated with 0.3 ml supernatant from N-P-S (GA) incubation of Experiment 2. In Experiments 4 and 6, cells were incubated with 0.3 ml supernatant of N-P-S (GA) obtained by centrifugation of the Sephrose preparation used in the same experiment.

[d]Significantly different from control (P < 0.01, standard t test).

and unaffected by four antihistamines, all at 10^{-4} M. None of the four batches of Sepharose, activated Sepharose, R-S, P-S, H-R-S, H-R-S (GA), H-P-S, H-P-S (GA), Iso-R-S, or N-P-S stimulated cyclic AMP accumulation. Two preparations of PGE_2-P-S, but not PGE_2-R-S, stimulated cyclic AMP accumulation. Some of these results are shown in Table 3.

Although the results with various batches of N-P-S (GA) were not consistent, the unique augmentation of cyclic AMP accumulation by N-P-S (GA) suggests that the glutaraldehyde linking procedure had produced chemical and pharmacologic properties different from those resulting from the ECDI procedure. The glutaraldehyde procedure may have preserved some of the reactivity of the amino group, which in other systems is essential for adrenergic agonist activity. The likely structure of N-P-S (GA) (Fig. 1) probably preserves the basicity of the amine group. In addition, alkyl substitution of norepinephrine (producing a secondary amine) clearly enhances beta rather than alpha agonist activity in other systems (Innes and Nickerson, 1965). The proposed structure (Fig. 1) for N-P-S (GA) provides just such an alkyl substitution, producing a catecholamine similar to epinephrine. Incidentally, further substituion (with ECDI or glutaraldehyde) of the amine group of isoproterenol and epinephrine (which are already secondary amines) would be expected to decrease agonist activity by decreasing the basicity of the nitrogen atom (ECDI) or introducing steric hindrance (GA) at a site that is critical for intrinsic activity of the agonist (Innes and Nickerson, 1965). Iso-R-S and H-R-S both failed to stimulate leukocyte cyclic AMP accumulation (Table 3).

E. Sepharose-Drug Columns

Column experiments were carried out for three purposes: (1) to further define the character of cell binding and the distribution of what might be specific receptors in the total population of cells, (2) to try to quantitate the number of cells bound to any Sepharose rather than to depend on visual scoring of binding, and (3) to devise a method by which cells with specific receptors could be recovered.

Using H-R-S, we found that the columns retained $82 \pm 7\%$ of the cells, or 31% more than was retained by the R-S (Table 4). Further, when the H-R-S column was washed extensively with phosphate buffer, pH 8.1, almost all the retained cells were eluted from the column. More than 95% of the eluted cells excluded methylene blue dye and, therefore, were presumably viable. The recovery from H-R-S columns by elution with an antihistamine, diphenhydramine, was 29%. We have no way at present to determine whether some of the cells bound by R-S had histamine receptors or whether the great majority of cells with histamine receptors were specifically eluted by diphenydramine. However, the similarity of 31% additional retention by H-RSA over RSA and the 29%

Table 4. Determination of Distribution
of Receptors to Catecholamine and Histamine

Sequential drug or RSA column		Column				Mean % over nonspecific (RSA) binding	% cells with single receptors after 2nd column
Column 1	Column 2						
RSA[a]	—	51 ± 13				—	—
Histamine	—	82 ± 7[b]				31	—
N-R-S (GA)	—	79 ± 2[c]				38	—
		Experiment					
		1	*2*	*3*	*4*		
RSA	RSA	42	21	20	—	—	—
	Histamine	80	63	65	—	40 ± 2[d]	—
	N-R-S (GA)	75	59	52	—	34 ± 3[e]	—
Histamine	RSA	0	16	0	0	—	—
	Histamine	42	26	13	40	26 ± 14	—
	N-R-S (GA)	74	56	42	53	56 ± 13[e]	30
N-R-S (GA)	RSA	18	36	25	30	—	—
	Histamine	73	70	87	85	51 ± 12[f]	21
	N-R-S (GA)	40	49	65	75	30 ± 15	—

[a]Mean and SD in single column experiments are taken from 10 separate experiments.
[b]P values < 0.001 vs control (calculated by student t test).
[c]P values < 0.01 vs. control (calculated by student t test).
[d]P values < 0.001 vs. control (calculated by student t test).
[e]P values < 0.01 vs. control (calculated by student t test).
[f]P values < 0.02 vs. control (calculated by student t test).

elution of cells from H-R-S by an antihistamine is interesting. Propranolol $(1 \times 10^{-3}\text{M})$ did not displace cells from the H-R-S columns.

When columns were constructed from H-R-S (GA), pH changes did not increase elution of cells over that obtained by washing at pH 7.1. Diphenhydramine did elute about the same percentage of cells from H-R-S as from H-R-S (GA). Adrenergic antagonists had no effect, and because histamine was relatively inefficient in displacement of cells from beads, it was not tried in the experiments with columns.

N-R-S (GA) columns retained significantly more cells than R-S and about the same number as H-R-S. Only 15% of the total number were eluted by a gradient elution of 10^{-5} to 10^{-2} M propranolol. Surprisingly, only 8% of the total were eluted by washing with PBS at pH 8.5. Likewise, there was no effect of the antihistamine pyrilamine on elution of cells from these columns.

Determinations of the "distribution of receptors" in the population of cells were approached by sequential column experiments in which the unbound cells of the first column were passed (recycled) through a second column (Table 4). About 50% of cells were retained by R-S. We consider this retention nonspecific because retention by Sepharose, activated Sepharose, and RSA-Sepharose columns was equivalent. We have not been able to determine precisely how much retention is due to the RSA, to the contiguous surface of beads, to the sponges or column material, or is a function of the volumes of fluid we use on the columns. When cells passed through the R-S were subsequently reexposed to new batches of R-S, substantially fewer cells were retained. Regardless of which Sepharose-drug conjugate was used first, a lower percentage of cells was retained when the nonbound cells were rechromatographed on fresh batches of the same substance or R-S. In the four experiments in which the cells excluded from columns of H-R-S were chromatographed on R-S, practically no additional retention was seen. This seems to indicate that RSA may contribute to binding of cells that are also bound by H-R-S, but once cells that are bound by histamine-Sepharose columns are removed from the population, R-S will have little more effect.

The binding of cells excluded by a H-R-S column was only significant on N-R-S (GA). Conversely, the binding of cells excluded by N-R-S (GA) columns was only significant on H-R-S (Table 4). This may indicate that a maximum of 21 to 30% of cells from the second column (4 to 6% of the total cell population) has receptors either to histamine alone or to catecholamine alone, but that the majority of cells bound probably have multiple receptors on them.

Columns constructed of PGE_2-R-S were used in a similar set of sequential column experiments. These columns did not retain significantly more cells than R-S (57 ± 5 vs. 51 ± 2%, respectively). However, when cells excluded from an R-S column were exposed to a second column composed of PGE_2-R-S, there was a small increase of binding on the PGE_2-R-S compared to recycling on R-S. Preliminary experiments indicated that when PGE_2-R-S served as the first column, a second PGE_2-R-S column retained only a small additional percentage of cells (about 18%), whereas an H-R-S column retained 65% of the cells that were not bound by the prostaglandin column. These experiments suggest that the column procedure (in contrast to direct visualization, Table 1) may detect binding of cells to beads coated with prostaglandin. Since there is presently no specific procedure for displacing cells from PGE_2-R-S, the presence of receptors to prostaglandins on cell surfaces must be established by more direct methods.

In summary, receptors for small endogenous hormones on human leukocytes were studied by insolubilizing the hormones and incubating them with the cells. Histamine, norepinephrine, and prostaglandin E_2 (PGE_2) were conjugated to either of two types of carrier: (bovine or rabbit) serum albumin or a random copolymer of DL-alanine and L-tyrosine. The conjugates were linked to agarose beads (Sepharose), and the resultant drug-conjugate beads were incubated with leukocytes. Norepinephrine (when linked to its carrier via glutaraldehyde) and histamine preparations bound the majority of leukocytes. The binding appeared to be specific for the hormones tested. For example, the binding by histamine-RSA-Sepharose was prevented or reversed by high concentrations of histamine and histamine antagonists, but not by catecholamines or their pharmacologic antagonists. Similarly, binding of cells to the norepinephrine conjugate was inhibited by catecholamines and propranolol, but not by histamine or histamine antagonists. Conjugates of norepinephrine linked via carbodiimide did not bind cells. The protein or copolymer carriers did not contribute to binding per se. The hormone-protein conjugates bound more cells than the hormone-polymer conjugates. The former (unlike the free amines) failed to stimulate accumulation of cyclic AMP in leukocytes. The norepinephrine linked to polymer via glutaraldehyde, however, did stimulate leukocyte cyclic AMP accumulation, possibly because of the flexibility of the polymer. Columns of the various Sepharoses were used to determine the distribution of receptors to each hormone in mixed leukocyte populations. The majority of cells appeared to have receptors for both histamine and norepinephrine (bound through glutaraldehyde). Receptors to prostaglandins may have been detected by the column procedure, but their distribution could not be quantitated. The approach described provides a means to separate leukocytes on the basis of what are likely to be preformed receptors to small endogenous hormones, and to study the physiologic importance and function of the receptors.

III. Separation of Plaque-Forming Cells by Hormone-RSA-Sepharose Columns

The facts that we seemed to be able to separate cells on the basis of physiologic receptors for endogenous biogenic amines and that the biogenic amines inhibited release of antibody from lymphocytes formed the basis of an additional experimental approach. We have used the binding by columns of hormone-carrier-Sepharose to separate cells with different immunologic functions. This section will discuss the ability to subtract plaque-forming cells from a suspension of murine spleen cells. The following chapter will discuss the possibility that cells expressing receptors for histamine can also function as regulators of humoral antibody responses (Shearer et al., 1973). The relationship of plaque formation to cyclic AMP production has already been stressed in the last chapter (Bourne et al., 1973). The possibility that biogenic amines and prostaglandins may modulate other immunologic processes is suggested by the studies in this chapter and the next. The feasibility of discovering possible relationships between cyclic AMP and immunologic functions other than suppression of antibody release, cytolysis, and histamine release is increased with the approaches we describe.

A. Methods

Three strains of mice [C57 BL/6, Balb/c, and (Balb/c × C57 BL/6)F$_1$ = Balb/BL] were immunized by ip injection of 3 × 10^8 sRBC, and the spleens were removed on either day 6, 9, or 13 after immunization. The cells were were teased from the spleens, strained through stainless steel mesh (200 mesh), and suspended in Eagle's minimal essential media pH 7.4 (MEM) (from Microbiological Associates, Israel). The hemolytic plaque assays were done using a modification of the Jerne agar plate technique (Jerne et al., 1963; Shearer et al., 1968).

The assays were performed in agarose (L'Industrie Biologique Francaise S.A., Gennevilliers, France) on 60-mm plastic disposable petri dishes (Shearer et al., 1968). The plates were incubated at 37° for 1 hr, and then for an additional hour in the presence of 1 ml of 1:10 diluted lyophilized guinea pig complement (Grand Island Biol. Co., Grand Island, N.Y.). Direct PFC were counted after the 2-hr incubation period. Indirect PFC were developed on the same plates (after the direct PFC had been marked) by incubation for two additional hours in the presence of 0.1 ml 1:100 diluted rabbit antimouse IgG (Wortis et al., 1966). Triplicate plates were made from each sample and read by two investigators at different times. Rosette formation was performed by the method of Shearer and Cudkowicz (1968).

1. Cell Incubates and Hormone-RSA-Sepharose Columns

The Sepharose preparations [0.6 ml of 25% (w/v) of hormone-RSA-Sepharose in MEM] were incubated in plastic tubes with 10 to 200 × 10^6 cell spleen cells at 37°C for 15 min with gentle intermittent shaking. The cell-Sepharose mixture was then poured into a plastic column, and the unbound cells collected after washing the column with two 1-ml aliquots of MEM (Weinstein et al., 1973). Control preparations of cells were made with RSA-Sepharose, activated Sepharose, or Sepharose alone. Equal numbers of control and experimental cells were then used for the rosette, or plaque assay with sheep red blood cells.

B. Separation of Plaque-Forming Cells

In separate experiments using spleens from Balb/BL mice, cells excluded from columns of histamine-RSA-Sepharose produced between 56 and 84% fewer 19S or 7S plaques than control cells not put over columns (Table 5). The high cell/Sepharose ratio and the considerable variability of the effects of different batches of histamine-RSA-Sepharose in part accounted for the lower mean effect of the columns represented in Table 6. In all three experiments tried, the histamine-RSA-Sepharose did not subtract rosette-forming cells from the same spleens in which plaque-forming cells were bound. That the attraction of the insolubilized hormones for the plaque-forming cells was specific for the hormones and not the carriers is illustrated by (1) the fact that plaque-forming cells from

Table 5. Simultaneous Determination of the Effects of Two Sepharose
Preparations on 19S and 7S Plaques and Rosettes

Column composition[a]	Percent decrease of plaques[b]		
	19S	7S	Rosettes
RSA-Sepharose			
Exp. 1	22	18	4
Exp. 2	19	18	3
Exp. 3	12	8	0
Histamine-RSA-Sepharose			
Exp. 1	75	84	2
Exp. 2	56	67	0
Exp. 3	62	58	0

[a]Experiment number indicates same experiment on same day done nine days after immunization; e.g., Experiment 1 was done on same day. Cell experiments were done using Balb/BL spleen cells.
[b]Each point represents the average of two separate determinations done in two cell concentrations in triplicate plates (plaques) or tubes (rosettes).

the same animal attached to the histamine column, but were not bound by RSA-Sepharose (Table 5, upper portion), and (2) in 10 experiments conducted using each of the Sepharose conjugates represented in Table 6, only the histamine, epinephrine, isoproterenol, and prostaglandin E_2 preparations significantly altered the plaque-forming population of plated cells. None of the control column preparations (Sepharose, RSA-Sepharose, or activated Sepharose) selectively subtracted substantial numbers of plaque-forming cells. Neither did Sepharose preparations

Table 6. Effect of Hormone-RSA-Sepharose Columns
on Plaque Formation

Column composition	Percent decrease of plaques[a]	
	19S	7S
Sepharose	3 ± 1	+ 2 ± 6
Activated Sepharose	5 ± 3	6 ± 2
RSA-Sepharose	14 ± 12	+2.7 ± 5
Histamine-RSA-S	54 ± 12	64 ± 19
Epinephrine-RSA-S	32 ± 7	50 ± 23
Isoproterenol-RSA-S	43 ± 21	29 ± 17
Norepinephrine-RSA-S	7 ± 2	6 ± 4
Prostaglandin E_2-RSA-S	44 ± 17	28 ± 5
Prostaglandin $F_2\alpha$-RSA-S	4 ± 8	7 ± 10

[a]Each data point represents the mean ± SD of 10 separate experiments done 6 days (for 19S) and 13 days (for 7S) after immunization. In each experiment, duplicate points were averaged as explained in Table 1. Balb/BL animals were used. Inhibition of plaques by histamine, epinephrine, isoproterenol, and prostaglandin E_2 preparations was significantly different from any control ($P < 0.001$ by student t test).

made with norepinephrine and prostaglandin $F_{2\alpha}$ alter the plaque-forming ability of excluded cells. There was no significant difference between a substance's ability to subtract direct or indirect plaque-forming cells from the mixed population. If a Sepharose preparation retained 19S plaque-forming cells (PFC), it also removed 7S plaque formers. No preparation bound only one of the populations of antibody-forming cells. None of the Sepharose preparations used in this study attracted the population of cells capable of forming rosettes.

In order to determine whether the Sepharose preparations were able to subtract plaque-forming cells equally in various strains of mice, spleen cells from the two parental strains of Balb/BL were studied (Table 7). Rosette-forming cells were not removed by filtration over histamine-RSA-Sepharose in any of the three strains. The pattern of PFC retention by histamine-RSA-Sepharose was similar in the Balb/c and Balb/BL strains. In contrast, spleen cells from C57 BL/6 mice were not bound by any of the Sepharose preparations.

None of the Sepharose preparations used in these studies was pharmacologically active. None stimulated leukocyte adenyl cyclase.

C. Conclusions

Not all of the plaque-forming cells were subtracted by any one Sepharose preparation. There are two major possibilities and a number of technical considerations that may explain why the depletion of antibody-forming cells was incomplete:

(1) A number of the antibody-producing cells may not have receptors to any single hormone tested; or,

(2) A substantial number of cells may have receptors to each hormone, but technical aspects of the experimental procedure do not allow adequate binding of all cells with receptors.

Perhaps analogies can be drawn to human leukocytes where the majority of cells have multiple receptors to both histamine and epinephrine and only 6 to 10% of cells have receptors to only one hormone or another (see above). The data in the present studies indicate that some mouse spleen leukocytes have receptors for a number of biologically active substances. A number of insolubilized hormones attracted plaque forming cells, and each substance that bound antibody-forming cells did so with about equal effectiveness. It will be of interest to determine whether the majority of PFC can be removed from the population of mouse spleen cells by passing those that do not bind to one hormone column over a series of columns containing other hormones.

Additional explanations also should be considered—that the average depletion of PFC by columns was only about 50% of the total is likely to be due in part to the experimental conditions. For purposes of adequate

Table 7. Effect of Drug-Conjugate-Sepharose Column on 19S Plaque
Formation of Three Strains of Mice

Column material	Percent inhibition of plaques[a]		
	In C57 BL	In Balb/BL	In Balb/c
RSA	8 ± 14	14 ± 12	5 ± 6
Histamine-RSA	0 ± 7	54 ± 12	45 ± 13
Epinephrine-RSA	2 ± 9	32 ± 7	31 ± 6
Prostaglandin E_2-RSA	6 ± 3	44 ± 17	58 ± 3

[a]Each data point represents the mean ± SD of 10 experiments in Balb/BL and 5 experiments in each of C57 BL and Balb/c mouse spleen cells. Assays were done six
days after immunization. C57 BL had significantly less effect than the other two
strains tested ($P < 0.01$ — student t test). Data for each experiment were handled as
described in the footnote to Table 1.

yield of cells from the column within a reasonable time, we worked at
ratios of cells/Sepharose beads that produced submaximal depleting
effects. Using most preparations of histamine-RSA-Sepharose, even
under optimal conditions, the cells that were not bound to the column
could still respond to histamine and prostaglandin stimuli. Certainly in
most experiments, not all cells with receptors were removed from the
mixture.

The data in Table 7 indicate that strain differences may be associated with alterations in the affinity of cells for the insolubilized hormones. However, despite our inability to subtract PFC from C57 BL
mouse spleen cells, the cells did respond to free histamine, epinephrine,
and prostaglandin E_2 (Melmon et al., 1973b). They, therefore, have
receptors to the biogenic substances used here. The increase in cyclic
AMP content of the cells and inhibition of plaque formation during such
an increase was similar to those obtained in cells from the related strains
of mice used in these experiments. The pharmacologic effects of any
drug seemed to be equivalent in each strain listed.

Experiments reported elsewhere (Melmon et al., 1973b) have shown
that using histamine-RSA-Sepharose, some plaque-forming cells could be
removed from cell mixtures without removing all cells that are able to
respond to free histamine. These results indicate that it may be possible
to have a spectrum of affinity of cells from a given spleen with receptors
for the same hormone. We cannot determine whether all cells that produce antibody have receptors to the endogenous hormones tested or
whether the ability to separate antibody-producing cells on the basis of
their hormone receptors is dependent on the antigen used for immunization. Using antigen other than sRBC, it will be of interest to determine
whether antibody-forming cells can be removed from the population of
mouse spleen cells by the same column used in these experiments.

Each hormone-RSA-Sepharose that bound antibody-forming cells
was made of agonists for which corresponding leukocytic pharmacologic
receptors have been found. Thus, histamine, beta catecholamines, but

not alpha mimetic catecholamines, and the E series prostaglandins, but not the F series, stimulate separate receptors to increase cellular cyclic AMP (Bourne et al., 1973). None of the Sepharose preparations used in this study was pharmacologically active or noxious to the cells. Therefore, the effects of the columns must have been on the basis of simple physical separation of cells.

There are two lines of evidence that receptors to the hormones investigated here do not develop or are not exposed in the precursors of antibody-forming cells. They seem to appear sometime after the cell has been committed to production of antibody, i.e., after immunization. Either spleen cells from primed or unprimed donor mice have been passed over a histamine-RSA-Sepharose column and the unbound cells transferred to irradiated, syngeneic recipients that are simultaneously immunized with sRBC. These transferred cells generated enhanced PFC responses when compared with transfers of equal numbers of control cells or cells excluded from a RSA-Sepharose column (Shearer et al., 1972). Such transfers would have been expected to result not in enhancement, but depression of the antibody response if the precursors expressed histamine receptors and would, therefore, have been bound to histamine-RSA-Sepharose columns before transfer. The second line of evidence is more tenuous. One would have expected some binding effect of the hormone-RSA-Sepharose on rosette formers if (1) rosette formers express receptors to sRBC antigen, and some portion of these also are antigen-sensitive cells, and (2) if the hormone receptors are expressed at the precursor cell level (Brody, 1970). There was no effect of the columns on rosette formers (RFC). Therefore, the evidence implies that the receptors develop or are exposed during the differentiation processes that occur between antigenic stimulation and antibody production. The observation that hormone-Sepharose columns retained PFC but not RFC supports other published results that suggested that PFC and RFC represent distinct populations of immunocompetent cells (Shearer et al., 1968; Brody, 1970; Wilson, 1971).

In summary, we have shown that cells from immunized sRBC mouse spleens can be separated on the basis of their physiologic receptors to small endogenous hormones. Histamine, catecholamines (epinephrine, isoproterenol, and norepinephrine), and prostaglandin E_2 and $F_{2\alpha}$ were insolubilized by linking them to a carrier protein and Sepharose beads. Columns containing those Sepharose preparations for which there are corresponding pharmacologic receptors on leukocytes, bound and separated a portion of the cells from the general population. The cells that were not bound to (i.e., passed through) histamine, epinephrine, isoproterenol, and prostaglandin E_2 columns had significantly less plaque-forming ability per 10^6 plated cells than control cells or cells that did not adhere to RSA-Sepharose or norepinephrine and prostaglandin $F_{2\alpha}$-RSA-Sepharose columns. Chromatographing cells by the histamine, beta catecholamines, and E_2 prostaglandin columns decreased both 19S and 7S plaque-forming cells. None of the columns retained rosette-forming cells, and none stimulated cellular adenyl cyclase. Evidence is provided that the cellular receptors to the hormones are not expressed on precursors to plaque-forming cells, but develop or become exposed only after the cell has been committed to the production of antibody.

The methods used in the study allow separation of cells from a heterogeneous group on the basis of their receptors. The separation is not quantitative, but makes possible substantial concentrations of cells on the basis of their physiologic receptors to small hormones. The full function of such receptors on these cells remains to be seen. However, we think that our approach might provide a better understanding of the possible biochemical basis for a relationship between mediators of inflammation (histamine and prostaglandin) and modulation of an immune process.

Discussion of Dr. Melmon et al.'s Paper

DR. HADDEN: Since phytohemagglutinins that are bound to agarose beads behave differently than the native mitogens, I wonder what functional studies you have done with your Sepharose bead-bound amines. Do they have an antiproliferative effect in PHA-stimulated lymphocytes?

DR. MELMON: The amines insolubilized via water-soluble carbodiimides and carried on rabbit albumin have no effect on adenylate cyclase. The preparation was used simply to subtract cells. The cells that could be eluted from these columns were viable and had the capacity to function after transfer to irradiated animals. When you use the copolymer as the carrier and bind the amine with glutaraldehyde, the amino function of the different hormones is protected, and some preparations have pharmacologic effects like those of the free drug. Stimulation of cyclic AMP accumulation with the histamine polymer is relatively poor, whereas stimulation by norepinephrine is quite good. Norepinephrine effects on cyclic AMP are appropriately blocked by propranolol and not by phentolamine or any of the antihistamines. Thus, procedures may be chosen that either retain or inactivate the function of these amines. We think that the difference between the effects of the copolymer and the other carriers is relative to increased polymer flexibility as well as to its univalence.

DR. HADDEN: I am particularly interested in the effect of bound catecholamines on the proliferation system. In the material presented by Dr. Parker this morning, an effect of epinephrine on nuclear cyclic AMP content was shown. I would be very interested to know if your bound catecholamines have an effect on PHA-induced proliferation.

DR. MELMON: If we use an insolubilized hormone preparation, the amine will not penetrate to the nucleus.

DR. HADDEN: That is why I am asking.

DR. MELMON: We have not done the experiment.

DR. PARKER: Quantitatively, how much of a response do you get to insolubilized norepinephrine? Is it commensurate with the response obtained when cells are stimulated with soluble catecholamine?

DR. MELMON: If you calculate the amount of norepinephrine that is bound on a given pharmacologically active hormone-carrier preparation and select a density of beads that can be used in experiments with cells, you come up with a maximal concentration of about 10^{-4} M. However, 10^{-4} M bound catecholamine is not equivalent to 10^{-4} M free drug. One has no way of anticipating what degree of response he should see. This depends upon the accessibility of Sepharose-bound hormone to the cells and upon whether the drug retains its binding and hormonal activity. We usually get stimulation approaching that seen with 10^{-4} M free hormone. The responses do not appear to be due to solubilized drugs. We have not been able to demonstrate release of radioactive drug in our insolubilized preparations either in the presence or absence of cells. We have, moreover, incubated the bead and cell supernatants with fresh cells and not seen response.

DR. PARKER: Are your cyclic AMP values based on total cells or on cells that are adhering to the beads?

DR. MELMON: Our cyclic AMP values are for total cells in the incubation mixture. Of course, a number of those cells did not bind to beads.

DR. PARKER: Have you taken the cells that have been exposed to the beads and added high concentrations of free isoproterenol, norepinephrine, and histamine to see whether further stimulation will occur?

DR. MELMON: That has not been done. Another interesting experiment would be to incubate cells with beads that bind but do not stimulate and see whether the response to the corresponding free amine would be blocked.

References

Bourne, H. R., Melmon, K. L., Weinstein Y., and Shearer, G. M. "Pharmacologic Regulation of Antibody Release in Vitro: Effects of Vasoactive Amines and Cyclic AMP," Proc. of Conf. on Cyclic Nucleotides, Immune Responses, and Tumor Growth (1973).

Brody, T. "Identification of Two Cell Populations Required for Mouse Immunocompetence," *J. Immunol.* **105**:126 (1970).

Henney, C. S., Bourne, H. R., and Lichtenstein, L. M. "The Role of Cyclic 3',5'-Adenosine Monophosphate in the Specific Cytolytic Activity of Lymphocytes," *J. Immunol.* **108**:1526 (1972).

Innes, I. R., and Nickersen, M. "Drugs Acting on Postganglionic Adrenergic Nerve Endings and Structures Innervated by Them (Sympathomimetic Drugs)." In *The Pharmacologic Basis of Medical Practice,* L. S. Goodman and A. Gilman, eds., 3rd ed. New York: MacMillan Co., p. 477 (1965).

Jerne, W. K., Nordin, A. A., and Henney, C. "The Agar Plaque Technique for Recognizing Antibody Producing Cells." In *Cell Bound Antibodies*, B. Amos and H. Koprowski, eds. Philadelphia: Wistar Institute Press, p. 109 (1963).

Melmon, K. L., Bourne, H. R., Weinstein, Y., and Sela, M. "Receptors for Histamine Can be Detected on the Surface of Selected Leukocytes," *Science* 177:707 (1972).

Melmon, K. L., Bourne, H. R., Weinstein, Y., Shearer, G. M., and Bauminger, S. "Release of Antibody from Leukocytes in Vitro: Control by Vasoactive Hormones and Cyclic AMP" (submitted for publication—1973a).

Melmon, K. L., Weinstein, Y., Shearer, G. M., Bourne, H. R., and Bauminger, S. "Separation of Specific Antibody-Forming Mouse Spleen Cells by Their Receptors to Endogenous Hormones" (submitted for publication—1973b).

Porath, J., Axen, R., and Ernback, S. "Chemical Coupling of Proteins to Agarose," *Nature* 215:491 (1967).

Sela, M., and Fuchs, S. "Preparation of Synthetic Polypeptides As Antigens." In *Methods in Immunology and Immunochemistry*, C. A. Williams and M. W. Chase, eds., vol 1. New York: Academic Press, p. 167 (1967).

Shearer, G. M., Cudkowicz, G., Connell, M. J., and Priore, R. L. "Cellular Differentiation of the Immune System of Mice. I. Separate Splenic Antigen-Sensitive Units for Different Types of Antisheep Antibody-Forming Cells," *J. Exp. Med.* 128: 437 (1968).

Shearer, G. M., Melmon, K. L., Weinstein, Y., and Sela, M. "Regulation of Antibody Response By Cells Possessing Histamine Receptors," *J. Exp. Med.* 136:1302 (1972).

Shearer, G. M., Weinstein, Y., Melmon, K. L., and Bourne, H. R. "Separation of Leukocytes by Their Amine Receptors: Subsequent Immunologic Functions," *Proc of Conf. on Cyclic Nucleotides, Immune Responses, and Tumor Growth* (1973).

Singhal, S. K., and Wizzell, H. "In Vitro Induction of Specific Unresponsiveness of Immunologically Reactive, Normal Bone Marrow Cells," *J. Exp. Med.* 131:149 (1970).

Venter, Y. C., Dixon, Y. E., Maroko, P. R., and Kaplan, J. O. "Biologically Active Catecholamines Covalently Bound to Glass Beads," *Proc. Nat. Acad. Sci.* 69:1141 (1972).

Weinstein, Y., Melmon, K. L., Bourne, H. R., and Sela, M. "Specific Leukocyte Receptors for Small Endogenous Hormones: Detection by Cell Binding to Insolubilized Hormone Preparations," *J. Clin. Invest.* (in press—1973).

Wilson, J. D. "The Relationship of Antibody-Forming Cells to Rosette-Forming Cells," *Immunol.* 21:233 (1971).

Wizzell, H., and Andersson, B. "Cell Separation on Antigen-Coated Columns," *J. Exp. Med.* 129:23 (1969).

Wofsy, L., Kimura, J., and Truffa-Bachi, P. "Cell Separation on Affinity Columns: The Preparation of Pure Populations of Antihapten Specific Lymphocytes," *J. Immunol.* 107:725 (1971).

Wortis, H. H., Taylor, R. B., and Dresser, D. W. "Antibody Production Studied by Means of the LHG Assay. I. The Splenic Response of CBA Mice to Sheep Erythrocytes," *Immunol.* 11:603 (1966).

SEPARATION OF LEUKOCYTES BY THEIR AMINE RECEPTORS: SUBSEQUENT IMMUNOLOGIC FUNCTIONS

G. M. Shearer, Y. Weinstein, K. L. Melmon, and H. R. Bourne

I. Introduction

Lymphoid cell populations have been separated into various functional categories by empirical procedures such as albumin gradient centrifugation (Dutton et al., 1970; Haskill and Marbrook, 1971; Miller and Cudkowicz, 1971), velocity sedimentation (Miller and Phillips, 1969; Haskill and Moore, 1970; Haskill and Axelrod, 1972), and by virtue of antibodylike receptors on immunocompetent cells (Wigzell and Andersson, 1969; Wofsy et al., 1971; Singhal and Wigzell, 1970). Affinity immunocyte column chromatography has been achieved using antigenic molecules attached to inert beads (Wigzell and Andersson, 1969), or to nylon threads (Rutishauser et al., 1972). Immunocytes possessing functional properties characteristic of antibody-producing cells or their precursors were retained by such chromatography techniques, whereas cells presumed not to express the relevant immunoglobulinlike receptors passed through the columns.

It is possible that lymphoid cells possessing other types of membrane receptors could play a significant, although less obvious, role in modulation of immune processes. Endogenous hormones such as histamine, the beta catecholamines, and prostaglandins E_1 and E_2 have been shown to inhibit the cytotoxic effects of lymphocytes sensitized to fibroblasts (Henny and Lichtenstein, 1971), as well as the IgE-mediated release of histamine from human leukocytes (Bourne et al., 1972). Chromatographic methods, which can be used to separate cells based on their surface receptors to these hormones, have been developed recently (Weinstein et al., 1973). Leukocytes expressing receptors for histamine were separated from human peripheral blood cells by fractionation over columns of histamine-RSA-Sepharose (Melmon et al., 1972). Cells expressing histamine surface receptors, which appear to function as regulators of humoral antibody responses, have been removed from mouse spleen cell suspension (Shearer et al., 1972). Thus, cell surface receptors for biogenic amines seem to play a significant role in regulation of antibody production.

II. Enhancement of PFC Responses by Passing Mouse Spleen Cells over Amine-RSA-Sepharose Columns

The last chapter presented evidence that at least some direct and indirect plaque-forming cells (PFC) for sheep erythrocytes (sRBC) express surface receptors for certain biogenic amines. Thus, PFC were specifically retained by columns of histamine, epinephrine, isoproterenol, and prostaglandin E_2, insolubilized by attachment to RSA-Sepharose beads. Since PFC descend from bone-marrow-derived precursors by proliferation and differentiation, it was of interest to determine whether the precursor of PFC (P-PFC) would also exhibit similar receptors. Were P-PFC to express such receptors, columns of the same amine-RSA-Sepharose composition that retained PFC should also bind P-PFC. This would result in a reduced immune response potential of the cells that pass through the columns, due to retention of the P-PFC by the amine-RSA-Sepharose beads. Alternatively, if the amine receptors were not present on the antibody-forming cell precursors, but were to develop during the process of antigen-induced differentiation from the progenitor to the antibody-producing cell, then the cell fractionation procedure should have no effect on the immunocompetent potential of the cells that pass through the columns. A third possibility is that the cells that pass through the amine columns would display enhanced immune potential. This could result if removal of PFC by the columns affected antibody-feedback mechanisms, or if cells that function to regulate immunity express surface receptors for these amines.

The immunocompetent potential of murine lymphoid cells can be analyzed by cell transfer into heavily irradiated, syngeneic recipients accompanied by immunization. The donor-derived PFC responses are subsequently assayed using cell suspensions from the repopulated recipient spleens (Shearer et al., 1968). The direct PFC responses generated following spleen cell transfer: (1) 5×10^6 unfiltered cells, (2) 5×10^6 cells passed over RSA-Sepharose control columns, or (3) 5×10^6 spleen cells passed over histamine-RSA-Sepharose columns were compared in irradiated, syngeneic recipients immunized simultaneously with 3×10^8 sRBC. In the first experiments, the spleen cell suspensions used were prepared from donors that had been immunized with 3×10^8 sRBC one week earlier, since the appearance of amine receptors could have depended on immunization. The seven-day PFC responses from five independent experiments are shown in Table 1. The immunocompetence of spleen cells passed over histamine-RSA-Sepharose columns was enhanced from 2.2- to 5.4-fold when compared with the responses of an equal number of unfiltered cells and between 1.6- and 4.4-fold when compared with responses generated by cells passed over RSA-Sepharose control columns (Shearer et al., 1972).

Indirect PFC responses obtained on days 9, 13, and 23 after transfer of 5×10^6 unfiltered spleen cells or cells passed over histamine-RSA-Sepharose are summarized in Table 2. On day 13, the peak of the indirect PFC response, 5.8-fold more indirect PFC were generated by the fractionated than by the unfractionated cells. By day 23, the responses of the two groups were similar.

The data presented thus far indicate that the removal of a cell population from the spleens of immunized mice results in elevated 19S and 7S humoral antibody response potential of the cells not retained by the columns. This enhancement could be due either to the exclusion of a cell population expressing histamine receptors, which functions to regulate the number of PFC responding, or to the removal of PFC, which might have affected the level of responsiveness via removal of the source of antibody feedback (Uhr and Moller, 1968). In order to test the latter

Table 1. Direct PFC Responses in Irradiated Mice Injected with Unfiltered Spleen Cells, Spleen Cells Passed over RSA-Sepharose Columns, or Spleen Cells Passed over H-RSA Sepharose Columns [a]

	Mean No. direct PFC per spleen		
Exp. No.	No col.	RSA-S	H-RSA-S
1	1060 ± 204	1310 ± 224	4830 ± 873
2	3550 ± 410	4370 ± 793	9300 ± 2470
3	2370 ± 820	4340 ± 610	6940 ± 1680
4	1960 ± 836	1890 ± 302	4320 ± 680
5	1350 ± 302	1870 ± 844	3040 ± 690

	Ratio of direct PFC per spleen		
	RSA-S	H-RSA-S	H-RSA-S
Exp. No.	No col.	RSA-S	No col.
1	1.2	3.7	4.6
2	1.2	4.4	5.4
3	1.8	1.6	2.9
4	0.97	2.3	2.2
5	1.4	1.6	2.3

[a](Balb/c × C57 BL/6)F_1 mice exposed to 750 rad of ^{60}Co gamma irradiation were injected with 5×10^6 spleen cells (mixed with 3×10^8 sRBC) from synergeneic donors immunized seven days before with sRBC. The donor spleen cells were not passed over columns or were passed over either RSA-Sepharose or H-RSA-Sepharose columns before transfer into recipients. Direct PFC assays were made seven days after cell transfer; ± indicates standard error.

Table 2. Indirect PFC Responses in Irradiated Mice Injected with
Unfiltered Spleen Cells or with Spleen Cells Passed over H-RSA-
Sepharose Columns[a]

Day of assay	Indirect PFC per spleen		Ratio of indirect PFC per spleen $\frac{\text{H-RSA-S}}{\text{No col.}}$
	No col.	H-RSA-S	
9	1010 ± 143	3170 ± 910	2.9
13	3950 ± 580	$23{,}100 \pm 4950$	5.8
23	1440 ± 169	1940 ± 400	1.3

[a](Balb/c × C57 BL/6)F_1 mice exposed to 750 rad of ^{60}Co gamma irradiation were injected with 5×10^6 spleen cells (mixed with 3×10^8 sRBC) from syngeneic donors
immunized seven days before with sRBC. The donor spleen cells were not passed
over columns or were passed over H-RSA-Sepharose columns before transfer into
recipients; ± indicates standard error.

possibility and to establish whether spleen cells from unimmunized mice
show enhanced immune response potential after fractionation over
histamine-RSA-Sepharose columns, the PFC results of transfer experiments were compared using unfiltered spleen cells, cells passed over
RSA-Sepharose control columns, and cells passed over histamine-RSA-
Sepharose columns, prepared from normal donors. The results of six independent experiments are shown in Table 3. The direct PFC responses
generated by 5×10^6 spleen cells passed over the histamine-bead columns was from 3.1- to 12.5-fold greater than that resulting from the
transfer of an equal number of unfiltered spleen cells (Shearer et al.,
1972). These results, using spleen cells from unimmunized donors, indicate:

(1) that removal of the antibody-producing cells and hence the
source of antibody feedback control of the antibody response cannot account for the enhanced immune potential of cells that pass through the
columns;

(2) that the enhancement is likely to be due to the removal of a cell
population other than antibody-producing cells, which modulates the
number of PFC appearing in the recipients; and

(3) that these regulator cells exist in the lymphoid system prior to
active immunization.

In order to determine whether enhanced immune responses would
result from cell fractionation using columns composed of other amines
which stimulate cyclic AMP production in soluble form (Bourne et al.,
1972) and which retain PFC when insolubilized on RSA-Sepharose beads
(the last chapter), irradiated mice were injected with 3×10^8 sRBC and
5×10^6 unfiltered spleen cells or an equal number of cells that passed

Table 3. Direct PFC Responses in Irradiated Mice Injected with Unfiltered Spleen Cells, Spleen Cells Passed over RSA-Sepharose Columns, or Spleen Cells Passed over H-RSA-Sepharose Columns[a]

Exp. No.	Mean No. direct PFC per spleen			Ratio direct PFC per spleen		
				RSA-S	H-RSA-S	H-RSA-S
	No col.	RSA-S	H-RSA-S	No col.	RSA-S	No col.
1	690 ± 122	810 ± 255	2160 ± 610	1.2	2.7	3.1
2	592 ± 190	1170 ± 570	2490 ± 950	2.0	2.1	4.2
3	537 ± 411	NT	1810 ± 486	—	—	3.4
4	188 ± 71	NT	755 ± 38	—	—	4.0
5	109 ± 29	NT	1380 ± 342	—	—	12.5
6	270 ± 74	NT	1190 ± 640	—	—	4.4

[a](Balb/c × C57 BL/6)F_1 mice exposed to 750 rad ^{60}CO gamma irradiation were injected with 5 × 10^6 spleen cells (mixed with 3 × 10^8 sRBC) from unimmunized syngeneic donors. The donor spleen cells were not passed over columns or were passed over either RSA-Sepharose or H-RSA-Sepharose columns before transfer into recipients. Direct PFC assays were made seven days after cell transfer; ± indicates standard error. NT indicates not tested.

over histamine-RSA-Sepharose, epinephrine-RSA-Sepharose, or isoproterenol-RSA-Sepharose columns. The results of these preliminary studies (data not shown) indicate that PFC enhancement similar to that observed for histamine-bead columns can be obtained using epinephrine- or isoproterenol-bead columns. These observations raise the possibility that the immunoregulatory process under study may be associated with cyclic AMP production by lymphoid cells since (1) the same amines found to modulate humoral antibody production by column fractionation have also been shown to stimulate cyclic AMP production in human leukocytes and to inhibit antigen-induced release of histamine from cells of allergic donors (Bourne et al., 1972), and (2) depletion of immune splenic PFC by columns identical to those used in this study resulted in a concomitant reduction in the ability of the remaining cells to accumulate cyclic AMP after stimulation with either histamine or prostaglandin E_2 as discussed in the preceding chapter.

III. Kinetics of PFC Response After Passing Spleen Cells over Histamine-RSA-Sepharose Columns

Although the PFC responses generated by 5 × 10^6 histamine-RSA-Sepharose filtered cells were greater than those produced by an equal number of unfiltered cells seven days after cell transfer, the possibility was not excluded that the observed differences in PFC number could have been due to a shift in the kinetics of PFC production. Thus, it was important to compare the responses of fractionated and unfractionated cells on various days after transfer. Results of the kinetic studies are summarized graphically in Fig. 1. The patterns of direct PFC production in

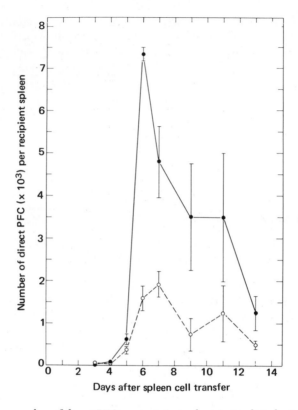

Fig. 1. Mean number of direct PFC per recipient spleen in irradiated mice injected with 5 × 10⁶ unfiltered spleen cells(o) or 5 × 10⁶ spleen cells passed over H-RSA-S columns (●), and immunized with 3 × 10⁸ sRBC, as a function of the time after cell transfer. (I) indicates standard deviations. p values: days 6, 7 <0.001; day 9 <0.05.

the two groups of recipients were similar to each other and to those reported earlier (Shearer et al., 1968). However, 4.5-fold (p < 0.001) and 2.5-fold (p < 0.001) differences were detected between the number of direct PFC generated by the filtered and unfiltered cells on days 6 and 7 after transfer. Significant differences (p < 0.05) were also observed between the two groups throughout the first two weeks of the response. By day 23, the responses of the two groups were indistinguishable from each other, although both were considerably above background.

The PFC data shown thus far are expressed per recipient spleen, and therefore compare the response potentials of a fixed number of donor-derived spleen cells. However, the differences in the responses might also be reflected in the number of direct PFC detected per 10⁶ cells from the recipient spleen. The kinetics of the direct PFC responses per 10⁶ recipient cells are shown in Fig. 2. A twofold difference (p < 0.05) was

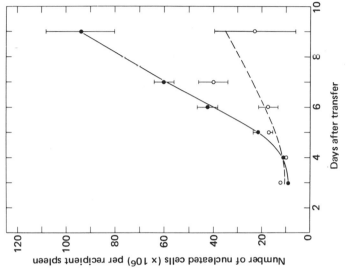

Fig. 3. Mean number of nucleated cells per recipient spleen in irradiated mice injected with 5×10^6 unfiltered spleen cells (○) or 5×10^6 spleen cells passed over H-RSA-S columns (●), and immunized with 3×10^6 sRBC, as a function of time after cell transfer. (I) indicates standard deviations; p values days 6, 7 <0.001; day 9 <0.01.

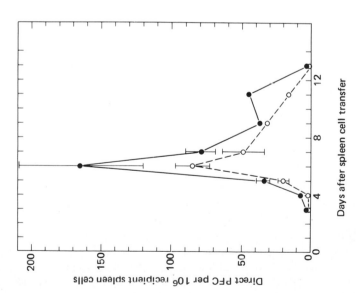

Fig. 2. Mean number of direct PFC per 10^6 recipient spleen cells in irradiated mice injected with 5×10^6 unfiltered spleen cells (○) or 5×10^6 spleen cells passed over H-RSA-S columns (●), and immunized with 3×10^8 sRBC, as a function of time after cell transfer. (I) indicates standard deviation. p values: day 6 <0.05; day 7 <0.1.

detected at the peak of the response on day 6 between the number of PFC per 10^6 cells generated by histamine-RSA-Sepharose filtered versus the unfiltered spleen cells. The fact that enhancement of the response was only partially accounted for in terms of PFC per 10^6 spleen cells suggested that a part of the overall 4.5-fold enhancement could have been due to differences in the total number of nucleated cells in the spleens of the two groups of recipients. The mean number of nucleated cells per spleen are shown for days 3 to 9 in Fig. 3. Significantly higher cell counts ranging from 1.5- to 4.5-fold were obtained on days 6 ($p < 0.01$), day 7 ($p < 0.001$), and day 9 ($p < 0.01$) in the spleens of recipients injected with 5×10^6 histamine-RSA-Sepharose filtered cells. Thus, the enhancement in the PFC responses obtained by cell fractionation over histamine-bead columns can be partially accounted for by an overall increase in total number of recipient spleen cells, a portion of which is anti-sRBC PFC, and to some extent by a selective increase in the number of PFC themselves.

IV. Effect of Inoculum Size on Enhancement of PFC Response After Passing Cells over Histamine-RSA-Sepharose Columns

The PFC results of using inocula of 5, 10, 15 $\times$ 10^6 filtered and unfiltered spleen cells are compared in Table 4 along with the ratios of the responses. The lower the cell inoculum, the greater the differences in the responses generated by the two groups. The regulatory effects of the cells retained by the histamine-bead columns seem to be expressed over a limited range of cell inocula. Haskill and Axelrod (1972) have observed a similar cell dose-dependent regulatory effect on in vitro PFC responses using empirical cell fractionation techniques.

V. Suppression of PFC Response by Spleen Cells Retained by Histamine-RSA-Sepharose Columns

If the enhanced immune potential of spleen cells that pass through amine-RSA-Sepharose columns is due to the removal of a regulator cell population by virtue of its amine receptors, then the regulatory function of this population should be verifiable by addition of the cells eluted from histamine-RSA-Sepharose beads to the cells that pass through the columns. In other words, the enhanced immunocompetence exhibited by the cells that pass through the columns should be reduced to the level observed for unfractionated spleen cells by the addition of relatively small numbers of eluted cells. In order to test this possibility, 1×10^6 spleen cells eluted from histamine-RSA-Sepharose beads by agitation

Table 4. Comparison of Direct PFC Responses in Irradiated Mice Injected with Different Numbers of Unfiltered Spleen Cells with Those Obtained in Mice Injected with Equal Numbers of Spleen Cells Passed over H-RSA-Sepharose Columns[a]

No. cells injected ($\times 10^6$)	Mean No. direct PFC per spleen		Ratio of direct PFC per spleen $\dfrac{\text{H-RSA-S}}{\text{No Col.}}$
	No Col.	H-RSA-S	
5	109 ± 29	1380 ± 339	13.0
10	711 ± 194	2620 ± 194	3.0
15	2690 ± 620	6690 ± 1120	2.5

[a](balb/c × C57 BL/6)F_1 mice exposed to 750 rad ^{60}Co gamma irradiation were injected with spleen cells (mixed with 3×10^8 sRBC) from unimmunized, syngeneic donors. The donor spleen cells were not passed over columns or were passed over H-RSA-Sepharose columns before transfer into recipients. Direct PFC assays were made seven days after cell transfer; ± indicates standard error.

were mixed with 5×10^6 spleen cells that passed through the histamine-bead columns and injected into irradiated, syngeneic recipients. The donor-derived direct PFC responses were accessed seven days later. Eluted cells were also centrifuged, the supernatant removed, and the pellet of eluted cells was resuspended in fresh media. The immunosuppressive effect of the supernatant as well as that of the resuspended pellet was tested in a similar transfer experiment. (The results are not summarized in the tables.) The number of direct PFC resulting from the transfer of 5×10^6 spleen cells that passed through histamine-RSA-Sepharose columns was reduced to the number generated by 5×10^6 unfiltered cells as a result of adding either 1×10^6 cells eluted from histamine-bead columns or 1×10^6 eluted, centrifuged cells. Addition of the supernatant of the 1×10^6 eluted, centrifuged cells also suppressed the response, but not as much as the eluted cells themselves. These results indicate that cells that function to regulate the level of humoral antibody responses by suppression of the number of antibody-forming cells produced can be eluted from histamine-RSA-Sepharose columns. These findings also suggest that a soluble product of the eluted cells is involved in the regulatory process. This immunoregulatory process may be amenable to detailed biochemical, pharmacological, and cellular analysis.

It remains to be established whether the regulatory cells expressing receptors for biogenic amines are antigen specific, whether they can account for other known immunosuppressive phenomena such as tolerance or antigenic competition, whether these regulatory mechanisms are modulated via cAMP-mediated events, whether they function for other types of immunity such as cell-mediated reactions, and if so, whether they could account for the weak reactivity toward minor histocompatibility and tumor-specific antigens.

Discussion of Dr. Shearer et al.'s Paper

DR. HENNEY: Do you have any indication of whether the immunoregulatory cells you are studying belong to a T or B class?

DR. SHEARER: This will be one of our next experiments. We plan to add antitheta antiserum and also look at cells derived directly from bone marrow and the thymus.

DR. HENNEY: The elution cell procedure you described was mechanical. Are you not able to elute with histamine?

DR. SHEARER: No, not by histamine nor by a change in pH. These are approaches that may well work, but so far we have only been successful with mechanical elution.

DR. WEBB: Have you considered using a T cell independent antigen in your system?

DR. SHEARER: We have considered, but have not done the experiment you mention. I agree that the results would be very interesting.

DR. AUSTEN: Can we be absolutely certain that the cells that passed through the column and then used to reconstitute recipients were not, in fact, stimulated through the histamine receptor, but simply did not have the necessary binding affinity to adhere to the column?

DR. MELMON: The answer is as sure as we know how to be sure. The insolubilized hormone preparations did not stimulate cyclic AMP accumulation in either the bound or free cells even though the cells are viable. The insoluble derivatives that do stimulate cyclic AMP accumulation were not used to separate cells for either the transfer experiments or the plaque-forming cell experiments.

DR. AUSTEN: Are you saying that the ones that are trapped are not stimulated?

DR. MELMON: That is correct.

DR. AUSTEN: How do you account for that?

DR. MELMON: By virtue of nullification of the basic charge of the amino group without major losses in binding affinity for histamine receptors. We know, for instance, that certain antihistamines incapable of stimulating cells have a much higher affinity for the cell than histamine. Indeed, it is by such a mechanism that they are able to function as competitive antagonists. So the modified hormone cannot stimulate the cell metabolically, but retains its cell-binding capability. In a sense, we may have made an insoluble antihistamine that binds to cells in part through histamine receptors.

DR. AUSTEN: Let me ask my question in another way. If you take the two populations of cells, those that are retarded and those that are not, and expose them to histamine in its simple nonconjugated form, will they both respond with an increase in cyclic AMP? And if not, how do they now differ?

DR. MELMON: If the appropriate column conditions are used, it is possible to show that the cells that come through will not respond to the corresponding amine. If the adherent cells are mechanically eluted from the column, they do respond, but we have not done many of these experiments.

DR. WEISSMANN: Lymphocytes of the T and B series display ultrastructural differences. Have you looked by ultrastructural techniques to see what the morphologic differences are between the cells that stick and the cells that do not stick on the histamine column? One would expect that the sticky cells might very well be cells rich in Golgi material with a large nuclear hof area that Allison has described as a third type of lymphocyte.

DR. MELMON: Ultrastructural studies are in progress.

DR. PLESCIA: Are the cells that have been filtered through the column completely devoid of cells that bind to beads or did you pass them through only once?

DR. SHEARER: In the transfer experiments, they were only passed once.

DR. PLESCIA: I think it would be of interest to completely remove the cells that specifically bind and then do transfer experiments.

DR. MELMON: The problem is that the columns have a lot of nonspecific binding, and total depletion of specifically binding cells would result in a very low overall cell yield, limiting the cells available for transfer.

DR. MORSE: Other than antihistamines, are there ways of or treatment of the cells with enzymes such as trypsin and neuraminidase.

DR. SHEARER: pH changes have been used to elute cells. The optimal pH for binding is around 6.8 to 7.4. At about pH 8.0, cells start to elute from the column.

DR. MELMON: If cells are incubated in culture medium, a substance is obtained in the supernatant that will bind to the beads. Perhaps this substance is soluble receptor material. If cells are incubated with trypsin, they very quickly lose their ability to adhere to histamine columns. These are very preliminary studies, and they are in no way as sophisticated as the experiments of Roth and Cuatrecasas and their colleagues with insulin and ACTH.

References

Bourne, H. R., Lichtenstein, L. M., and Melmon, K. L. "Pharmacologic Control of Allergic Histamine Release in Vitro: Evidence for an Inhibitory Role of 3', 5'—Adenosine Monophosphate in Human Leukocytes," *J. Immunol.* **108**:695–705 (1972).

Dutton, R. W., McCarthy, M. M., Mishell, R. I., and Raidt, D. J. "Cell Components in the Immune Response. IV. Relationships and Possible Interactions," *Cell. Immunol.* **1**:196 (1970).

Haskill, J. S., and Axelrod, M. A. "Cell Mediated Control of an Antibody Response," *Nature (New Biol.)* **237**:251–252 (1972).

Haskill, J. S., and Marbrook, J. "The in Vitro Immune Response to Sheep Erythrocytes by Fractionated Spleen Cells: Biological and Immunological Differences," *J. Immunol. Methods* **1**:43–54 (1971).

Haskill, J. S., and Moore, M. A. S. "Two-Dimensional Cell Separation: Comparison of Embryonic and Adult Hemopoietic Stem Cells," *Nature* **226**:853–854 (1970).

Henny, C., and Lichtenstein, L. M. "The Role of Cyclic AMP in the Cytolytic Activity of Lymphocytes," *J. Immunol.* **107**:610–615 (1971).

Melmon, K. L., Bourne, H. R., Weinstein, J., and Sela, M. "Receptors for Histamine Can Be Detected on the Surface of Selected Leukocytes," *Science* **177**:707–709 (1972). Miller, H. C., and Cudkowicz, G. "Density Gradient Separation of Marrow Cells Restricted for Antibody Class," *Science* **171**:913–915 (1971).

Miller, R. G., and Phillips, R. A. "Separation of Cells by Velocity Sedimentation," *J. Cell. Physiol.* **73**:191–201 (1969).

Rutishauser, U., Millette, C. F., and Edleman, G. M. "Specific Fractionation of Immune Cell Populations," *Proc. Nat. Acad. Sci.* (U.S.A.) **69**:1596–1600 (1972).

Shearer, G. M., Cudkowicz, G., Connell, M. S. J., and Priore, R. L. "Cellular Differentiation of the Immune System of Mice. I. Separate Splenic Antigen-Sensitive Units of Different Types for Antisheep Antibody-Forming Cells," *J. Exp. Med.* **128**:437–457 (1968).

Shearer, G. M., Melmon, K. L., Weinstein, Y., and Sela, M. "Regulation of Antibody Response by Cells Expressing Histamine Receptors," *J. Exp. Med.* **136**:1302–1307 (1972).

Singhal, S. K., and Wigzell, H. "In Vitro Induction of Specific Unresponsiveness of Immunologically Reactive, Normal Bone Marrow Cells," *J. Exp. Med.* **131**:149–164 (1970).

Uhr, J. W., and Moller, G. "Regulatory Effect of Antibody on the Immune Response," *Advan. Immunol.* **8**:81–127 (1968).

Weinstein, Y., Melmon, K. L., Bourne, H. R., and Sela, M. "Specific Leukocyte Receptors to Small Endogenous Hormones: Detection by Cell Binding to Insolubilized Hormone Preparation," *J. Clin. Invest.* (in press—1973).

Wigzell, H., and Andersson, B. "Cell Separation on Antigen-Coated Columns," *J. Exp. Med.* **129**:23–36 (1969).

Wofsy, L., Kimura, J., and Truffa-Bachi, P. "Cell Separation on Affinity Columns: the Preparation of Pure Populations of Antihapten Specific Lymphocytes," *J. Immunol.* **107**:725–729 (1971).

THE ROLE OF THE CYCLIC AMP SYSTEM IN INFLAMMATION: AN INTRODUCTION

LAWRENCE M. LICHTENSTEIN

The first suggestion that agents that could increase the intracellular level of cyclic AMP were capable of inhibiting inflammatory responses was in 1968 with the demonstration that the catecholamines and methylxanthines could inhibit antigen-induced, IgE-mediated histamine release from human leukocyte preparations (1). It had been shown, to be sure, as early as 1937, that epinephrine could inhibit histamine release from the lung of a sensitized guinea pig (2). These experiments, however, were carried out long before the cyclic AMP system was discovered by Sutherland and his colleagues. It was their research, demonstrating the effects of the cyclic AMP system in secretory reactions, that led us to undertake this experimentation. A number of different lines of evidence, generated in the early sixties, convinced us that IgE-mediated histamine release from human basophils, and similar reactions in other systems, represented secretory reactions (3, 4). It was natural at that point to ask whether cyclic AMP controlled this reaction as it did other secretory phenomenon. The fact that it appeared to do so was not a surprise; what was surprising was that an increased level of cyclic AMP turned off the response in this system rather than facilitating it, which was the common experience at that time. Obviously, this is no longer so novel.

These earlier observations provided, at best, circumstantial evidence that cyclic AMP was involved. That is, we were able to show that the catecholamines and methylxanthines, which were capable, in other tissues, of increasing cyclic AMP levels, inhibited IgE-mediated histamine release. Moreover, as expected, isoproterenol and theophylline had a synergistic effect, and the dibutyryl derivative of cyclic AMP was also inhibitory. These observations, with respect to the reactions of immediate hypersensitivity, have been amply confirmed by others; notably by Austen and his collaborators in experiments studying

This work was supported by Research Career Development Award AI-42373 from the National Institute of Allergy and Infectious Diseases, National Institutes of Health. This is publication 86 of the O'Neill Research Laboratories.

IgE-mediated histamine release from monkey and human lung (5,6). Moreover, the level of circumstantial evidence linking cyclic AMP to the inhibition of the allergic response has now been increased by actual measurements of the drug-induced changes in the cyclic AMP levels of leukocyte or lung preparations (7,8). This type of evidence is shown in Fig. 1. There is a satisfying parallelism between the ability of a series of catecholamines to inhibit IgE-mediated histamine release and to increase the cyclic AMP levels of the leukocyte preparations. It seems evident from these data, incidentally, that the catecholamine receptor on the basophil leukocyte, and on the cells producing the increased level of cyclic AMP, is of the beta adrenergic type—the activity of the series decreases from isoproterenol, which is most active and is largely a beta adrenergic agonist, to phenylnephrine, which has little or no activity and is predominately an alpha adrenergic agonist.

A similar parallel is seen in Fig. 2, which compares the effects of the various prostaglandins with respect to their ability either to inhibit leukocyte histamine release or to increase the intracellular cyclic AMP levels in the same cells (9). In fact, using a variety of different agonists and antagonists, we have never shown a discrepancy between the change in cyclic AMP level that is induced and the inhibition of histamine release (10,11). Nonetheless, this evidence remains circumstantial. We are

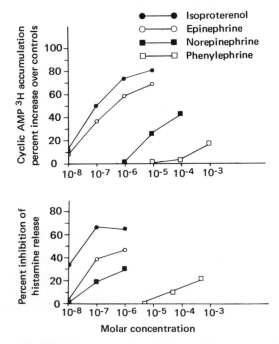

Fig. 1. The effects of adrenergic agonists on leukocyte cyclic AMP levels and histamine release.

measuring histamine release from the subpopulation of leukocytes that contain histamine, the basophils, which constitute only 1% of the total, while we are measuring increases in cyclic AMP in a much larger number of cells.

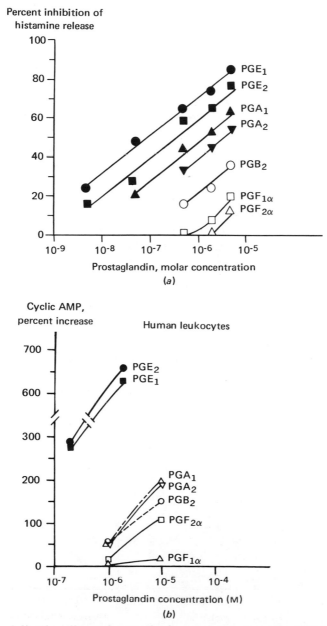

Fig. 2 (*a* and *b*). The effects of prostaglandins on leukocyte cyclic AMP levels and histamine release.

The pharmacologic construction of receptors on the cell surface that respond to an agonist and are blocked by an antagonist is readily demonstrated in this system. This, the beta adrenergic antagonist, propranolol, blocks the action of the catecholamines, but has no effect on the response to the prostaglandins or any other agonist. Moreover, propranolol's antagonism of catecholamine-induced increases in cyclic AMP coincides with the drug's ability to inhibit histamine release (8). Experiments of this sort, with histamine, led to the discovery that a number of effector cells, which contribute to the inflammatory response, have what are called H_2 receptors. In collaboration with Dr. Henry Bourne, we showed that histamine in the range of 10^{-7} to 10^{-6} was an efficient inhibitor of IgE-mediated histamine release (12). The ability of histamine to inhibit histamine release coincides in a dose-response fashion with its ability to increase leukocyte cyclic AMP levels (Fig. 3). In seeking to demonstrate formally that there was a histamine receptor on the basophil, we attempted to block the response with various antihistamines. We tried the five classes of antihistamines at usual concentrations (10^{-7} to 10^{-4} M) and uniformly failed to block the histamine-induced inhibition of histamine release or the resulting increase in cellular cyclic AMP levels. At higher concentrations, the antihistamines seemed to have agonist activity themselves or were cytotoxic, and thus could not be studied.

In retrospect, this failure was not surprising, since it is well established that antihistamines can block only certain of the actions of histamine. Included are many that we normally think of as inflammatory, such as increased vascular permeability, smooth muscle contraction, and the like. In contrast, the histamine-stimulated secretion of acid by the stomach and the histamine-induced relaxation of the rat uterus are not

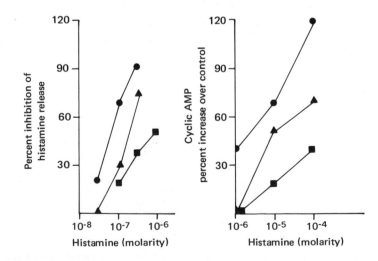

Fig. 3. The effects of histamine on leukocyte histamine release and cyclic AMP levels.

blocked by the standard antihistamines. Because of these observations, Schild had some years ago postulated that there must be at least two classes of histamine receptors (13). Following this, Black et al. recently described a new antihistamine, burimamide, which blocked the actions of histamine that are not antagonized by the standard antihistamines, without interfering with those that are (14). Black, therefore, has defined two classes of histamine receptors, which he calls H_1 and H_2, a terminology which unfortunately will be confusing for the immunologists. Using this H_2 antihistamine, we were able to effectively block histamine's inhibition of histamine release and the induced rise in the cellular cyclic AMP levels (Fig. 4). Again, the action of burimamide was specific for histamine in that it antagonized neither the catecholamine nor prostaglandin effects we have cited. Thus, the histamine receptor on the basophil is of the H_2 type; as I shall mention shortly, this receptor is also on the mouse lymphocyte, where it exercises a similar role.

The next agent I shall consider in this context, cholera enterotoxin, is hardly likely to be of interest in a physiologic or pharmacologic sense. We feel, however, that it will be of use in studying some of the questions that are being raised at this meeting. That is, we suggest that this compound is an ideal agent with which to question whether control by the cyclic AMP system exists in each of the many systems being studied. While some cell types have receptors for prostaglandins or for

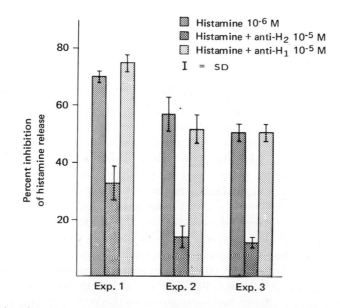

Fig. 4(*a*). The antagonism of histamine inhibition of histamine release by burimamide (anti-H_2) and diphenhydramine (anti-H_1).

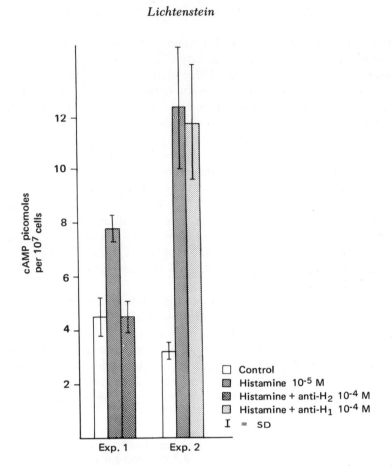

Fig. 4(*b*). The antagonism of histamine-stimulated increases in leukocyte cyclic AMP by burimamide (anti-H$_2$) and diphenhydramine (anti-H$_1$).

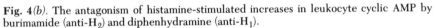

cathecholamines, others do not. On the other hand, all cell types that have been studied to date seem to have a receptor for cholera enterotoxin: All respond with an increased cyclic AMP level (15). The action of cholera enterotoxin has a unique time course; the increased level of cyclic AMP occurs only after a considerable lag period, and the level remains elevated for an extremely long periods. Moreover, cholera enterotoxin has two specific antagonists: an antitoxin and a toxoid. Finally, and perhaps most important, cholera enterotoxin seems to have only one effect on cells: the prolonged elevation of cyclic AMP levels.

The effect of cholera enterotoxin on the human leukocyte system is shown in Fig. 5. Cholera enterotoxin in subnanogram concentrations inhibits histamine release from the leukocyte preparations and markedly increases the cyclic AMP level. Its effect on both of these activities is not readily apparent for 30 min and increases over the 90 to 120 min in which it is being studied (10).

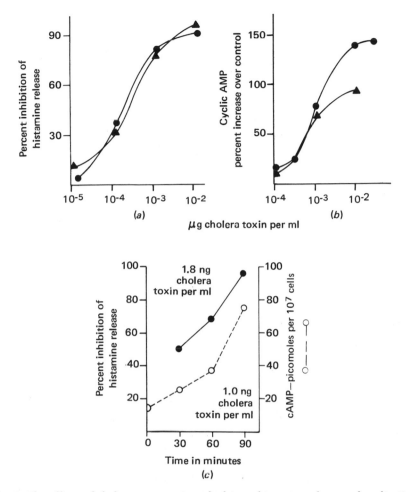

Fig. 5. The effects of cholera enterotoxin on leukocyte histamine release and cyclic AMP levels. In (*a*) and (*b*), the leukocytes were incubated with the stipulated doses of toxin for 60 min before antigen was added. In (*c*), the time of preincubation of cells with cholera enterotoxin before antigen addition or cyclic AMP measurements is the variable.

The level of circumstantial evidence, which suggests that alterations in cyclic AMP levels control the release of histamine, is not only supported by the similarity in dose-response curves. As suggested by the cholera data just presented, there is also supporting evidence from kinetic considerations. Figure 6 illustrates three different patterns of change in cellular cyclic AMP levels induced by the agents discussed above. With isoproterenol, for example, the cyclic AMP level of the leukocyte preparation increases rapidly and, even while the isoproterenol remains in contact with the cells, falls rapidly so that it is back to base line levels in approximately 20 to 30 min. The prostaglan-

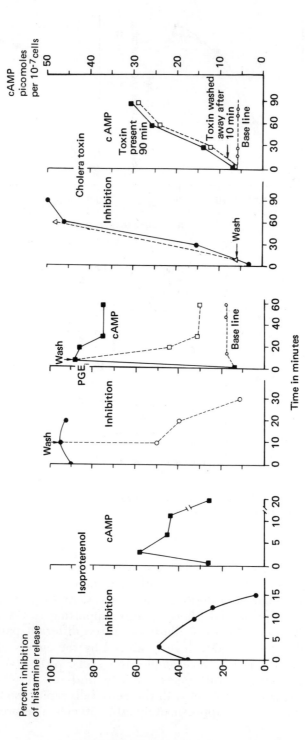

Fig. 6. Kinetic patterns of leukocyte cyclic AMP levels and histamine release after exposure to three different agonists.

dins similarly increase the cyclic AMP level rapidly, but this level is maintained over the period of study, up to 90 min, if the cells are not washed. Washing the leukocyte preparations free of the prostaglandin, however, leads to a prompt fall of the cyclic AMP levels to base line. Finally, as noted, cholera enterotoxin requires a relatively protected period to induce the cyclic AMP levels to rise; but once the cells have been exposed to the toxin for a few minutes, washing the cells free of the cholera enterotoxin has no effect on the subsequent rise in the intracellular levels of cyclic AMP. As can be seen by reference to Fig. 6, each of these time-course patterns for changes in the cyclic AMP levels is paralleled by the time-course activity of these agents with respect to their ability to inhibit IgE-mediated histamine release (11).

While this type of evidence certainly cannot prove the hypothesis in question, it tends to strengthen it. Another reason for carrying out these experiments, however, is that if a discrepancy is found, it must cast into question whether cyclic AMP is involved in the system in question. I shall consider shortly a system where this problem exists.

Before turning to other systems, however, I would like to briefly comment on the question of cholinergic receptors and changes in cyclic GMP levels, a subject which has been among the most fascinating topics discussed in this book. Dr. Frank Austen and his collaborators have made the observation that stimulation of cholinergic receptors by acetylcholine analogues leads to enhancement of histamine and SRS-A release from the human lung. This work is presented fully in the paper from Dr. Austen's laboratory (16). Following their lead, we attempted to obtain cholinergic enhancement in the leukocyte system. As can be seen in Table 1, a considerable series of experiments with both acetylcholine bromide and methacholine (mecholyl) yielded completely negative results; we see neither enhancement nor inhibition. This is the most significant discrepancy between the human basophil and the human lung systems that has been observed to date. Whether we must conclude from these data that the human basophil and the human mast cell have different types of receptors or whether the observed differences are due to the fact that the basophils are studied as isolated cells while the mast cells are studied in tissue fragments, which contain many other cell types, remains to be seen.

A similar discrepancy with respect to cholinergic enhancement may be found in Dr. Henney and Dr. Strom's papers elsewhere in this book (17,18).

These studies of the immediate component of the inflammatory response led us to question whether other reactions that participate in inflammation are basically secretory in nature and would, therefore, be similarly controlled by the cyclic AMP system. In this endeavor, it seems reasonable to look at cell types responding to specific stimuli by an event which could be viewed as consisting of the secretion of products which

Table 1. Effects of Carbachol and Acetylcholine Bromide (AcBr) on the Release of Histamine in the Leukocyte System

Dose (M)	% histamine released after carbachol					% histamine released after AcBr				
	Experiment No.					Experiment No.				
	1	2	3	4	5	1	2	3	4	5
Control	19	43	78	29	59	51	28	34	17	41
10^{-5}	19	42	—ᵃ	32	—	—	—	—	—	35
10^{-6}	16	44	—	—	62	—	—	—	—	41
10^{-7}	21	37	72	—	—	51	—	33	19	35
10^{-8}	20	38	—	29	—	51	—	—	—	30
10^{-9}	17	37	64	—	59	—	26	34	17	30
10^{-10}	19	37	—	—	—	—	—	—	16	—
10^{-11}	17	37	71	38	—	51	—	34	—	—
10^{-12}	—	—	—	—	62	—	27	—	17	—
10^{-13}	—	—	72	—	—	54	—	34	—	—
10^{-14}	—	—	—	29	—	—	—	—	17	—
10^{-15}	—	—	72	—	53	51	26	31	20	—
10^{-16}	—	—	—	—	—	54	27	—	—	—
10^{-17}	—	—	—	29	—	54	27	35	18	—
10^{-18}	—	—	—	—	56	—	30	—	—	—
Avg. of drug contained in tubes ± SD	18.4 ± 1.8	38.9 ± 2.9	70.2 ± 3.5	31.4 ± 3.9	58.4 ± 3.9	52.5 ± 2.3	27.2 ± 1.5	33.5 ± 1.4	17.8 ± 1.5	34.2 ± 4.6

ᵃDashes indicate no experiment performed with this dosage.

lead to inflammation. A likely candidate was the system first described by Brunner, which involves the killing of target cells by sensitized T lymphocytes (19–21). Dr. Henney has described the results of this work in a separate report which is included in this book (17). For the present purposes, it is sufficient to point out that each of the phenomena expected from the data presented above was observed in this model of cellular immunity. That is, in the putative secretory act by which a sensitized T cell delivers lymphokines to, and thereby kills, the target cells, the killing was blocked by the catecholamines and prostaglandins, as well as by histamine and cholera enterotoxin. In each instance, the dose-response curves relating to the inhibition of cellular immunity and to the increased cyclic AMP levels were similar, and, finally, the kinetics of the changes in lymphocyte cyclic AMP levels and the duration of inhibition of target cell killing show striking similarities. In this system, too, the evidence linking cyclic AMP to the observed abrogation of the immune response is circumstantial. Although in Dr. Henney's system, most (and if necessary all) of the cells are splenic lymphocytes, it is clear that only a few percent of these cells are sensitized and acting as killer cells. Thus, again, we are studying the biologic effect, in this instance killing, of a few percent of the cell population, while the cyclic AMP levels are being measured in a larger population. Nevertheless, the concordance between the two systems is gratifying.

Attempts to extend this type of observation to another system were not so successful. In this instance, I am referring to the studies carried out by Dr. Gerald Weissmann and his collaborators, which are presented elsewhere in this book (22). Well before our collaboration, Dr. Weissmann had developed evidence suggesting that cyclic AMP controlled the release of lysosomal enzymes by human polymorphonuclear cells stimulated by phagocytosis (23). In this instance, we asked whether cholera enterotoxin would be useful in determining whether or not cyclic AMP was involved in this response. In this series of experiments, it was found that cholera enterotoxin increased the cyclic AMP levels of the human leukocyte preparation, but did not interfere with or effect the release of lysosomal enzymes (11). Surprised by this result, Dr. Bourne and Dr. Weissmann pursued it further by fractionating the leukocytes into populations composed largely of lymphocytes or polymorphonuclear cells. Cholera enterotoxin was found to increase cyclic AMP levels in relatively purified preparations of human granulocytes. These fractionation experiments also led to the discovery of another discrepancy in this system. Isoproterenol, which Dr. Weissmann had previously shown could inhibit the release of lysosomal enzymes, was found not to stimulate the cyclic AMP levels of granulocytes. Thus, it would appear that human granulocytes have receptors for cholera enterotoxin (i.e., they respond with an increase in the cyclic AMP level), but this agent does not affect the release of lysosomal enzymes. On the other hand, the human granulocyte does not have isoproterenol receptors, but this agent is capable of depressing the release of lysosomal enzymes. These discrepancies, in my mind, raise the question of whether cyclic AMP is involved in this system, but certainly do not prove that such is not the case. Dr. Weissmann has evidence that perhaps the situation with the human granulocyte is more complex than we anticipated. His paper, in this book, deals with this subject fully (22).

I would like to close on a somewhat more speculative note. We have defined the effects of a number of hormones or drugs on several systems that are important in the inflammatory responses. All of these experiments, however, are in vitro, and one wonders about their in vivo significance. That these agents have the appropriate in vivo effects seems likely. For example, in studies carried out with Dr. Robert Perper (Table 2), we have been able to show that the methylxanthine, theophylline, and the catecholamine, salbutamol (chosen because of its long duration of action after oral administration), are able to significantly impair the wheal and flare response caused by an intradermal injection of ascaris antigen into the skin of sensitive monkeys. This is an IgE-mediated response caused by released histamine. It can be shown that the effect of the drugs is primarily in stopping the release of histamine, since they do not have any effect on the wheal and flare response caused by the injection of histamine.

The in vivo importance of the two types of histamine receptors remains to be studied fully. Whether, in fact, the H_2 receptor allows histamine to feed back and decrease the inflammatory response caused by antigens that release histamine remains to be seen, but the availability of an anti-H_2 type of antihistamine should allow this study to be carried out with relative ease.

Finally, in a more teleological sense, what is the importance of these types of receptors in the inflammatory response? Here, of course, we are on shaky ground. It does seem likely, however, that it will be found that these receptors have a profound influence on the control and development of the inflammatory response. For example, very recent experiments by Drs. Plaut and Henney have shown that the activated T lymphocyte, which causes target cell destruction, becomes more and

Table 2. Effects of Drugs on Ascaris Antigen
and Histamine Skin Reactions in Monkey

Drug	Dose (P.O.)	% inhibition of ascaris reactivity	% inhibition of histamine reaction
Salbutamol	0.5 mg/kg	42[a]	11
Theophylline	12.0 mg/kg	94[a]	9
Mepyramine	10.0 mg/kg	4	37[a]
Cyproheptadine	0.5 mg/kg	6	44[a]

[a] $p < 0.05$ versus group of animals receiving placebo drug.

more sensitive to histamine as the time after immunization proceeds. That is, there is an increase in the histamine receptors on these developing cells as they reach a fully active state; shortly thereafter, the histamine receptor level on the cell peaks and the ability of the cells to cause cytolysis decrease. Whether there exists in vivo a cause-and-effect relationship—i.e., endogenously occurring levels of histamine which, at the outset of the reaction, cannot stimulate the insensitive T cells become adequate to do so as the receptor concentration on the cells increases and thereby shuts off the response—remains to be seen.

In other experiments with Dr. Henney and Dr. Gillespie, it was demonstrated that agents that increase the cyclic AMP levels of splenic lymphocytes, in vivo, can have profound inhibitory effects on the subsequent expression of cellular immunity long after the cyclic AMP levels have returned to normal levels. These results are similar to work carried out by Dr. Werner Braun, which is presented elsewhere in this book (24). It is this kind of admittedly fragmentary evidence that suggests that we are just beginning to understand the role of hormones such as histamine, epinephrine, and the prostaglandins in controlling the development and expression of the inflammatory response. Once this understanding is increased, it will obviously open the door for more effective pharmacologic control of these types of responses.

Discussion of Dr. Lichtenstein's Paper

DR. J. HADDEN: I am very interested in your demonstration and discussion of the H_2 receptor for histamine on the lymphocyte. We have done some preliminary work, which would indicate that histamine suppresses DNA and protein synthesis in PHA-stimulated lymphocytes. The inhibition may be analogous to that of other agents that raise cyclic AMP in lymphocytes.

DR. MELMON: The data on burimamide and the inability of diphenhydramine to block histamine effects on leukocytes are of considerable interest. The picture being developed may be analogous to that in the beta adrenergic system in which isoproterenol stimulates a different receptor in the lung and the heart. As far as the difference in your results and ours, it is possible that the apparent conflict might be resolved if we were to look at the relative potency of inhibition of cell binding by burimamide versus the other antihistamines.

DR. LICHTENSTEIN: And you have not done that yet?

DR. MELMON: No, our results on histamine stimulation of cyclic AMP accumulation and its inhibition by standard antihistamines have varied. There are experiments in which very high concentrations of the more classical antihistamines will inhibit the response of lymphocytes to histamine. There are other experiments in which it is very clear that burimamide is the more potent inhibitor.

DR. AUSTEN: If I understood you correctly, the inhibition of cytotoxicity by histamine increased with time after immunization, whereas the capacity of the cells to accumulate cyclic AMP in response to histamine was constant throughout. How do you interpret these observations?

DR. LICHTENSTEIN: Dr. Henney has calculated that only about 1 to 4% of the total cell population is involved in cytotoxicity. We are measuring cyclic AMP in the total lymphocyte population. We believe that cells that are differentiating into "killer" lymphocytes are also getting new receptors for histamine, making them more susceptible to inhibition. But this change in a few percent of the cells would not be reflected in cyclic AMP measurements of the whole lymphocyte population.

DR. HENNEY: The overall cytolytic activity of the various populations varies with time after immunization. To ascertain the role of total cytolytic activity in susceptibility to histamine inhibition, we can dilute sensitized cells with normal lymphocytes. When this is done, the 11-day effector cells are only inhibited some 8% by 10^{-5} M histamine, whereas the 17-day effector cells are inhibited almost 60% by the same amount of histamine. Thus, at a time when cytotoxicity is maximal (11 days), little inhibition is obtained.

DR. AUSTEN: Does the inhibition of cytotoxicity by histamine involve H_1 or H_2 receptors?

DR. LICHTENSTEIN: It appears to be an H_2 receptor as does the receptor involved in cyclic AMP accumulation.

DR. HENNEY: Can I just parenthetically add a plea that we not use the term H_2 for the type 2 histamine receptor. We already have a perfectly adequate H-2.

DR. BLOOM: The terminology is already complex enough.

References

1.   Lichtenstein, L. M., and Margolis, S. "Histamine Release in Vitro: Inhibition by Catecholamines and Methylxanthines," *Science* **161**: 902–903 (1968).
2.   Schild, H. "Histamine Release in Anaphylactic Shock in Isolated Lungs of Guinea Pigs," *Quart. J. Exp. Physiol.* **26**:165–179 (1937).
3.   Lichtenstein, L. M., and Osler, A. G. "Studies on the Mechanisms of Hypersensitivity Phenomena. IX. Histamine Release from Human Leukocytes by Ragweed Pollen Antigen," *J. Exp. Med.* **120**:507–530 (1964).
4.   Austen, K. F., and Humphrey, J. H. "In Vitro Studies of the Mechanism of Anaphylaxis," *Advan. Immunol.* **3**:1 (1963).
5.   Assem, E. S. K., and Schild, H. O. "Inhibition by Sympathomimetic Amines of Histamine Release Induced by Antigen in Passively Sensitized Human Lung," *Nature* (London) **224**:1028 (1969).
6.   Ishizaka, T., Ishizaka, K., Orange, R. P., and Austen, F. K. "Pharmacologic Inhibition of the Antigen-Induced Release of Histamine and Slow-Reacting Substance of Anaphylaxis (SRS-A) from Monkey Lung Tissues Mediated by Human IgE," *J. Immunol.* **106**:1267 (1971).

7. Orange, R. P., Austen, W. G., and Austen, K. F. "Immunologic Release of Histamine and Slow-Reacting Substance of Anaphylaxis from Human Lung. I. Modulation by Agents Influencing Cellular Levels of Cyclic 3',5'-Adenosine Monophosphate," *J. Exp. Med.* **134**:136S (1971).
8. Bourne, H. R., Lichtenstein, L. M., and Melmon, K. L. "Pharmacologic Control of Allergic Histamine Release in Vitro: Evidence for an Inhibitory Role of 3',5'-Adenosine Monophosphate in Human Leukocytes," *J. Immunol.* **108**:695–705 (1972).
9. Lichtenstein, L. M., Gillespie, E., Bourne, H. R., and Henney, C. S. "The Effects of a Series of Prostaglandins on in Vitro Models of the Allergic Response and Cellular Immunity," *Prostaglandins* **2**:519–528 (1972).
10. Lichtenstein, L. M., Henney, C. S., Bourne, H. R., and Greenough, W. B. "Effects of Cholera Toxin on in Vitro Models of Immediate and Delayed Hypersensitivity: Further Evidence for the Role of cAMP," *J. Clin. Invest.* **52**:691–697 (1973).
11. Bourne, H. R., Lehrer, R. I., Lichtenstein, L. M., Weissmann, G., and Zurier, R. "Effects of Cholera Enterotoxin on Adenosine 3',5'-Monophosphate and Neutrophil Function: Comparison with Other Compounds Which Stimulate Leukocyte Adenyl Cyclase," *J. Clin. Invest.* **52**:698–708 (1973).
12. Bourne, H. R., Melmon, K. L., and Lichtenstein, L. M. "Histamine Augments Leukocyte Cyclic AMP and Blocks Antigenic Histamine Release," *Science* **173**:743–745 (1971).
13. Ash, A. S. F., and Schild, H. O. "Receptors Mediating Some Actions of Histamine," *Brit. J. Pharmacol. Chemother.* **27**:427 (1966).
14. Black, J. W., Duncan, W. A. M., Durant, C. J., Ganellen, C. R., and Parsons, E. M. "Definition and Antagonism of Histamine H_2-Receptors," *Nature* **236**:385 (1972).
15. Pierce, N. F., Greenough, W. B., III., and Carpenter, C. C. J. "*Vibrio Cholerae* Enterotoxin and Its Mode of Action," *Bact. Rev.* **35**:1 (1971).
16. Kaliner, M., and Austen, K. F. "Hormonal Control of the Immunologic Release of Histamine and Slow-Reacting Substance of Anaphylaxis from Human Lung" (in this book, 1974).
17. Henney, C. S. "Relationships Between the Cytolytic Activity of Thymus-Derived Lymphocytes and Cellular Cyclic Nucleotide Concentrations" (in this book, 1974).
18. Strom, T. B., Deisseroth, A., Morganroth, J., Carpenter, C. B., and Merrill, J. P. "Modulation of Cytotoxic T Lymphocyte Function by Cyclic 3',5'-Mononucleotides" (in this book, 1974).
19. Brunner, K. T., J. Mauel, Cerottini, J. C., and Chapuis, B. "Quantitative Assay of the Lytic Action of Immune Lymphoid Cells on ^{51}Cr-Labeled Allogeneic Target Cells in Vitro: Inhibition by Isoantibody and by Drugs," *Immunol.* **14**:181 (1968).
20. Henney, C. S. "Quantitation of the Cell-Mediated Immune Response. I. The Number of Cytolytically Active Mouse Lymphoid Cells Induced by Aimmunization with Allogeneic Mastocytoma Cells," *J. Immunol.* **107**:1558 (1971).
21. Henney, C. S., Bourne, H. R., and Lichtenstein, L. M. "The Role of Cyclic 3',5'-Adenosine Monophosphate in the Specific Cytolytic Activity of Lymphocytes," *J. Immunol.* **108**:1526–1534 (1972).

22. Weissmann, G., Zurier, R. B., and Hoffstein, S. "Leucocytes as Secretory Organs of Inflammation: Control by Cyclic Nucleotides and Autonomic Agonists" (in this book, 1974).
23. Weissmann, G., Dukor, P., and Zurier, R. B. "Effect of Cyclic AMP on Release of Lysosomal Enzymes from Phagocytes," *Nature* (New Biol.) **231**:131 (1971).
24. Braun, W. "Regulatory Factors in the Immune Response: Analysis and Perspective" (in this book, 1974).

HORMONAL CONTROL OF THE IMMUNOLOGIC RELEASE OF HISTAMINE AND SLOW-REACTING SUBSTANCE OF ANAPHYLAXIS FROM HUMAN LUNG

Michael Kaliner and K. Frank Austen

The antigen-induced secretion of the chemical mediators histamine and slow-reacting substance of anaphylaxis (SRS-A) has been detected in human lung tissue from asthmatic patients (Brocklehurst, 1960). These same chemical mediators are also released from human lung tissue obtained from normal individuals after passive sensitization with sera from allergic patients and challenge with specific antigen (Sheard et al., 1967; Parish, 1967). The antigen-antibody initiated secretory mechanism is modulated by a variety of hormones, which affect the intracellular levels of cyclic nucleotides (Orange et al., 1971a; Kaliner et al., 1972; Tauber et al., submitted for publication). These findings suggest that the subpopulation of target cells involved in the immunologic release of chemical mediators have specific receptors for alpha and beta adrenergic, cholinergic, and prostaglandin hormones as well as for IgE antibody. Although the hormones noted above can themselves stimulate secretion in some tissues, in the reaction to be considered, they modulate the secretory process stimulated by antigen-antibody interaction, instead.

I. IgE–Target Cell Interaction

Evidence implicating IgE as the immunoglobulin that prepares the target cells in human lung fragments in vitro for the subsequent antigen-induced release of histamine, SRS-A, and eosinophil chemotactic factor of anaphylaxis (ECF-A) includes the following observations:

(1) The capacity of reaginic serum to sensitize human lung tissue is lost after absorption of that serum with monospecific anti-IgE (Orange et al., 1971a; Kay and Austen, 1971).

This work was supported by grants AI-07722 and RR-05669 from the National Institutes of Health and a grant from the John A. Hartford Foundation, Inc.

(2) IgE myeloma protein, but not myeloma protein, of other immunoglobulin classes inhibits sensitization, presumably by competing for receptor sites with IgE antibody from allergic serum (Orange et al., 1971c).

(3) anti-IgE antibody, but not antibodies, directed against the other immunoglobulin classes induces mediator release from unsensitized lung, presumably by interacting with endogenous IgE on the appropriate receptor sites (Kay and Austen, 1971).

The target cells in human lung tissue have not been identified, but studies employing monkey lung fragments and implicating human IgE as the sensitizing immunoglobulin by the same three criteria noted above (Ishizaka et al., 1970a) have localized IgE by radioautography to the tissue mast cell (Tomioka et al., 1972). IgE has also been indentified on the human basophil by radioautography (Ishizaka et al., 1970b) and by electron microscopy employing hybrid antibodies (Sullivan et al., 1971). The detection of histamine in the granules of mast cells and basophilic leukocytes of many species (Riley and West, 1953; Graham et al., 1955) and of ECF-A in the rat mast cell (Wasserman et al., manuscript in preparation) as well as the stimulatious immunologic release of histamine and ECF-A from purified human basophilic leukocytes (Parish, in press) localize these preformed chemical mediators to cells known to be sensitized by IgE. In contrast, SRS-A is not preformed, but evolves as a consequence of the antigen-antibody interaction and appears to be derived from cells in addition to mast cells.

II. The Antigen-Induced Secretory Process

Studies in a variety of in vitro model systems, reviewed in two sequential symposia (Austen and Becker, 1968 and 1971) have supported the formulation that mediator release is noncytotoxic and is essentially a form of immunologically induced secretion. This view, although substantially supported by the findings relating to the pharmacologic controls of mediator release, was originally based upon biochemical and morphologic considerations. From a biochemical viewpoint, the available studies in human lung, although limited, are in essential agreement with the findings in guinea pig lung prepared with IgG_1 or the IgE-dependent mixed cell systems of human peripheral blood or the rat peritoneal cavity. Histamine release from human lung requires the activation of a serine esterase, inhibited by diisopropylfluorophosphate (DFP); the availability of a glycolytic pathway as revealed by inhibiton with iodoacetic acid or with 2-deoxyglucose (2-DG) when glucose is limited; and calcium ions as indicated by inhibition with ethylene-diaminetetraacetate (EDTA) and reversal upon restoration of the free ion concentration (Orange et al., 1971c; Orange and Austen, in press).

When sensitized lung fragments are incubated with DFP, EDTA, or 2-DG and these agents are removed before antigen challenge, no impairment of the subsequent antigen-activated secretory pathway results. If the lung tissue is challenged with antigen in the presence of these agents and both the antigen and the inhibitor are removed a few minutes afterward, reversal of the inhibition occurs as manifested by mediator release without the necessity for additional antigen.

Evidence that a cation requirement precedes the activation of a serine esterase in the secretory process initiated by antigen is as follows: Antigen challenge in the presence of DFP for 10 min permits little histamine release upon removal of the DFP, while antigen challenge in the presence of DFP and in the absence of divalent cations yields full histamine release upon removal of DFP and restoration of divalent cations. It is suggested that the presence of divalent cations is required for the generation of a serine esterase, which has an autocatalytic effect on the reaction. An energy-requiring step appears to precede a second divalent cation-requiring step in the immunologic release of mediators; inhibition of mediator release by 2-DG is reversed by removal of the 2-DG within 3 min after antigen challenge; the addition of EDTA after removal of 2-DG inhibits mediator release, and the subsequent removal of EDTA and restoration of divalent cations permit the release of histamine and SRS-A. Although EDTA can inhibit mediator release after reversal of 2-DG inhibition, 2-DG does not inhibit mediator release after cation reversal of EDTA inhibition. It is suggested, therefore, that the interaction of specific antigen with tissue-fixed IgE antibody initiates the divalent cation-requiring activation of a DFP-sensitive serine esterase, which precedes an energy-requiring step leading to a second divalent cation-requiring step for the release of the chemical mediators histamine and SRS-A from human lung tissue in vitro (Kaliner and Austen, in press).

III. Modulation of Mediator Release

A. Beta Adrenergic

Hormones with beta adrenergic activity such as epinephrine and norepinephrine or the synthetic catecholamine isoproterenol produce dose-dependent inhibition of the immunologic release of chemical mediators (Orange et al., 1971a) and have been found to produce concomitant increases in the tissue concentrations of cyclic adenosine 3',5'-monophosphate (cyclic AMP) (Orange et al., 1971b; Orange et al., 1971c; Kaliner et al., 1972). That the suppression of mediator release from a subpopulation of target cells is, in fact, related to the elevation of cyclic AMP observed in the entire lung fragment is supported by the kinetic relation of the two events (Orange et al., 1971b; Orange et al., 1971c). Further, the combination of beta adrenergic stimulants with a methylxanthine capable of protecting cyclic AMP from breakdown through competitive inhibition of phosphodiesterase yields both a synergistic inhibition of mediator release and accumulation of cyclic AMP. Finally, the relative capacity of isoproterenol, norepinephrine, and the prostaglandins E_1 and $F_{2\alpha}$ to inhibit mediator release is in accord with the relative potency of these agents to increase cyclic AMP concentrations (Tauber et al., submitted for publication). The beta adrenergic receptor site involved is of the $beta_2$ type as judged by the capacity of the beta adrenergic antagonist propranolol to prevent both the accumulation of cyclic AMP and inhibition of mediator release otherwise produced by norepinephrine, while practolol, a $beta_1$ adrenergic antagonist, is ineffective (Table 1).

B. Alpha Adrenergic

Alpha adrenergic stimulation achieved with the relatively pure synthetic catecholamine phenylephrine or the hormone norepinephrine either alone (Fig. 1) or in combination with propranolol (Table 1) consistently produced a depletion of the total tissue cyclic AMP concentration with an associated augmentation of the antigen-induced release of histamine and SRS-A (Orange et al., 1971b; Kaliner et al., 1972). In 13 experiments performed to study alpha adrenergic effects upon human lung tissue, the mean depletion of cyclic AMP levels was 32% (range of decrease 17 to 68%), while the mean enhancement of the immunologic release of histamine was 37% (range of increase 7 to 70%) and of SRS-A was 84% (range of increase 33 to 180%). Thus, there exists an inverse relationship between the tissue levels of cyclic AMP and the immunologic release of chemical mediators.

C. Cholinergic

The cholinergic neurohormone acetylcholine and the synthetic cholinesterase-resistant cholinomimetic agent carbamylcholine (Carbachol) in picomolar concentrations markedly enhance the IgE-dependent secretion of histamine and SRS-A without eliciting a measurable change in the tissue concentrations of cyclic AMP (Fig. 2) (Kaliner et al., 1972). In 16 experiments with 10^{-10} M Carbachol, the mean enhancement of the immunologic release of histamine was 44% (range of increase 8 to 181%) and of SRS-A was 110% (range of increase 23 to 270%), while the mean change in the tissue levels of cyclic AMP was less than 1% (range of change −4 to +32). Atropine prevention of cholinergic enhancement indicates that the receptor site involved is of the muscarinic type.

Evidence that the mechanisms of alpha adrenergic and cholinergic enhancement of the immunologic release of chemical mediators are discrete includes (1) the consistent association of alpha adrenergic stimulation with depletions in cyclic AMP levels, while cholinergic stimulation is not associated with changes in the concentration of this nucleotide; (2)

the additive enhancement produced by the combination of alpha adrenergic and cholinergic agents; and (3) the capacity of atropine to selectively prevent cholinergic, but not alpha adrenergic, enhancement.

Table 1. Comparison of the Ability of Propranolol and Practolol to Prevent the Beta Adrenergic Effects of Norinephrine on the Immunologic Release of Histamine and SRS-A from Human Lung and the Concomitant Changes in the Tissue Levels of Cyclic AMP

Drug (10^{-6} M)	Histamine release[a] (% control)	SRS-A release[a] (% control)	Cyclic AMP[b]	
			(% control)	($\mu\mu$Moles/g)[c]
Experiment 1				
Norepinephrine	−40	− 80	+128	450
Propranolol	+ 1	0	− 4	190
Norepinephrine and propranolol	+11	+100	− 25	150
Experiment 2				
Norepinephrine	−39	− 68	+200	ND
Practolol	+ 6	0	− 0.5	ND
Norepinephrine and practolol	−40	− 68	+214	ND

[a]Preparation of human lung for the immunologic release of histamine and SRS-A. Human lung tissue obtained at time of surgery was dissected free of pleura, bronchi, blood vessels, and grossly diseased areas; washed; fragmented and replicated into 200-mg samples; sensitized in 1.0 ml of a 1:3 dilution of atopic serum for 2 to 4 hr at 37°C or overnight at room temperature; washed; incubated with the agent(s) under study for varying time periods; and challenged with 0.2 μg/ml ragweed antigen E (Research Resource Branch, NIAMD) for 15 min at 37°C inducing the release of the chemical mediators histamine and SRS-A, which were quantitated by bioassay. The residual tissue histamine was extracted by boiling the fragments for 8 min (Orange et al., 1971a).

[b]Cyclic AMP assay, Experiment 1: Lung fragments were handled in a fashion parallel to sensitized fragments and at the time of antigen challenge were transferred to boiling distilled water for 30 min; homogenized employing a Potter-Elvehjem tissue homogenizer; centrifuged at 6000 × g for 20 min and the supernatant charged onto 0.5 × 2 cm Dowex AG-1 columns. The columns were washed with 10 ml distilled water and eluted with 10 ml 4 N formic acid. The eluate was evaporated to dryness; resuspended in 1.0 ml acetate buffer (50 mM, pH 4.0) and assayed as described (Kaliner et al., 1972) by the protein binding assay of Gilman (1970).

Experiment 2: 200 mg replicates of human lung tissue were incubated in 1.0 ml of a 1:3 dilution of atopic serum containing 0.2 μCi/ml ^{14}C-Adenine for 2 hr at 37°C; washed and incubated with the agent(s) under study as described above. At the time of antigen challenge, the replicates to be used for cyclic AMP assay were, instead, transferred to 1.0 ml boiling distilled water for 30 min to extract the ^{14}C-cyclic AMP, which was isolated from the other ^{14}C-labeled nucleotides by passage of the boiled tissue extracts over Dowex 50 H+ columns, which were eluted with 10 ml water. The eluates were evaporated to dryness, resuspended in 0.2 ml 50% methanol and streaked onto Whatman no. 40 paper. The chromatograms were developed in descending fashion for 15 to 18 hr using a solvent system of 95% ethanol:1.0 N ammonium acetate (7:3 v/v) and the segment containing ^{14}C-cyclic AMP identified using a superimposed reference marker of 15 μg cyclic AMP. The ^{14}C-cyclic AMP thereby isolated was determined by liquid scintillation counting as described (Kaliner et al., 1971).

[c]control = 200 $\mu\mu$Moles/g: ND = not done.

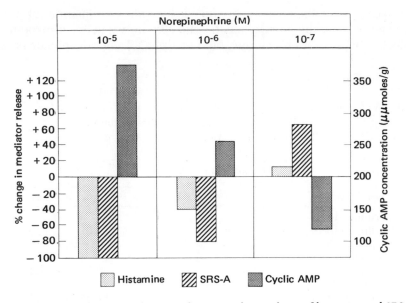

Fig. 1. The effects of norepinephrine on the immunologic release of histamine and SRS-A from human lung and the concomitant changes in the tissue levels of cyclic AMP.

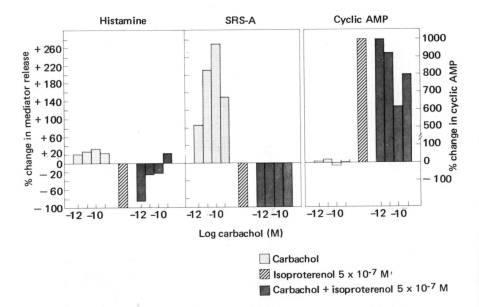

Fig. 2. The effects of combined cholinergic and beta adrenergic stimulation on the immunologic release of histamine and SRS-A from human lung and the concomitant changes in the tissue levels of cyclic AMP. The effects of isoproterenol (5×10^{-7} M) and Carbachol (10^{-12} to 10^{-9} M) were determined both alone and in combination.

That cholinergic enhancement may be mediated by cyclic guanosine 3′,5′-monophosphate (cyclic GMP) is suggested by the findings of others (George et al., 1970; Ferrendelli et al., 1970; Kuo et al., 1972) that cholinergic stimulation of several tissues selectively increases the concentration of cyclic GMP and by the demonstration that exogenous 8-bromo-cyclic GMP produces a dose-related enhancement of the immunologic release of histamine and SRS-A (Kaliner et al., 1972), while dibutyryl cyclic AMP inhibits mediator release (Orange et al., 1971a).

D. Prostaglandin

Because of (1) the presence of both prostaglandin E_1 (PGE_1) and $F_{2\alpha}$($PGF_{2\alpha}$) in relatively high concentrations in human lung (Karim et al., 1967), (2) the ease of formation and release of these ubiquitous hormones after a variety of stimuli (Piper and Walker, in press), (3) the detection of prostaglandins in the prefusate from immunologically challenged guinea pig lungs (Piper and Vane, 1969) and human lungs (Piper and Walker, in press), and (4) the capacity of PGE_1 to inhibit the immunologic release of histamine from human peripheral leukocytes (Bourne et al., 1972), PGE_2 and $PGF_{2\alpha}$ were studied for their effect on the immunologic release of histamine and on the levels of cyclic AMP in human lung (Tauber et al., in press; Tauber et al., submitted for publication). PGE_1 (5×10^{-4} to 5×10^{-6} M) produced both a dose-dependent inhibition of the antigen-induced release of histamine as determined by fluorometric assay (Shore et al., 1959) and an increase in cyclic AMP; the time course of both these effects was identical. That PGE_1 was interacting with a specific receptor site activating the enzyme adenylate cyclase was determined in two ways: A methylxanthine combined with PGE_1 produced a synergistic inhibiton of mediator release and increase in cyclic AMP, and propranolol failed to prevent the effects of PGE_1 while totally suppressing the actions of isoproterenol. PGE_2 was as potent as PGE_1, while $PGF_{2\alpha}$ was less potent in its ability to increase the tissue concentration of cyclic AMP and to inhibit the immunologic release of histamine, suggesting that the prostaglandin receptor site has at least part of its specificity on the keto of the cyclopentane ring of PGE_1.

Both PGE_1 and $PGF_{2\alpha}$ at concentrations of 5×10^{-8} M were found to consistently enhance the antigen-induced secretion of histamine and induce depletions of cyclic AMP concentrations—this effect was not prevented by phenoxybenzamine. PGE_1 examined at this concentration in four experiments produced a mean enhancement of histamine release of 41% in association with a 12% reduction in cyclic AMP, while $PGF_{2\alpha}$ in two experiments at this concentration elicited a 26% increase in the immunologic release of histamine along with a 30% reduction in cyclic AMP levels. The biphasic dose response appreciated with the prostaglandins is analogous to the findings with the adrenergic hormone norepinephrine (Fig. 1); at high concentrations, both norepinephrine and the prostaglan-

dins stimulate adenylate cyclase and suppress the immunologic release of
mediators, while at dilute concentrations, the tissue levels of cyclic AMP
are depleted with a concomitant enhancement of the release of mediators. The stimulation of phosphodiesterase with imidazole produced a
depletion of tissue cyclic AMP and enhancement of mediator release
(Fig. 3). The capacity for such a nonhormonal stimulus to reproduce the
enhancement observed with either alpha adrenergic or low-dose
prostaglandin stimulation supports the suggestion that these hormones
modulate mediator release by alterations in cyclic AMP concentrations.

E. Combined Cholinergic and Beta Adrenergic

The effect of 5×10^{-7} M isoproterenol on the suppression of mediator release and cyclic AMP levels was determined alone and in the presence of incremental concentrations of Carbachol (10^{-12} to 10^{-9} M) (Fig.
2). Carbachol alone had no effect on the level of cyclic AMP while
enhancing the release of both histamine and SRS-A at all concentrations.
The capacity of isoproterenol to increase cyclic AMP levels and inhibit
the release of SRS-A was unaffected by any dose of Carbachol, while the
inhibition of histamine release was progressively reversed by increasing
concentrations of Carbachol. These data again demonstrate that
cholinergic stimulation is independent of a measurable change in cyclic
AMP levels and suggest that the effects are due to an action antagonistic
to the cAMP-dependent protein kinase or subsequent phosphorylation
step. The failure of Carbachol to reverse the isoproterenol inhibition of
SRS-A release may relate to the necessity for formation of SRS-A before
release, thereby providing additional steps which may be cyclic AMP
modulated. In contrast, histamine is preformed, and only the secretory
process initiated by antigen-antibody action is available for modulation.

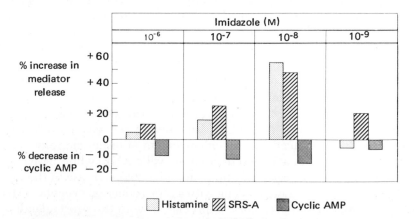

Fig. 3. The effects of imidazole on the immunologic release of histamine and SRS-A from
human lung and the concomitant changes in the tissue levels of cyclic AMP.

IV. Concluding Comments

The antigen-induced, IgE-dependent secretion of histamine and SRS-A from human lung in vitro is modulated by a variety of hormones, which demonstrate characteristic dose-related responses in terms of their capacity to increase or decrease the levels of cyclic nucleotides and influence mediator release. These hormones each interact with individual receptor sites as indicated by the ability of pharmacologic antagonists such as propranolol and atropine to prevent the effects of beta adrenergic and cholinergic stimulation without influencing alpha adrenergic or prostaglandin stimulation. The evidence that the hormonally induced changes in cyclic AMP levels are related to the modulation of mediator release includes (1) the kinetic relationship between increases in cyclic AMP and inhibition of mediator release after stimulation with isoproterenol or prostaglandins; (2) the parallel rank order of potency of isoproterenol $\simeq$ epinephrine $>$ nor epinephrine $>$ PGE$_1$ $>$ PGF$_{2\alpha}$ in terms of increasing cyclic AMP levels and inhibiting mediator release; (3) the synergism expressed between stimulators of adenylate cyclase, whether catecholamines or prostaglandins, and inhibitors of phosphodiesterase upon both phenomena; and (4) the association of depletions of cyclic AMP levels with enhancement of mediator release produced either by alpha adrenergic or low-dose prostaglandin hormonal stimulation or by the nonhormonal agent imidazole.

The sequence of biochemical events leading to mediator release from human lung appears to proceed from antigen interaction with tissue-fixed IgE to the divalent cation-dependent activation of a DFP-sensitive serine esterase, to an energy-requiring step which precedes a second divalent cation-requiring step. It appears that this second cation-requiring step is associated with the action of phosphodiesterase and may be the stage at which accumulations in cyclic AMP inhibit the release of mediators (Kaliner and Austen, in press). The requirement for divalent cations and energy in the immunologic release process conforms to the biochemical requirements for the secretion of hormones and enzymes in general, but the inhibitory role of cyclic AMP is an exception to the general finding that increases in cyclic AMP stimulate secretion. One could argue that the secretion of hormones and enzymes is designed to maintain homeostasis, while the release of the chemical mediators of inflammation is antihomeostasis; therefore, changes in the levels of cyclic AMP in the cells involved in mediator release, while producing opposite effects as far as secretion is concerned, are still designed to maintain homeostasis.

Discussion of Dr. Kaliner and Dr. Austen's Paper

DR. BOURNE: I have two questions. Firstly, what does histamine do to cyclic AMP and histamine release in the lung? Secondly, could you expand on this mysterious secondary process involving divalent cations in terms of what you think its relation is, if any, to the first and second stages that Larry Lichtenstein has worked out with human basophiles. Would you also comment on where you think the cholinergic agents act in relation to drugs that function through cyclic AMP. Which reaction occurs first?

DR. KALINER: We have not studied histamine affects on mediator release. Histamine has produced inconsistent changes in cyclic AMP in our system. Usually it does not alter cyclic AMP, but occasionally it has increased the concentration of cyclic AMP to a minor extent. In regard to the second question, we have used EDTA in the absence of divalent cations, and we find an increase in cyclic AMP concentrations in lung tissue. Synergistic effects on cyclic AMP accumulation are obtained with beta adrenergic agents, suggesting that EDTA may be influencing phosphodiesterase activity. The third question was concerned with the relative sequence of cyclic AMP and cyclic GMP induced events. I can't answer that question except by inference. In experiments combining Carbachol with isoproterenol, we find that high concentrations of Carbachol prevent the action of isoproterenol on histamine release. This would suggest that the metabolic consequences of cholinergic stimulation are antagonistic to one of the steps affected by beta adrenergic agents.

DR. LICHTENSTEIN: In the presence of calcium, isoproterenol is many logs less potent than it is without calcium.

DR. KALINER: We are suggesting that the absence of calcium in tris A buffer inhibits phosphodiesterase so that you simply are protecting the cAMP increases produced by beta adrenergic agents analogous to combining a PGE inhibitor and a beta agent.

DR. LICHTENSTEIN: But prostaglandin effects are not sensitive to calcium.

DR. KALINER: I can refer back to what Dr. Parker said before: There are subcellular distribution differences between prostaglandin induced and beta adrenergic induced increases in cyclic AMP. I think your findings fit in quite nicely with what Dr. Parker has done. There may be a marked difference in the activity of phosphodiesterase in the cytosol and the nucleus in terms of how they are affected by calcium.

DR. GOLDBERG: It is curious that because you have been so concerned with cyclic AMP and, finding a lowering of its concentrations, you have worked the system hard enough to establish that there may be a 10 or 12% drop in cyclic AMP concentration, and, because it is consistent with the current dogma, have chosen to place a great deal of significance on only a 10 to 12% change. We showed a while ago that imidazole is a more potent inhibitor of phosphodiesterase-promoted hydrolysis of cyclic GMP than it is a stimulator of cyclic AMP hydrolysis when micromolar concentrations of these cyclic nucleotides are used as substrate. I think you ought to reserve final comment on whether a decrease in cyclic AMP or an increase in cyclic GMP levels is really the more striking change that occurs with imidazole treatment.

DR. HARDMAN: There are a number of reports in the literature of cell types in which cyclic AMP levels are lowered by alpha adrenergic agents and of other cell types in which cyclic AMP levels are not lowered, or even elevated by alpha adrenergic agents. Since mast cells must be a very small percentage of the total mass of your lung fragments, your studies do not prove that a lowering of cyclic AMP is taking place in the mast cell.

DR. KALINER: There is no question that the cyclic AMP we measure is in the entire lung fragment population, which includes a small proportion of mast cells. In order to answer your question, we need to obtain a homogeneous target cell population which we can properly sensitize and study.

DR. HARDMAN: Isn't it conceivable that the alpha adrenergic effects here are through some mechanism other than a lowering of cyclic AMP?

DR. KALINER: Absolutely. There are a variety of plausible explanations for alpha adrenergic enhancement of mediator release. The reason we did the imidazole experiment was to see if nonhormonally induced changes in cyclic AMP would simulate hormonal enhancment. There is good evidence that imidazole activates phosphodiesterase, thereby reducing cyclic AMP without introducing the multiple metabolic consequences possible after hormonal stimulation.

DR. HARDMAN: However, it takes about 10^{-3} to 10^{-2} M imidazole to see stimulatory effects of imidazole on phosphodiesterase; this is several orders of magnitude more than the concentrations you have used.

DR. KALINER: The studies you cite are in broken cell preparations with crude or partially purified phosphodiesterase, whereas the studies reported here employ intact lung fragments. Quite possibly, very different imidazole concentrations are needed in the two situations.

References

Austen, K. F., and Becker, E. L., eds. *Biochemistry of the Acute Allergic Reactions.* Oxford, England: Blackwell Scientific (1968).

Austen, K. F., and Becker, E. L., eds. *Biochemistry of the Acute Allergic Reactions, 2nd Symposium.* Oxford, England: Blackwell Scientific (1971).

Bourne, H. R., Lichtenstein, L. M., and Melmon, K. L. "Pharmacologic Control of Allergic Histamine Release in Vitro: Evidence for an Inhibitory Role of 3′,5′-adenosine Monophosphate in Human Leukocytes," *J. Immunol.* **108**:695–705 (1972).

Brocklehurst, W. E. "The Release of Histamine and Formation of a Slow-Reacting Substance (SRS-A) During Anaphylactic Shock," J. Physiol. **151**:416–435 (1960).

Ferrendelli, J. A., Steiner, A. L., McDougal, D. R., and Kipnis, D. M. "The Effect of Oxotremorine and Atropine on cGMP and cAMP Levels in Mouse Cerebral Cortex and Cerebellum" *Biochem. Biophys. Res. Commun.* **41**:1061–1067 (1970).

George, W. J., Polson, J. B., O'Toole, A. G., and Goldberg, N. D. "Elevation of Guanosine 3',5'-Cyclic Phosphate in Rat Heart After Prefusion with Acetylcholine," *Proc. Nat. Acad. Sci.* (U.S.A.) **66**:398–403 (1970).

Gilman, A. G. "A Protein Binding Assay for Adenosine 3',5'-Cyclic Monophosphate," *Proc. Nat. Acad. Sci.* (U.S.A.) **67**:305–312 (1970).

Graham, H. T., Bowry, O. H., Wheelwright, B. F., Lenz, M. A. and Parish, H. H. Jr. "Distribution of Histamine Among Leukocytes and Platelets," *Blood J. Hematol.* **29**:467–481 (1955).

Ishizaka, T., Ishizaka, K., Orange, R. P., and Austen, K. F. "The Capacity of Human Immunoglobulin E to Mediate the Release of Histamine and Slow-Reacting Substance of Anaphylaxis (SRS-A) from Monkey Lung," *J. Immunol.* **104**:335–343 (1970a).

Ishizaka, K., Tomioka, H., and Ishizaka, T. "Mechanisms of Passive Sensitization. I. Presence of IgE and IgG Molecules on Human Leukocytes," *J. Immunol.* **105**:1459–1467 (1970b).

Kaliner, M. A., and Austen, K. F. "The Antigen Activated Biochemical Steps in the Release of Mediators from Human Lung," *Fed. Proc.* (in press).

Kaliner, M. A., Orange, R. P., Koopman, W. J., Austen, K. F., and LaRaia, P. J. "Cyclic Adenosine 3',5'-Monophosphate in Human Lung," *Biochem. Biophys. Acta* **252**:160–164 (1971).

Kaliner, M. A., Orange, R. P., and Austen, K. F. "Immunological Release of Histamine and Slow-Reacting Substance of Anaphylaxis from Human Lung. IV. Enhancement by Cholinergic and Alpha Adrenergic Stimulation," *J. Exp. Med.* **136**:556–567 (1972).

Karim, S. M. M., Sandler, M., and Williams, E. D. "Distribution of Prostaglandins in Human Tissues," *Brit. J. Pharmacol. Chemother.* **31**:340–344 (1967).

Kay, A. B., and Austen, K. F. "The IgE-Mediated Release of an Eosinophil Leukocyte Chemotactic Factor from Human Lung," *J. Immunol.* **107**:899–902 (1971).

Kuo, J., Lee, T., Reyes, P. L., Walton, K. G., Donnelly, T. E., Jr., and Greengard, P. "Cyclic Nucleotide-Dependent Protein Kinases. X. An Assay Method for the Measurement of Guanosine 3',5'-Monophosphate in Various Biological Materials and a Study of Agents Regulating Its Levels in Heart and Brain," *J. Biol. Chem.* **247**:16–22 (1972).

Orange, R. P., and Austen, K. F. "Immunologic and Pharmacologic Receptor Control of the Release of Chemical Mediators from Human Lung." In *The Biologic Role of the Immunoglobulin E System*, D. D. Dayton, ed. Washington, D.C.: U.S. Govt. Printing Office (in press).

Orange, R. P., Austen, W. G., and Austen, K. F. "Immunological Release of Histamine and Slow-Reacting Substance of Anaphylaxis from Human Lung. I. Modulation by Agents Influencing Cellular Levels of Cyclic 3',5'-Adenosine Monophosphate," *J. Exp. Med.* **134**:136s–148s (1971a).

Orange, R. P., Kaliner, M. A., LaRaia, P. J., and Austen, K. F. "Immunological Release of Histamine and Slow-Reacting Substance of Anaphylaxis from

Human Lung. II. Influence of Cellular Levels of Cyclic AMP," *Fed. Proc.* **30**:1725–1729 (1971b).

Orange, R. P., Kaliner, M. A., and Austen, K. F. "Immunological Release of Histamine and Slow-Reacting Substance of Anaphylaxis from Human Lung. III. Biochemical Control Mechanisms Involved in the Immunologic Release of the Chemical Mediators." In *Biochemistry of the Acute Allergic Reactions, 2nd Symposium,* K. F. Austen and E. L. Becker, eds. Oxford, England: Blackwell Scientific, pp. 189–202 (1971c).

Parish, W. E. "Antigen Induced Release of Histamine and SRS-A from Human Lung Passively Sensitized with Reaginic Serum," *Nature* (London) **215**:738–739 (1967).

———. "Reaginic and Nonreaginic Antibody Reactions on Anaphylactic Participating Cells." In *Control Mechanisms in Reagin Mediated Hypersensitivity,* L. Goodfriend and A. Sehon, eds. New York: Marcel Dekker, Inc. (in press).

Piper, P. J., and Vane, J. R. "Release of Additional Factors in Anaphylaxis and Their Antagonism by Antiinflammatory Drugs," *Nature* (London) **223**:29–35 (1969).

Piper, P. J., and Walker, J. L. "The Release of Spasmogenic Substances from Human Chopped Lung Tissue and Its Inhibition," *Brit. J. Pharmcol.* (in press).

Riley, J. F., and West, G. B. "Presence of Histamine in Tissue Mast Cells," *J. Physiol.* **120**:528–537 (1953).

Robison, G. A., Butcher, R. W. and Sutherland, E. W. *Cyclic AMP.* New York: Academic Press, pp. 17–46 (1971).

Sheard, P., Killingback, P. G., and Blair, A. M. J. N. "Release of Histamine and Slow-Reacting Substance with Mast Cell Changes After Challenge of Human Lung Sensitized Passively with Reagin in Vitro," *Nature* (London) **216**:283–284 (1967).

Shore, P. A., Burkhalter, A., and Cohen, Jr., V. H. "A Method for the Fluorometric Assay of Histamine in Tissues," *J. Pharmacol. Exp. Ther.* **127**:182–186 (1959).

Sullivan, A. K., Grimley, P. M., and Metzger, H. "Electron Microscopic Localization of Immunoglobulin E on the Surface Membrane of Human Basophils," *J. Exp. Med.* **134**:1403–1416 (1971).

Tauber, A. I., Kaliner, M. A., Stechschulte, D. J., and Austen, K. F. "Prostaglandins and the Immunological Release of Chemical Mediators from Human Lung," Symposium on Medical Aspects of Prostaglandins and Cyclic AMP, Univ. of Michigan, Ann Arbor (in press).

Tauber, A. I., Kaliner, M. A., Stechschulte, D. J., and Austen, K. F. "Immunological Release of Histamine and Slow-Reacting Substance of Anaphylaxis from Human Lung. V. Effects of Prostaglandins on Release of Histamine (submitted for publication).

Tomioka, H., and Ishizaka, K. "Mechanisms of Passive Sensitization. II. Presence of Receptors for IgE on Monkey Mast Cells," *J. Immunol.* **107**:971–978 (1971).

Wasserman, S. I., Goetzl, E. J., and Austen, K. F. (manuscript in preparation).

LEUCOCYTES AS SECRETORY ORGANS OF INFLAMMATION: CONTROL BY CYCLIC NUCLEOTIDES AND AUTONOMIC AGONISTS

GERALD WEISSMANN, ROBERT B. ZURIER, and SYLVIA HOFFSTEIN

I. Introduction

Although a series of studies has indicated that extracts of leucocyte lysosomes can provoke acute and chronic inflammation in experimental animals (Weissmann, 1972), a clear definition of discrete mechanisms, which account for release from inflammatory cells of such materials, is just beginning to emerge. Four separate circumstances are recognized under which substances ordinarily sequestered within lysosomes may gain access to the exterior of cells.

One mechanism has been termed "regurgitation during feeding" (Weissmann et al., 1971) and may be important to the propagation of joint inflammation in rheumatoid arthritis. When cells engage in phagocytosis (e.g., leucocytes which engulf immune complexes in the synovial fluid of patients with rheumatoid arthritis), they release a portion of their lysosomal hydrolases into the surrounding medium. This effect appears due to extrusion of lysosomal materials from incompletely closed phagosomes open at their external border to tissue space while joined at their internal border with granules discharging acid hydrolases into the vacuole (phagolysosome). Under such circumstances, lysosomal enzymes are selectively released to the outside of the cell without necessarily causing cytoplasmic damage; electron photomicrographs consistent with this mechanism have been published (Zucker-Franklin and Hirsch, 1964; Henson, 1971).

A second mechanism has been called "reverse endocytosis" (Weissmann et al., 1972), and may be pertinent to the pathogenesis of tissue injury in nephritis, vasculitis, and rheumatoid arthritis. When leucocytes encounter immune complexes which have been dispersed

This study was aided by grants from The National Institutes of Health (AM-11949), The New York Heart Association, The Glenn B. and Gertrude P. Warren Foundation, and The Whitehall Foundation.

The abbreviation MSU refers to monosodium urate.

along a nonphagocytosable surface such as a millipore filter or collagen membrane, there is similar, selective release of lysosomal enzymes directly to the outside of the cell (Hawkins, 1971; Henson, 1971a). Enzyme release may occur when leucocytes are in apposition to immune complexes in a blood vessel wall, in the glomerular basement membrane, or when pannus encounters articular cartilage.

Another mechanism for enzyme release—"perforation from within" (Weissmann and Dukor, 1970)—occurs when certain materials gain access to the vacuolar system wherein membrane damage results with release of cytoplasmic and lysosomal enzymes followed by cell and tissue death. The inflammatory episodes of acute gout appear due to this type of encounter between leucocytes and crystals of monosodium urate (McCarty and Hollander, 1961; Weissmann and Rita, 1972).

Finally, inflammatory substances may leak from cells simply as a result of cell death due to plasma membrane injury. A number of animal, bacterial, and chemical toxins, as well as synthetic detergents, may cause such lysis of the outer cell membrane (Weissman et al., 1969).

II. Inhibition of Enzyme Release

Since enzyme release under each of the above circumstance is crucial to the perpetuation of tissue injury, it may be that *reduction* of such enzyme release would prove beneficial. The effect on these four mechanisms of enzyme release was, therefore, studied of agents that regulate the secretion of stored proteins in tissues such as pancreas, salivary gland, and thyroid. Two types of compounds were studied: those that effect the function of microtubules and microfilaments and those that influence the accumulation within cells of cyclic nucleotides. The results suggest that selective enzyme release from human phagocytes is independent of cell death and that pharmacologic agents can reduce lysosomal enzyme release in three of four experimental challenges. When cells were killed by means of a lethal injury to the plasma membrane (Triton X-100), enzyme release could not be inhibited by pharmacologic means.

A. *Cells Exposed to Zymosan and to Immune Complexes in the Bulk Phase*

Exposure of human leucocytes to zymosan or to aIgG/IgM complexes resulted in the selective release of lysosomal enzymes with maintenance of cell viability (Weissmann et al., 1971). Release of enzyme, therefore, appeared to be closely related to phagocytosis and not to an event which was significantly delayed. However, when cells were exposed to smaller numbers of particles (particle/cell ratio 3:1), phagocytosis was completed by 30 min. The kinetics of enzyme release suggested two possibilities: (1) formation within cells of phagocytic vacuoles may stimulate movement of granules to the cell periphery, or (2) phagocytic vacuoles remain open to the outside during the entire 120 min period.

When cells were incubated with compounds that increased cellular levels of cyclic AMP or that interfere with microtubule integrity before their exposure to zymosan or immune complex, reduction in lysosomal enzyme release was observed (Weissmann et al., 1971). Cyclic AMP itself had no effect on extrusion of lysosomal enzymes. However, in combination with theophylline or 2-chloroadenosine, both of which inhibit phosphodiesterase, cyclic AMP caused a considerable reduction in enzyme release. Prostaglandin E_1 (PGE_1) also blocked selective hydrolase release. When theophylline or 2-chloroadenosine was added during preincubation with PGE_1, they acted further to reduce hydrolase release. Colchicine at a concentration of 10^{-3} M, but not at lower concentrations, also reduced enzyme release (Zurier et al., 1973).

PGE and PGA compounds stimulate the accumulation of cyclic AMP in mixed populations of human leucocytes, whereas prostaglandin F compounds have little or no effect on cyclic AMP levels (Scott, 1970; Bourne et al., 1971). Preincubation of leucocytes with prostaglandin E and A compounds consistently reduced selective extrusion of lysosomal enzymes from cells exposed to zymosan. However, pretreatment of cells with the same concentrations of prostaglandin $F_{1\beta}$ had little effect on enzyme release. Prostaglandin $F_{2\alpha}$ enhanced enzyme release by virtue of causing cell injury. Similar results were obtained when cells were exposed to the immune complex.

The concentrations of prostaglandins employed were quite large. Although optimal results were obtained with 2.8×10^{-4} M concentrations, PGE_1 and PGA_2 retarded hydrolase release from phagocytes at concentrations as low as 2.8×10^{-6} M (1 μg/ml).

In inital experiments, PGE_2 did not reduce enzyme release. It has been demonstrated that PGE_2 reduces the deformability of red blood cells, and it has been suggested that erythrocytes may be one of the primary receptors for PGE_2 (Allen and Rasmussen, 1972). When contaminating erythrocytes were removed from cell suspensions by means of hypotonic lysis, the effect of PGE_2 was comparable to PGE_1

B. Cells Exposed to Nonphagocytosable Stimuli

At the concentration used (10^{-3} M), colchicine significantly retarded particle uptake by phagocytic cells, whereas the cyclic nucleotides and prostaglandin E and A compounds also depressed C-1 glucose oxidation, a metabolic concomitant of phagocytosis (Zurier et al., 1973). This suggested that reduced enzyme release might simply be a consequence of reduced particle uptake. Nonetheless, the experiments did not exclude additional effects of these compounds on intracellular events subsequent to particle ingestion. Therefore, experiments were designed so that enzyme release might be studied without concern for the engulfment phase of phagocytosis. Henson has demonstrated that the encounter between neutrophiles and immune complexes dispersed along a non-phagocytosable surface results in selective release of lysosomal enzymes (Henson, 1971b). The technique was adapted to our studies.

Human leucocytes were allowed to settle onto millipore filters and incubated at 37°C for 2 hr. Under these circumstances, in the absence of phagocytosis, there was selective release of lysosomal enzymes occurring chiefly during the first 60 min. When cells were incubated with the dibutyryl analogue of cAMP, with prostaglandin E and A compounds, and with 10^{-6} M colchicine before being exposed to immune complex on filters, there was reduction of induced enzyme release (Zurier et al., 1973). $PGF_{1\beta}$ again failed to retard enzyme release (Zurier et al., 1973).

C. Effect of Serum Inactivation

It has been shown that heat-inactivated serum supports particle uptake, C-1 oxidation, and enzyme release almost as well as fresh serum when cells are exposed to immune complexes or MSU crystals (Weissmann et al., 1971). However, when cells were challenged with zymosan, modest but significant increments in phagocytosis and its metabolic concomitants were observed with fresh serum. The effect on enzyme release of heat inactivation was also studied. The presence of complement components (fresh serum) augmented only slightly the release of lysosomal enzymes from cells that had encountered immune complexes on the millipore filter. Despite the evidence that phagocytosis of complexes does not occur in this system, it cannot be inferred that phagocytic vacuoles were not formed in neutrophiles adherent to the millipore filter. Hawkins has utilized ferritin-antiferritin complexes on the filters and, in electron micrographs, has observed phagosomes which contain clusters of ferritin (Hawkins, 1972).

D. The Effect of Inhibiting Phagocytosis by Means of Cytochalasin B

We therefore studied the morphological and biochemical consequences of exposing leucocytes to nonphagocytosable particles. This was accomplished by using cytochalasin B, a fungal metabolite known to disrupt microfilaments and inhibit ingestion of particles by human leucocytes (Carter, 1967; Wessells et al., 1971; Allison et al., 1971). Incubation of cells with 5 μg/ml cytochalasin B before exposure of cells to zymosan resulted in complete inhibition of zymosan uptake. Zymosan particles adhered to the plasma membrane and induced enzyme release, but were not ingested (Fig. 1). Phagocytic vacuoles within neutrophils were not seen on any of many electron microscopic grids observed, enzyme appeared to be released directly to the outside of the cell (Fig. 1), and since treatment of cells with cytochalasin B did not enhance enzyme release, this system converts the "regurgitation during feeding"

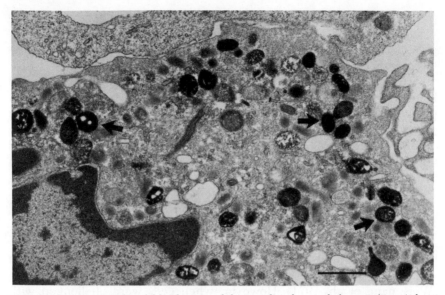

Fig. 1(a). Human peripheral blood neutrophil treated with cytochalasin B (5 μg/ml, 15 min), then fixed and stained for myeloperoxidase (Graham and Karnovsky, 1966). Numerous peroxidase-positive lysosomes (arrows) are apparent in the peripheral cytoplasm with only a few in the interior of the cell. Magnification 18,000 X. Line equals 1 μ.

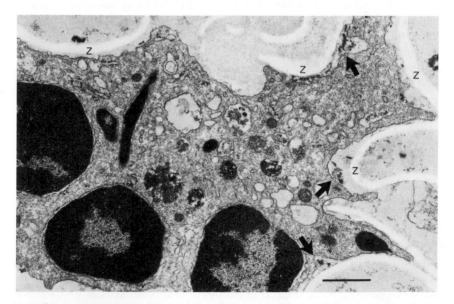

Fig. 1(b). Human peripheral blood neutrophil treated with cytochalasin B (as above), then exposed to zymosan for 60 min. Only a few peroxidasepositive lysosomes remain undischarged, mainly in the interior of the cell, whereas peroxidase-positive deposits (arrows) can be seen extruded into the space between the cell membrane and zymosan particles (Z). Magnification 18,000 X. Line equals 1 μ.

model into the "reverse endocytosis" model, in which release of lysosomal enzyme is independent of particle uptake and phagosome formation. It was, therefore, of interest to study the effect on enzyme release of colchicine and of agents that increase cyclic AMP concentration. Incubation of cells with 10^{-5} M colchicine, 5×10^{-5} M vinblastine, or with dibutyryl cAMP and PGE_1 before treatment of cells with cytochalasin B and exposure to zymosan reduced extrusion of lysosomal enzymes. Zurier et al., 1973c).

E. Cells Exposed to Membrane-Lytic Agents

Cell suspensions incubated with monosodium urate (MSU) crystals release cytoplasmic as well as lysosomal enzymes with subsequent death of the cells (Weissmann et al., 1971). However, the kinetics of enzyme release from cells exposed to the membrane-lytic detergent Triton X-100 differed from those observed after cells ingested MSU crystals (Zurier et al., 1973). In the first instance, there was rapid release of large proportions of total enzyme, whereas enzyme release was delayed in the latter. In addition, incubation of cells with MSU crystals resulted in normal increments of C-1 oxidation (a metabolic consequence of phagocytosis) prior to enzyme release, whereas the addition of Triton X-100 prevented C-1 oxidation, and led to immediate release of enzymes. It may, be inferred therefore, that enzyme release *follows* phagocytosis of crystals and is due to "perforation from within" rather than lysis of plasma membrane by crystals in the media.

There is no evidence that agents that affect the intracellular level of cyclic nucleotides prevent rupture of lysosomal membrane by MSU. However, if merger of granules with phagocytic vacuoles were to be impeded, or if phagocytosis were inhibited, reduced release from cells of acid hydrolase would be expected. Leucocytes were, therefore, incubated with PGE_1, $PGF_{1\beta}$, and dibutyryl cAMP before challenge by MSU crystals. Dibutyryl cAMP and PGE_1 (but not $PGF_{1\beta}$) did, in fact, reduce release of lysosomal enzymes, whereas such agents had no effect on enzyme release from cells incubated with Triton X-100.

III. Influence of Autonomic Agonists

Intracellular movements of lysosomes, and of secretory granules in a variety of cell types, appear to be accomplished by an interaction among cyclic nucleotides, microtubules, and microfilaments. In order for degranulation and enzyme release to proceed in response to perturbation of the cell membrane, microtubules must remain under appropriate controls. Although microtubules are in a dynamic state of assembly and disassembly, it is probably in their aggregated state that microtubules exert their influence on cell mechanics (Weisenberg et al., 1968). Cyclic AMP and colchicine both favor disassembly of microtubules (Gillespie, 1971), and both reduce selective enzyme release. Cyclic AMP and compounds that increase its cellular concentration also reduce antigen-induced histamine release from sensitized leucocytes and lung tissue (Lichtenstein and DeBernado, 1971; Orange et al., 1971). In contrast, deuterium oxide (D_2O) favors formation of microtubules (Marsland et al., 1971) and augments histamine release from leucocytes (Gillespie and Lichtenstein, 1972). In addition, it has recently been reported that beta adrenergic agents (which increase cyclic AMP levels) reduce, whereas cholinergic agonists enhance, the immunologic release of histamine and SRS-A from human lung fragments (Kaliner et al., 1972). Cholinergic stimulation of heart and brain preparations has produced an increase in levels of cyclic guanosine 3'5'-monophosphate (cyclic GMP) (Kuo et al., 1972), and the introduction of 8-bromo-cyclic GMP to sensitized lung tissue (Kaliner et al., 1972) and of cyclic GMP to leucocyte suspensions (Zurier et al., 1973) was associated with enhancement of antigen-induced release of histamine and SRS-A in the first instance and enhancement of lysosomal enzyme release in the latter circumstance. Thus, it appears that, in human leucocytes, granule movement and acid hydrolase release may be modulated through changes in concentrations of at least two cyclic nucleotides, cyclic AMP and cyclic GMP. It is therefore possible that the release of inflammatory substances from phagocytes can be controlled by humoral means and that many of these may effect microtubule function. Consequently, following the experimental protocol suggested by Kaliner et al. (1972), we exposed human leucocytes pretreated with cytochalasin B to a series of autonomic agonists. Since, as we have previously mentioned, cells treated with cytochalasin B selectively merge their lysosomes with the plasma membrane as if the latter was a large phagocytic vacuole, and since phagocytosis is not a factor under these circumstances, it is possible directly to measure the influence of these autonomic agonists on membrane fusion. When isoproterenol at 5×10^{-6} M (Fig. 2) was exhibited, enzyme release was significantly diminished. In contrast, when the beta adrenergic blocker, propanolol, was added at equal concentrations, inhibition was overcome. Similar findings were observed with epinephrine. In direct contrast, the alpha adrenergic agent, phenylephrine, inhibited acid hydrolase release only minimally, due perhaps to its modest beta adrenergic effect, which was overcome by the simultaneous administration of propanolol. It, therefore, appears that human leucocytes possess beta adrenergic receptors, the stimulation of which causes the inhibition of lysosomal hydrolase release. When alpha adrenergic agents are exhibited, no such action is seen in the presence of beta adrenergic blockers. It is, therefore, probable that these effects are mediated by the elevations of cyclic AMP that others have reported in mixed suspensions of human leucocytes following the administration of beta adrenergic agents (Bourne et al., 1971). Indeed, inhibition of

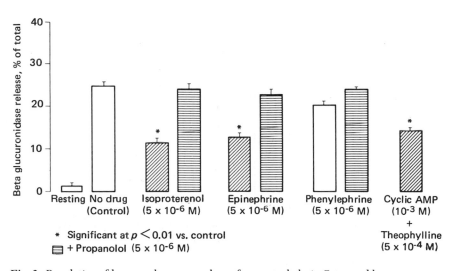

Fig. 2. Regulation of lysosomal enzyme release from cytochalasin-B-treated leucocytes exposed to zymosan (adrenergic agonists). Human leucocytes (4×10^6) treated with cytochalasin B ($5 \mu g/ml \times 10$ min), incubated with test compounds 20 min, then exposed to zymosan (particle/cell ratio 15:1) for 20 min. All incubations at 37°C. Beta-glucuronidase release expressed as percent of total (100%) enzyme release by 0.2% Triton X-100. Total activity = 23.7 ± 1.1 μg phenolphthalein per hour ($N = 4$).

enzyme release was demonstrated when cyclic AMP and theophylline were added to leucocytes in our experiments (Fig. 2).

To further test the hypothesis that autonomic agonists can regulate mediator release from the leucocyte, we studied the effect of the cholinergic agonist carbamylcholine at 5×10^{-6} M. Significant enhancement of lysosomal hydrolase release was observed in cells that had been treated with this agent. This enhanced release consequent to the cholinergic agonist could be reversed by 10^{-8} M of atropine. Moreover, cyclic GMP at 5×10^{-6} M had an effect similar to that of carbamylcholine, suggesting that the effects of carbamylcholine in this system, as in others, were due to the elevation within the appropriate cells of cyclic GMP. To prove that such cells were still intact, similar suspensions were treated by cyclic AMP and theophylline and enzyme release was, as expected, significantly inhibited (Fig. 3).

Such experiments would suggest that there is a sensitive balance in human polymorphonuclear leucocytes between secretion and retardation of secretion of lysosomal enzymes and that this balance may be set by autonomic agonists. If autonomic agonists in this system work as they do in other systems by virtue of an effect upon cyclic nucleotides, we can begin to put this sequence of events into proper perspective. A tentative scheme indicating this phenomenon is indicated in Table 1. It is clear that in leucocytes, secretion of acid hydrolases is enhanced by autonomic agonists. It may be mimicked by exhibition of cyclic GMP, but not other

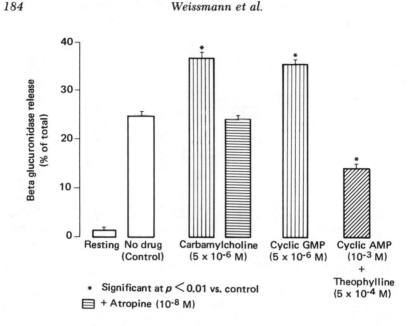

Fig. 3. Regulation of lysosomal enzyme release from cytochalasin-B-treated leucocytes exposed to zymosan (cholinergic agonists). Human leucocytes (4×10^6) treated with cytochalasin B (5 μg/ml $\times$ 10 min), incubated with carbamylcholine or cGMP for 2.5 min (cAMP + theophylline $\times$ 20 min), then exposed to zymosan (particle/cell ratio 15:1) for 20 min. All incubations at 37°C. Beta glucuronidase release expressed as percent of total (100%) enzyme released by 0.2% Triton X-100. Total activity = 25.6 $\pm$ 1.4 μg phenolphthalein per hour ($N = 4$).

cyclic nucleotides. Moreover, secretion in other systems, specifically chopped lung fragments and mast cells, appears to be *enhanced* when tubulin is in the aggregated state such as when D_2O is exhibited. In direct contrast, *diminished* secretion of inflammatory mediators (using beta glucuronidase as an indicator) is brought about when beta adrenergic stimuli are applied. These act via elevations of cyclic AMP, and indeed these effects may be mimicked by exhibition of cyclic AMP, especially in the presence of theophylline. Finally, tubulin in the disaggregated state may not provide the appropriate pathway for the flow of secretory granules (lysosomes) to the phagocytic vacuole or the cell periphery. Since colchicine and vinblastine mimic the effects of beta adrenergic agents, cyclic AMP, and prostaglandin E_1, it is reasonable to suspect that each of these agents acts at the same site, i.e., on the disaggregation of microtubules. We have found a unique protein kinase (Tsung et al., 1972), which in Hypaque-Ficoll preparations of polymorphonuclear leucocytes appears to be specifically activated by cyclic AMP. We have not yet determined whether tubulin prepared from polymorphonuclear leucocytes is the preferred substrate for this protein kinase. Should, however, further experiments indicate that tubulin is *disaggregated* in

Table 1. Controls of Mediator Secretion from Leukocytes

Mediator	Secretion increased	Secretion reduced
Autonomic agonist	Cholinergic	Adrenergic (β in PMN)
Cyclic nucleotide	Cyclic GMP	Cyclic AMP
Tubulin	Aggregated, = D^2O effect	Disaggregated, = colchicine, vinblastine
Enzyme	Unknown	Protein kinase (Ca^{2+} inhibits)

the phosphorylated form and *aggregated* in the dephosphorylated form, then secretion from leucocytes may be controlled by the phosphorylation of tubulin, a process regulated by the cyclic nucleotides.

IV. Summary

In order to study mechanisms underlying selective enzyme release from human leucocytes during phagocytosis, the effects were studied of compounds that affect microtubule integrity or the accumulation of cyclic nucleotides. Human leucocytes selectively extrude lysosomal enzymes (beta glucuronidase) from viable cells during phagocytosis of zymosan or immune complexes, or upon encounter with immune complexes dispersed along a nonphagocytosable surface such as a millipore filter. In each circumstance, lysosomal enzyme release was reduced by prior treatment of cells with pharmacological doses of drugs that disrupt microtubules (e.g., 10^{-3} to 10^{-6} M colchicine) or with agents that affect accumulation of cyclic nucleotides (e.g., 10^{-3} M cyclic nucleotides and 2.8×10^{-4} to 2.8×10^{-6} M prostaglandin E and A compounds). Preincubation of cells with 5 μg/ml cytochalasin B resulted in complete inhibition of zymosan ingestion, adherence of zymosan particles to plasma membrane, and selective enzyme release at the cell-zymosan interface. In this system, in which enzyme release was independent of particle uptake, preincubation of cells with colchicine, vinblastine, dibutyryl cAMP, or PGE_1 also reduced extrusion of lysosomal enzymes. The effects were also studied of adrenergic and cholinergic agonists on hydrolase release. Beta adrenergic stimulation with isoproterenol or epinephrine produced a dose-related reduction in beta glucuronidase release (50% inhibition at 5×10^{-6} M). This reduction in enzyme release was prevented by beta adrenergic blockade with 10^{-6} M propanolol. Alpha adrenergic stimulation (10^{-5} to 10^{-12} M phenylephrine or norepinephrine 10^{-5} to 10^{-8} M + propanolol) did not significantly alter enzyme release. The cholinergic agonist carbamylcholine (10^{-6} M Carbachol) increased beta glucuronidase release 41.6%. Atropine (10^{-8} M) prevented Carbachol-induced enhancement of hydrolase release. Incubation of cells with 10^{-6} M cGMP caused a 35.2% increase in beta glucuronidase release. The data suggest that in human leucocytes, granule movement and acid hydrolase release may be modulated by adrenergic and cholinergic agents, perhaps through changes in concentrations of cyclic nucleotides, acting on the aggregation of microtubules.

Discussion of Dr. Weissman et al.'s Paper

DR. PARKER: What is the evidence that isolated human polymorphonuclear leukocytes actually respond to isoproterenol? In most of our own preparations, the response is quite unimpressive. Do you have direct evidence that substantial changes in cyclic AMP concentration are taking place?

DR. WEISSMANN: We have not determined cyclic AMP in Ficoll-Hypaque purified human polymorphonuclear leukocytes. I think, however, the response of these cells to the usual spectrum of agents indicates that elevations of cyclic AMP may mediate inhibiton of hydrolase release.

DR. BOURNE: What does exogenous cyclic "I" do to hydrolase release. If the protein kinase is important and if cyclic IMP activates it, the nucleotide should inhibit hydrolase release.

DR. WEISSMANN: Exactly the same concentration of cyclic IMP as cyclic AMP inhibits hydrolase release. Cyclic GMP does not.

DR. BOURNE: I also have a comment about the Sutherland catechism and the measurement of drug effects on cyclic AMP. Like Dr. Parker, we have exposed purified (98%) neutrophils (capable of killing microorganisms and releasing hydrolases) to isoproterenol and have not seen an increase in cyclic AMP. By contrast, cholera toxin produces marked increases in cyclic AMP in these cells. Yet you have shown that isoproterenol inhibits hydrolase release, whereas cholera toxin at concentrations up to 1800 times those required for maximal stimulation of cyclic AMP accumulation does not inhibit hydrolase release. So, on the one hand, you have an agent that inhibits hydrolase release, but does not cause a rise in cyclic AMP (isoproterenol) and, on the other, an agent that increases cyclic AMP, but does not inhibit enzyme release (cholera toxin). It seems to me there is some contradiction there.

DR. WEISSMANN: I think it is possible that in mixed-cell populations, cyclic nucleotides might be secreted from other cells in the mixture and secondarily affect polymorphonuclear leukocytes. I am not suggesting that cyclic AMP synthesized in polymorphonuclear leukocytes necessarily produces the inhibition. I would also point out that we have found a cAMP-dependent protein kinase in the polymorphonuclear leukocyte, and I don't think it is there by whim. I think its function is to respond to cyclic AMP.

DR. BOURNE: Then why doesn't the cholera toxin inhibit hydrolase release with mixed cells?

DR. WEISSMANN: I do not know, but I think the answer will be fascinating. I think it is remarkable that so far the effects of nucleotides on hydrolase release have closely mimicked all the other mediator-release systems with respect to autonomic agonists, prostaglandins, and other agents. Clearly, this generality is probably more interesting than the specific exceptions. Although there are great gaps in the story, it is coherent for a number of cell types that secrete mediators of inflammation. Exogenous or perhaps endogenous cyclic AMP inhibits secretion, whereas exogenous or perhaps endogenous cyclic GMP enhances secretion.

DR. ROBISON: Isn't cholera toxin unstable? Did you check whether the cholera toxin used in your experiments really had cholera-toxin-like activity?

DR. WEISSMANN: We got our cholera toxin from Dr. Bourne and Dr. Lichtenstein.

DR. LICHTENSTEIN: The toxin was shipped back and forth across the country, but it was assayed after Dr. Weissmann's experiments had been completed and shown to be still active.

References

Allen, J. E., and Rasmussen, H. In *Prostaglandins in Cellular Biology*, Alza Conf. Series, Vol. 1. New York: Plenum Press, 1972, p. 27.

Allison, A. C., Davies, P., and dePetris, S. *Nature (New Biol.)* **232**:153 (1971).

Bourne, H. R., Lehrer, R. I., Cline, M. J., and Melmon, K. J. *J. Clin. Invest.* **50**:920 (1971).

Carter, S. B. *Nature* **213**:261 (1967).

Gillespie, E. *J. Cell Biol.* **50**:544 (1971).

Graham, R. C., Jr., and Karnovsky, M. J. *J. Histochem. Cytolchem.* **14**:291 (1966).

Hawkins, D. *J. Immunol.* **107**:344 (1971).

Hawkins, D. *J. Immunol.* **108**:310 (1972).

Henson, P. M. *J. Exp. Med.* **134**:114s (1971a).

Henson, P. M. *J. Immunol.* **107**:1547 (1971b).

Kaliner, M., Orange, R. P., and Austen, K. F. *J. Exp. Med.* **136**:556 (1972).

Kuo, J., Lee, T., Reyes, P. L., and Walton, K. G. *J. Biol. Chem.* **247**:16 (1972).

Lichtenstein, L. M., and DeBernado, R. *J. Immunol.* **107**:1131 (1971).

Marsland, D., Tilney, L. G., and Hirshfield, M. *J. Cell. Physiol.* **77**:187 (1971).

McCarty, D. J., and Hollander, J. L. *Ann. Intern. Med.* **54**:452 (1961).

Orange, R. P., Austen, W. G., and Austen, K. F. *J. Exp. Med.* **134**:136s (1971).

Scott, R. E. *Blood* **35**:514 (1970).

Tsung, P.-K., Hermina, N., and Weissmann, G. *Biochem. Biophys. Res. Commun.* **49**:1657 (1972).

Weisenberg, R. C., Borisy, G. G., and Taylor, E. W. *Biochem.* **7**:4466 (1968).

Weissmann, G., Hirschhorn, R., and Krakauer, K. *Biochem. Pharmacol.* **18**:771 (1969).

Weissmann, G., and Dukor, P. *Advan. Immunol.* **12**:283 (1970).

Weissmann, G., Dukor, P., and Zurier, R. B. *Nature (New Biol.)* **231**:131 (1971).

Weissmann, G., Zurier, R. B., Spieler, P., and Goldstein, I. *J. Exp. Med.* **134**:149s (1971).

Weissmann, G. *New Eng. J. Med.* **286**:141 (1972).

Weissmann, G., Zurier, R. B., and Hoffstein, S. *Amer. J. Pathol.* **68**:539 (1972).

Weissmann, G., and Rita, G. A. *Nature (New Biol.)* **240**:167 (1972).

Wessells, N. K., and Spooner, B. S., Ash, J. F., Bradley, M. O., Luduena, M. A., Taylor, E. L., Wren, J. T., and Yamade, K. M. *Science* **171**:135 (1971).

Zucker-Franklin, D., and Hirsch, J. G. *J. Exp. Med.* **120**:569 (1964).

Zurier, R. B., Tynan, N., and Weissmann, G. *Fed. Proc.* (Abstract, in press— 1973a).

Zurier, R. B., Hoffstein, S., and Weissmann, G. *J. Cell Biol.* (submitted for publication—1973b).

Zurier, R. B., Hoffstein, S., and Weissmann, G. *Proc. Nat. Acad. Sci.* (in press— 1973c).

THE IN VIVO EFFECT OF DIBUTYRYL CYCLIC AMP AND IMMUNOSUPPRESSIVE DRUGS ON THE SECONDARY ANTIBODY PRODUCTION OF MICE TO OVALBUMIN

R. J. Perper, V. Blancuzzi, and A. Oronsky

I. Introduction

It has been shown that there is genetic control over the ability for certain inbred strains of mice to produce high titers of homocytotrophic antibody to extremely small amounts of ovalbumin in alum adjuvants (1). These high-responder mice also produce IgG type antibodies against the same antigen. Thus, this model is a useful method to test for the effect of various pharmacological agents on two types of antibody responses, to the same antigen, in a strain of animals in which all are capable of responding. In previously available models, only a small percentage of immunized animals produced homocytotrophic antibody, making it difficult to accurately access drug effects. Using the mouse system, we now show that antiproliferative and cytotoxic immunosuppressive agents have a suppressive effect on the anamnestic IgG and reagin antibody responses, whereas dibutyryl cAMP and theophylline suppress the production of homocytotrophic antibody without effect on IgG production.

II. Methods

Groups of seven female 18 to 20 g SWR mice (Jackson Laboratories) were immunized intraperitoneally with 0.5 ml of a saline solution of 0.1 μg twice crystallized ovalbumin in aluminum hydroxide (1). Thirty days later, each animal was given a second injection of the same preparation, and they were sacrificed seven days later. Five separate experiments were performed, in which four to five groups of seven mice were treated with various agents given either before or after the second antigen injection. In each experiment, a controlled group of mice was included, and these were treated with vehicle only. Serums from each group of animals were pooled, and homocytotrophic antibody was assayed by injecting 0.05 ml of various dilutions of serum intradermally to each of three to five CFW 20 to 25 g male mice. Two days later, each recipient received an intravenous injection of 0.2 ml of a 1% saline solution of Evans blue dye containing 1 mg/ml ovalbumin. The reaction size at each dilution was measured, and the titer was expressed as the arithmetic mean of the highest dilution of antigen giving a positive reaction in the three to five recipients.

Each serum pool was assayed for IgG antibody content by passive hemagglutination techniques, using serial dilutions of serum and ovalbumin-coated tanned sheet erythrocytes. The pharmacological agents used were cyclophosphamide (Mead-Johnson Laboratories), theophylline (Nutritional Biochemicals), dibutyryl adenosine 3',5'-cyclic monophosphoric acid (Sigma Chemical Company), and rabbit antimouse thymocyte serum (ATS) made by the Freunds adjuvant method (2).

III. Results and Discussion

The effects of drug treatment on the secondary antibody responses of ovalbumin-immunized mice is given in Tables 1 to 3. The antibody titers and the PCA reaction size of the untreated control animals, which were included in the same experiment, are given with the corresponding values for the treated animals. Cyclophosphamide, 20 mg/kg P.O., given daily starting two days prior to the second antigen injection and continued until six days after (1 day prior to sacrifice), totally inhibited the production of both the IgG and homocytotrophic antibodies as compared to untreated controls (Table 1). When given by the same protocol at 5 mg/kg P.O., approximately a 50% inhibition of both responses occurred. Greater than a 50% reduction was observed even when the drug was given after the second antigen injection at 10 mg/kg. ATS given subcutaneously at 0.25 ml/day, from day −1 to day +2 in relation to the second antigen injection, totally inhibited homocytotrophic antibody production and also inhibited the IgG response. When given at the same dose after the second antigen injection (day +1 → +4), ATS also inhibited both responses; however, the IgG responses were effected to a greater extent.

It was apparent from these results that conventional type immunosuppressive agents were capable of inhibiting both IgG and homocytotrophic antibody production. It was not possible to test for the effect of these agents on the induction on the primary response, since the homocytotrophic antibody titers were less than 1:20 prior to the second antigen injection, a value too low for accurate quantitation. In order to test for the specificity of the 48-hr PCA reaction, serums were heated at 60°C for 90 min, which in all cases abolished the skin reaction. Alternatively heating the serum or treating it with 0.1 M mercaptoethanol did not affect the IgG titers.

Table 1. Drug Effect on 2° Antibody Response of Mice to Ovalbumin

Drug and route	Daily dose mg/kg	Days treated[a]	PCA reaction				Passive hemagglutination	
			Titer^{-1}		Reaction size (mm²)		Titer (1/log$_2$)	
			Treated	Control	Treated	Control	Treated	Control
Cyclophosphamide (P.O.)	20	−2 to +6	0[b]	60	0	125[c]	0.0	7.5
	5	−2 to +6	20	60	35	125	6.0	7.5
	10	+2 to +6	10	60	27	125	5.5	11.0
ATS (S.C.)	0.25 ml	−1 to +2	0[b]	120	0	90[d]	7.0	9.5
	0.25 ml	+1 to +4	120	240	10	48	4.5	7.0

[a]In relation to second antigen injection given 30 days after 1°.
[b]No positive reaction in three recipients at 1:10.
[c]Reaction at 1:20 dilution.
[d]Reaction at 1:60 dilution.

Table 2. Drug Effect on 2° Antibody Response of Mice to Ovalbumin

Drug and route	Daily dose mg/kg	Days treated[a]	PCA reaction				Passive hemagglutination	
			Titer^{-1}		Reaction size[b] (mm²)		Titer (1/log$_2$)	
			Treated	Control	Treated	Control	Treated	Control
Dibutyryl cAMP (S.C.)	10	−2 to +6	10	120	0	90	9.5	9.5
	15	+1 to +6	120	120	83	90	6.0	7.0
Theophylline (P.O.)	50		120	480	77	122	10.5	11.5
	150		7	50	3	28	–	–
	150	−2 to +6	13	120	0	64	9.5	10.0

[a]In relation to second antigen injection given 30 days after 1°.
[b]At 1:60 antiserum dilution.

Table 3. Antibody Content of "Pool" of Six Serums Compared with Average Titer of the Individual Samples[a]

Treatment	Passive cutaneous anaphylaxis		Passive hemagglutination
	Reciprocal titer[a]	x reaction (mm^2) 1:20 + 1:40	Titer (1/log^2)
Vehicle			
Pool	160	64	10
Individual	120 ± 25	48 ± 13	8.8 ± 0.7
Theoplylline (150 mg/kg P.O., −2 to +6 days)			
Pool	80	23	9
Individual	56 ± 14[b]	29 ± 11	7.8 ± 0.9

[a]Average titer of the six serums each assayed in at least three mice at dilutions of 1:20, 1:40, 1:80, and 1:160.
[b]$p < 0.01$ compared with controls.

It has been shown that IgE antibody production may not be regulated by the same mechanism as IgG responses are, since in the rat they can be inhibited selectively (3), and their induction in vivo has different cellular requirements (4). Since Braun and Ishizuka (5) had demonstrated that theophylline, possibly working through the cAMP system, can modulate antibody responses in vivo, it was of interest to determine the effect of these compounds on the homocytotrophic and IgG responses in the mouse. The results of these experiments given in Table 2 clearly indicate that treatment from day −2 → +6 with either dibutyryl cAMP or theophylline selectively reduced homocytotrophic antibody production without effects on IgG production. When given only after the second antigen injection, cAMP had no effect on either response, and in experiments not shown, theophylline acted similarly.

In order to determine whether the homocytotrophic antibody titers of the serum pools accurately reflected the individual responses, in one experiment the individual serums and the pool were each tested separately in groups of three to four recipients. The data given in Table 3 indicate that the titers of the pools accurately reflected the individual responses and reconfirmed the selective effect of theophylline on the homocytotrophic antibody response.

The finding that cAMP was ineffective when given only after the second antigen injection indicated that residual cAMP in the treated animal serum was not responsible for the selective inhibition of homocytotrophic antibody observed when both pretreatment and posttreatment schedules were employed. In order to confirm that residual cyclic cAMP in the serum was not responsible for inhibiting the PCA reactions in the recipients, experiments illustrated in Table 4 were performed. It can be seen that the control reaction was not inhibited when serum from cAMP-treated animals was mixed with control serum.

Table 4. Test for Residual cAMP in Serum of Treated Animals

Group number	Treatment of ovalbumin immunized serum donors[a]	Intradermal injection (ml)	PCA reaction[b]	
			Titer (reciprocal)	Size (mm^2) 1:40 dilution
I	0	0.05	120	125
II	8 daily injections dibutyryl cAMP. Last, 1 day prior to sacrifice (10 mg/kg)	0.05	40	20
III	Serum from I	0.05	120	160
	+	⟩0.01		
	Serum from II	0.05	120	160

[a]Serum pool from seven immunized animals seven days after 2nd ovalbumin injection.
[b]Average of three to four recipients.

In conclusion, we have demonstrated that cAMP and theophylline have a selective effect in reducing the secondary homocytotrophic antibody response of mice to ovalbumin without effecting the IgG titers to the same antigen. Standard immunosuppressive agents, however, inhibit both the responses approximately to the same extent. The results indicate that "IgE"-like antibodies are produced by different cell types since they can be selectively inhibited, which is in agreement with the findings of Ishizaka et al. (4). Since it has been demonstrated that cAMP has profound effects on the active export of many types of intracellular products, it is tempting to postulate that IgE antibody export is a cAMP-controlled event, which differs from that of IgG release. The fact that IgE-secreting cells are concentrated in intestinal and alveolar linings indicates a teleological function for such a mechanism; however, such a hypothesis can only be validated when the results of further experiments are available.

Discussion of Dr. Perper et al.'s Paper

DR. BLOOM: I think it is difficult to interpret results of drug inhibition in assays of different sensitivities, where the drug inhibits one and not the other. There is a system in which the idea that the cyclic AMP effect is on secretion rather than synthesis might be tested, namely, a human lymphoid blastoid cell line which both secretes and has membrane immunoglobulin. One could study that line for an effect on secretion relative to synthesis as manifested by surface immunoglobulin.

DR. PERPER: The relative sensitivity of the two assays was considered, and dose-response studies with cyclophosphamide were performed in order to determine how this nonspecific immunosuppressive agent affected IgG and IgE antibody titers. An equivalent effect on IgE and IgG levels was found in contrast to the selective effect of dibutyryl cyclic AMP. I agree that testing the effect of dibutyryl cAMP on the secretion of either IgE or IgG from synchronized myeloma cell lines would represent an ideal model to test our hypothesis, and we would like to try this experiment.

References

1. Vaz, N., Phillips, J., Quagliata, J., Levine, B., and Vaz, E. *J. Exp. Med.* **134**: 1335 (1971).
2. Perper, R. J., Lyster, S. C., Monovich, R. E., and Bowersox, B. E. *Transplantation* **9**: 447 (1970).
3. Taniguchi, M., and Tada, T. *J. Immunol.* **107**: 579 (1971).
4. Tadamitsu, T., and Ishizaka, K. *J. Immunol.* **109**: 1163 (1972)
5. Braun, W., and Ishizuka, M. *J. Immunol.* **107**: 1036 (1971).

RELATIONSHIPS BETWEEN THE CYTOLYTIC ACTIVITY OF THYMUS-DERIVED LYMPHOCYTES AND CELLULAR CYCLIC NUCLEOTIDE CONCENTRATIONS

I. Introduction

A proportion of the thymus-derived (T) lymphocytes from immune mice exhibits the capacity to specifically bind homologous antigen. When the antigen is cell associated, such binding may be followed by cytolysis of the antigen-bearing (target) cell. This phenomenon, called lymphocyte-mediated cytolysis, has been shown to play a central role in allograft rejection (1,2) and is of additional importance in vivo as a defense against neoplastic growth. While the mechanism of cytolysis remains poorly understood, extensive in vitro studies have established a number of basic requirements for the lytic process (3– 7).

The primary event in the cytolytic pathway involves the combination between antigen and a "sensitized"° T lymphocyte (3). The lymphocyte population must be viable (4) and capable of protein synthesis (1,7). Indeed, in the presence of pactamycin, an inhibitor of the initiation of protein synthetic pathways, no cytolysis is demonstrable (7). These findings suggest that the effector lymphocyte undergoes protein synthesis as a result of its combination with antigen, and that the protein thus synthesized is intimately associated with the lytic process. It seems reasonable, as a working hypothesis, to consider such protein as a soluble mediator, analogous to other mediators, e.g., migration inhibitory factor, synthesized by sensitized lymphocytes as a result of antigenic stimulation (8). If this is correct, then the delivery of mediator should be an essential feature of cytolysis, and it would be expected that agents that interfere

---

This work was supported by a Research Career Development Award and by grants from the National Institute of Allergy and Infectious Diseases and the National Science Foundation. This is communication 80 from the O'Neill Memorial Laboratories.

°The term "sensitized" is used here to describe lymphocytes bearing antigen-specific recognition units. In this context, "sensitized" cells are synonymous with the cytolytically active effector cells.

with cellular secretion would suppress target cell destruction. Studies employing cytochalasin B (9,10) and colchicine (10) have offered evidence strongly supportive of the concept that cytolysis involves a secretory event. These agents, which affect microtubular development and microfilament function and which suppress the secretory capacity of a number of cell types (11,12), were both found to be potent inhibitors of cytolysis.

Cyclic nucleotides, and in particular cyclic adenosine 3',5'-monophosphate (cAMP), have also been widely implicated in cellular secretory events (13,14). It was considered of interest, therefore, to modulate levels of cyclic AMP and of its guanosine analogue (cGMP) in an effector lymphocyte population, and to follow the effects of such changes on the cytolytic activity of the cells. An excellent inverse correlation between cAMP levels and specific cytolytic activity was demonstrable (15,16). On the other hand, modulation of cGMP levels did not affect the lytic capacity of effector cell populations.

II. Methods

A. In Vitro Cytolysis

Mouse mastocytoma cells (P815 of the DBA/2 strain) have been used in these studies both as antigen and as target cells in cytolytic assays. The methodology has previously been described in detail (1,17) and will only briefly be outlined here. Ten million viable mastocytoma cells were injected intraperitoneally into C57 Bl/6 mice, and 10 days later, spleen cell suspensions from these animals were prepared. Splenic lymphocytes (routinely 10^7 cells) were incubated for 2 to 6 hr at 37°C in an atmosphere of 5% CO_2 95% air with 10^5 ^{51}Cr-labeled mastocytoma cells. The culture medium employed throughout was Eagle's minimal essential medium containing 10% inactivated fetal calf serum and 100 units/ml of both penicillin and streptomycin. Following incubation, the amount of extracellular ^{51}Cr was measured and recorded as a percentage of that originally cell associated. The percent release of ^{51}Cr was equated with percent cytolysis (1,17). The percentage-specific cytolysis was obtained by subtracting the ^{51}Cr released in the presence of 10^7 normal C57 Bl/6 spleen cells (17). The percentage inhibition of specific cytolysis in the presence of various drugs was calculated relative to control cultures without drugs (16).

B. Measurement of cAMP and cGMP

Cyclic AMP was measured by Bourne's modification (18) of the competitive-binding assay of Gilman (19). Cyclic GMP was measured by radioimmunoassay according to the method of Steiner et al. (20).

Between 10^7 and 3×10^7 lymphocytes in 1 ml Eagle's medium were incubated with drugs for various periods of time at 37°C. Following in-

cubation, the cells were spun down, and the pellet resuspended in 1 ml
5% trichloracetic acid before nucleotide estimations were made.

III. Results

A. Relationship Between Lymphocyte cAMP Levels and Cytolytic Activity

Adult C57 B1 mice were immunized with 10^7 DBA/2 mastocytoma
cells intraperitoneally. Eleven days later, spleens were removed, and a
pooled lymphocyte population prepared. Such lymphocyte populations
were highly cytotoxic in vitro toward the DBA/2 mastocytoma cells. Ten
million viable lymphocytes, when incubated with 10^5 ^{51}Cr-mastocytoma
cells caused >50% release of ^{51}Cr from the target cells within a 6-hr incubation period. Incubation of the same target cells with 10^7 normal C57
B1 lymphocytes resulted in less than 10% of the ^{51}Cr being released over
the same time period. For a detailed kinetic analysis of cytolysis in this
system, see Henney's paper (17).

1. The Effect of Isoproterenol and Theophylline

In the presence of 10^{-5} M isoproterenol, inhibition of cytolysis was
marked in the early stages of the reaction. Thus, during an interaction
period of 30 min, 10^{-5} M isoproterenol inhibited specific cytolysis approximately 60%. This inhibitory activity was fairly short-lived, however, and
after a 3-hr incubation, only about a 5% inhibition of specific cytolysis
was observed. Longer periods of incubation (up to 9 hr) likewise resulted
in very little inhibition. A similar time dependence on the inhibitory activity of isoproterenol was observed in the presence of higher concentrations of the drug (e.g., 10^{-3} M).

In order to obtain an optimal dosage for the inhibitory activity of
isoproterenol, a relatively short (2-hr) incubation period of lymphocytes
with target cells was thus selected. Cytolytic assays were performed in
the presence of isoproterenol concentrations ranging from 10^{-8} to 10^{-4} M.
The inhibitory activity of the drug was calculated relative to the specific
cytolysis observed in its absence. As can be seen (Fig. 1), significant inhibition of cytolysis was observed at isoproterenol concentrations as low
as 10^{-7} M. Increasing drug concentrations above this resulted in a linear,
yet somewhat modest increase in inhibition.

Unlike isoproterenol, theophylline showed no time dependence on
its activity; a given concentration inhibited cytolysis reproducibly
regardless of the time of incubation. Concentrations of theophylline in
excess of 10^{-4} M were necessary for inhibition, but at concentrations approaching 10^{-3} M, cytolysis was almost completely inhibited (Fig. 2).

The inhibitory effect of combinations of theophylline and
isoproterenol was always at least additive to the effects observed with
the drugs singly. Typical results are shown in Fig. 3.

The results obtained from measurement of intracellular cyclic AMP levels in the presence of isoproterenol bore a striking resemblance to the inhibition studies, inferring an inverse relationship between cAMP levels

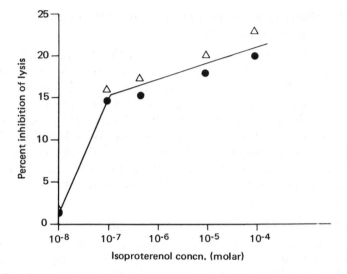

Fig. 1. The inhibition of specific lymphocyte-mediated cytolysis by isoproterenol. 10^7 C57 B1 lymphocytes, obtained 11 days after immunization with 10^7 DBA/2 mastocytoma cells, were incubated for 2 hr with 10^5 ^{51}Cr-labeled DBA/2 cells in the presence of various concentrations of isoproterenol. Cytolysis was then evaluated, and the percent inhibition calculated relative to cultures containing no drug [reproduced from *J. Immunol.* **108**:1526 (1972) by kind permission of the Williams and Wilkens Company, Baltimore].

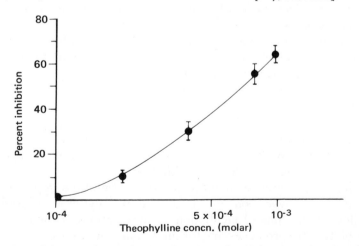

Fig. 2. The inhibition of specific lymphocyte-mediated cytolysis by theophylline. Lymphocyte and target cell populations were as in Fig. 1; incubation period 2 hr. The mean percentage inhibition by different concentrations of the drug and the SD of six observations are shown [reproduced from *J. Immunol.* **108**:1526 (1972) by kind permission of the Williams and Wilkens Company, Baltimore].

and cytolytic activity. Thus, isoproterenol at concentrations above 10^{-8} M caused increases in intracellular cyclic AMP levels (16). As suggested by the following observations, the augmentation of cAMP appeared to be only transient. Treatment of a lymphoid cell preparation with 10^{-5} M isoproterenol for 10 min at 37°C caused a fivefold increase in intracellular cAMP from 20 pmoles per 10^7 lymphocytes to more than 100 pmoles per 10^7 cells. After prolonged (60 min) treatment, however, the levels fell back again almost to those seen in the unstimulated state. These findings are directly in keeping with the time course for the inhibition of lymphocyte-mediated cytolysis seen in the presence of this drug. The augmentation of cAMP levels caused by isoproterenol was specifically reversed by addition of the beta blocker propranolol (Table 1).

2. *The Effect of cAMP and Dibutyryl cAMP*

Cyclic AMP and its dibutyryl derivative both markedly inhibited cytolysis, but 5'-AMP at comparable concentrations did not. Between 10^{-5} and 10^{-4} M, cyclic AMP and dibutyryl cAMP were found to be equally active and inhibited cytolysis between 20 and 50%. At the same concentrations, 5'-AMP caused only 5 to 7% inhibition. At higher concentrations, 5'-AMP caused definite inhibition itself, though in every case this was lower than that observed with the cyclic compounds (16).

3. *The Prostaglandins*

Of several synthetic prostaglandins tested, the E prostaglandins were found to be the most potent inhibitors of cytolysis, followed on a molar

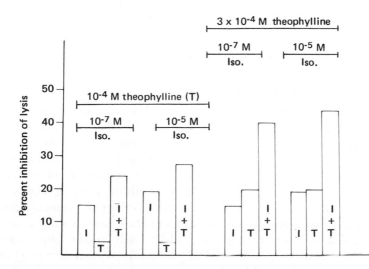

Fig. 3. The combined effects of theophylline and isoproterenol on specific lymphocyte-mediated cytolysis. Cell populations are as used in Fig. 1 [reproduced from *J. Immunol.* **108**:1526 (1972) by kind permission of the Williams and Wilkens Company, Baltimore].

Table 1. Effect of Propranolol on the Inhibition of Lymphocyte-Mediated
Cytolysis and the Augmentation of Lymphocyte cAMP Levels
Caused by Isoproterenol and by PGE_1

Drug	Percent inhibition of specific cytolysis[a]	cAMP pmoles per 10^7 Cells[b]
None	—	17
10^{-5} M propranolol	0	16
10^{-5} M isoproterenol	23.5	72
10^{-5} M isoproterenol plus 10^{-5} M propranolol	0	19
2×10^{-6} M PGE_1	45.0	45
2×10^{-6} M PGE_1 plus 10^{-5} M propranolol	45.8	54

[a]10^5 ^{51}Cr target cells + 10^7 "immune" lymphocytes; 2-hr incubation at 37°C. Inhibition in the presence of drug calculated relative to lysis observed in absence of drug. Values are means of nine assays derived from three similar experiments.
[b]10^7 C_{57}Bl/6 lymphocytes after a 10-min incubation with drug at 37°C. Each value is mean of eight determinations. The mean range of replicates was of the order of ± 10%.

basis by A_1, A_2, and B_2. Prostaglandins $F_{1\alpha}$ and $F_{2\alpha}$ were found to be only weakly inhibitory. The data from a number of experiments are summarized in Fig. 4.

The ability of the prostaglandins to increase the cyclic AMP content of lymphocyte preparations is shown in Fig. 5. As can be seen by comparison of Fig. 4 with Fig. 5 (see also Reference 21), there was shown to be virtually complete agreement between the cyclase-stimulating activity of the prostaglandins and their ability to inhibit lymphocyte-mediated cytotoxicity. The augmentation of cAMP levels caused by prostaglandin E_1 was, in contrast to that observed with isoproterenol, both long-lasting and unaffected by the presence of propranolol (Table 1).

4. The Effect of Cholera Enterotoxin

Cholera enterotoxin, which has previously been shown to be a potent adenyl cyclase stimulator in a number of systems (22), proved to be a potent inhibitor of cytolysis, with a prolonged action (23). Unlike the other inhibitors studied, however, the suppression of cytolysis was maximally expressed only if the effector cells were preincubated with cholera toxin for extended (up to 3 hr) periods prior to the addition of target cells. This is well illustrated in Fig. 6, which shows the effect of preincubation of effector cells with various concentrations of cholera toxin for 1, 2, or 3 hr prior to the addition of target cells. Additionally, unlike isoproterenol, theophylline, and the prostaglandins, the inhibitory activity of cholera enterotoxin was irreversible. Pretreatment of effector lymphoid cells with cholera toxin followed by extensive washing still resulted in the suppression of the subsequent cytolytic activity of these cells. Pretreatment of target cells with cholera toxin had no effect on their susceptibility to lysis by "sensitized" lymphocytes.

The inhibition of cytolysis observed with cholera enterotoxin was not due to nonspecific toxicity of this inhibitor. Trypan blue exclusion tests

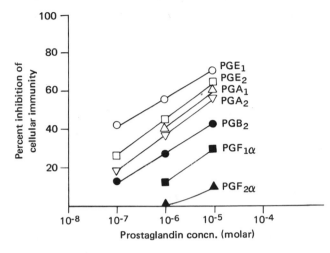

Fig. 4. Prostaglandin inhibition of lymphocyte-mediated cytolysis. Cell populations are as used in Fig. 1. Each determination was carried out in triplicate; the triplicates agreed within ±2%.

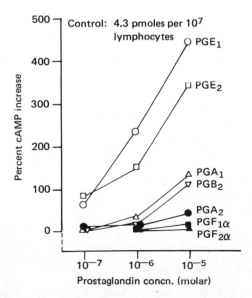

Fig. 5. The effect of prostaglandins on the intracellular cAMP level of mouse C57 Bl spleen cells. The level of cAMP in normal C57 Bl lymphocytes was 4.3 pmoles per 10^7 cells. Duplicate determinations were made of each point and agreed within ±10%.

showed no decrease in viability of the lymphoid cell preparations following prolonged (up to 24 hr) incubation with toxin.

Measurement of the cyclic AMP accumulation in mouse splenic lymphocytes stimulated by cholera enterotoxin is shown in Fig. 7. Again, the data are concordant with those derived from inhibition studies, demonstrating a considerable time lag before augmented cAMP levels were discernible.

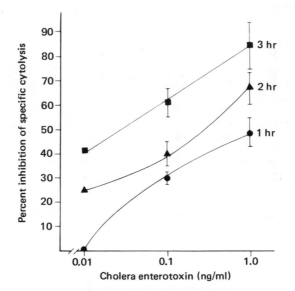

Fig. 6. Inhibition of lymphocyte-mediated cytolysis by cholera enterotoxin. Cell populations are as in Fig. 1. Lymphocytes were exposed to the various concentrations of toxin for 1, 2, or 3 hr before the addition of target cells; cytolysis was measured 4 hr later.

Table 2. Effect of Cholera Toxin on Lymphocyte-Mediated Cytolysis and Intracellular cAMP Levels[a]

Inhibitor	Percent inhibition of specific cytolysis[b]	cAMP pmoles per 10^7 lymphocytes
None	—	6.9
Toxin (1 ng/ml)	85.3	80.0
Toxoid (10 ng/ml)	0	9.2
Toxin plus Toxoid	25.0	16.0
Antitoxin (1:100)	<5	5.0
Toxin plus antitoxin	<5	17.2

[a]Effector cells were 10^7 "immune" C57 B1/6 splenic lymphocytes prepared from animals 11 days after immunization with 10^7 DBA/2 mastocytoma cells incubated for 4 hr with 10^5 ^{51}Cr target cells.
[b]Calculated relative to uninhibited controls. Lymphocytes and inhibitor were incubated together at 37°C for 3 hr prior to washing and the addition of target cells. Cytolysis was then evaluated 4 hr later. All values represent means of nine assays, derived from three similar experiments.

The augmentation of lymphocyte cyclic AMP levels and the inhibition of cytolysis caused by cholera enterotoxin were both specifically antagonized by a choleragenoid preparation and by a dog antitoxin antibody preparation (23). These results are shown in Table 2.

Following cholera enterotoxin treatment, we have recently demonstrated augmentation of lymphocyte cAMP levels for periods up to 24 hr in vitro and for 36 hr in vivo (24). These findings have led us to study the in vivo effects of cholera toxin on the development of immune responses to alloantigenic stimulation. Intraperitoneal administration of 1 μg cholera toxin to adult C57 Bl mice four days after immunization with 10^7 DBA mastocytoma cells resulted in 75 to 100% suppression in the number of cytolytically active lymphoid cells formed, and inhibited antibody formation to a similar degree (24). We are currently investigating the efficacy of cholera toxin treatment in prolongation of allograft survival. Preliminary experiments have demonstrated up to three-week prolongation of DBA/2 skin allograft survival in C57 Bl mice following administration of 2 μg cholera enterotoxin intraperitoneally on the day of grafting.

B. Relationship Between Lymphocyte cGMP Levels and Cytolytic Activity

The cytolytic activity of "immune" mouse spleen cells toward alloantigen-bearing target cells was unaffected by the presence of carbamyl choline chloride (Carbachol) over a range 10^{-14} to 10^{-3} M (Table 3). Such cholinergic stimulation also failed to stimulate cyclic cGMP levels in the

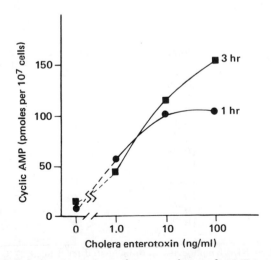

Fig. 7. Effect of cholera enterotoxin on the accumulation of cAMP in spleen cell suspensions. The lymphocytes were exposed to the noted concentration of toxin for 1 or 3 hr before the cAMP level was measured. All values are the mean of duplicate determinations differing by not more than 10%.

effector cell populations (Table 3). These findings, therefore, provide no evidence for lymphocyte cholinergic receptors functional either in the cGMP system or involved in the expression of cytolytic activity.

Incubation of effector lymphoid cell populations for short (2 to 5 min) periods with either cyclic GMP (10^{-6} to 10^{-3} M) of its dibutyryl derivative (10^{-9} to 10^{-3} M) caused marked increases in intracellular cyclic GMP levels. Such rises in cGMP concentrations, however, had no effect on the cytolytic activity of the lymphoid cells (Table 3). In further studies, to be reported in detail elsewhere (25), the concentration range of the reagents has been extended to 10^{-13} M, again with no effect on the cytolytic activity of the effector cells.

IV. Discussion

Our experiments provide convincing evidence that there is a direct correlation between the intracellular level of cyclic AMP and the immunologically specific cytolytic activity of mouse spleen cell suspensions toward allogeneic cells in vitro. The evidence may be summarized as follows:

(1) Cyclic AMP and dibutyryl cyclic AMP inhibit cytolytic activity. This effect was specific for the cyclic nucleotide, since 5'-AMP affected cytolysis only at much higher concentrations.

(2) At equivalent concentrations, drugs that augmented the cyclic

Table 3. Effect of Carbachol and of Dibutyryl cGMP on Lymphocyte-Mediated Cytolysis and on Intracellular cGMP Levels

Drug	Molar concn.	Percent change in cytolysis relative to no drug[a]	cGMP pmoles per 10^7 lymphocytes[b]
Carbachol (carbamyl	none	—	0.6
choline chloride)	10^{-5}	< 5	0.5
	10^{-7}	< 5	0.6
	10^{-9}	< 5	0.5
	10^{-11}	< 5	0.4
	10^{-12}	< 5	0.4
	10^{-14}	< 5	0.4
	none	< 5	0.5
Dibutyryl cGMP	10^{-3}	< 5	0.5 ± 0.1
	10^{-6}	< 5	600 ± 40
	10^{-9}	< 5	115 ± 35
			32 ± 9

[a]10^7 "sensitized" C57 Bl/6 lymphocytes were incubated for 4 hr at 37°C with 10^5 ^{51}Cr-DBA/2 mastocytoma cells in the presence of either Carbachol or of dibutyrl cGMP. The specific cytolysis resulting in the presence of these agents was compared with that seen in the absence of drug. Results are means of three experiments.
[b]Mean values (± SD) of eight assays from two experiments.

AMP content of spleen cell suspensions also consistently inhibited cytolysis. The specificity of this relationship is shown in three ways: (a) PGE$_1$ and PGE$_2$ both increased cyclic AMP content and inhibited cytolytic activity, while PGF$_{1\alpha}$ was inactive in both systems; (b) propranolol, a beta-adrenergic blocking agent, specifically prevented the effect of isoproterenol, a beta-adrenergic agonist, on both cyclic AMP and cytolytic activity, but did not prevent the effect of the prostaglandins in either system; (c) antagonists of cholera enterotoxin (toxoid and antibody) reversed both the augmentation of cAMP and the inhibition of cytolysis caused by the toxin.

(3) The time course of the increase of cyclic AMP content following exposure to drugs closely agreed with the time course of inhibition of cytolytic activity. The effect of isoproterenol on both cyclic AMP content and cytolytic activity was short-lived, while PGE$_1$ and cholera enterotoxin produced long-lasting effects on both systems.

(4) Theophylline, which inhibits enzymatic degradation of cyclic AMP in almost every tissue studied, caused consistent inhibition of cytolytic activity. As would be predicted from this mechanism of action, the inhibitory effect of theophylline was at least additive to that produced by isoproterenol, which elevates cyclic AMP by stimulation of adenyl cyclase.

We can find no evidence that implicates cGMP in T lymphocyte-mediated cytolysis. In contrast to the findings of Strom et al. (26), we have been unable to enhance cytolysis by cholinergic stimulation, and, moreover, we have not observed changes in cGMP following such stimulation as Strom et al. predicted would be the case (26). We can find no evidence either of active cholinergic receptors on the lymphocyte surface or of relationships between cholinergic receptors and the cGMP system. This conclusion is emphasized by the findings that the augmentation of intracellular cGMP levels by addition of dibutyryl cGMP to lymphocyte populations had no effect on the cytolytic activity of these cells.

Throughout this paper, the lymphocyte has been associated with both the cytolytic activity of the spleen cell suspensions and the changes observed in cyclic nucleotide levels following drug exposure. That the cytolytic activity is a lymphocyte function is unequivocal (27). That the changes in cAMP and cGMP levels observed in the spleen cell suspensions also reflect changes in the lymphocyte appears probable. Of the cell types present, erythrocytes lack cyclic AMP, and lymphocytes account for approximately 80% of the leucocytes. The presence of cAMP in lymphocytes and its augmentation by methylxathines and catecholamines have been established (28). It is recognized, however, that macrophages and other cell types present in the spleen populations contribute to the cyclic nucleotide measurements reported.

It is unlikely that the cAMP-active drugs prevented cytolytic activity by a protective effect on the target cells. If so, the protective effect was not exerted through an increase in cyclic AMP content; although the mastocytoma cell line employed did contain cyclic AMP (20 pmoles per 10^7 cells), the content of the nucleotide was not changed by exposure to isoproterenol (10^{-5} M) or to histamine (10^{-3} M) (results not shown). More conclusively, cholera enterotoxin treatment of effector cell populations prevented subsequent cytolytic activity, while pretreatment of target cells had no effect on their susceptibility to lysis.

Although the precise role of cAMP in lymphocyte-mediated cytolysis awaits definition, it has recently been established that the cAMP active drugs inhibit an event in the lytic pathway occurring considerably after antigen-lymphocyte binding (6). Such findings, coupled with the known effects of cAMP on the secretory capacity of many cell types, lead us tentatively to conclude that increased cAMP levels inhibit cytolysis by decreasing secretion of soluble mediator(s) from effector cells.

Acknowledgments

The author wishes to thank Dr. L. Lichtenstein for many helpful discussions, Dr. H. R. Bourne and Dr. E. Gillespie for carrying out the cAMP estimations, and Dr. M. Kurata for the cGMP assays. The excellent technical assistance of Greg Most and Anne Sobotka is gratefully acknowledged. The prostaglandins were kindly provided by Dr. J. Pike, Upjohn Company, and the cholera enterotoxin by Dr. W. Greenough, Johns Hopkins University.

Discussion of Dr. Henney's Paper

DR. GOLDBERG: Dr. Hardman and Dr. Gunter Schultz have shown a requirement for calcium in the maintenance and generation of cyclic GMP, and I believe we can show a dependence on calcium in the action of cyclic GMP. More generally, I think most people recognize that calcium is intimately involved in hormone action. If we assume that calcium and cyclic GMP work together, you are missing one of your partners in the scheme by just adding cyclic GMP. You may have gotten negative results because you didn't get the calcium into the cell. Translocating calcium and getting it into the right intracellular compartment present major problems.

DR. BOURNE: How do you get calcium into the cell?

DR. GOLDBERG: Nature has a special way of getting calcium in, and I don't know the secret.

DR. PARKER: Then I don't see how you can prove your point.

DR. HENNEY: I can't see any evidence to suggest that we have a calcium deficit either intra- or extracellularly.

DR. PARKER: Dr. Henney, you have shown very nicely when you take spleen cells from animals treated with cholera toxin and put them in your target cell system, you have a diminution in cell lysis. Do you have any evidence yet that cholera toxin can alter cellular immune responses in vivo? Can cholera toxin modify graft rejection? Is it possible that the cells might simply have been redistributed?

DR. HENNEY: More than redistribution is taking place, because we can suppress all antibody titers to the extent of about 95%. Effector cell function is suppressed to a similar extent. Whether cholera toxin prevents differentiation into effector cells or whether effector cells differentiate and are subsequently rendered inactive is currently under investigation. We have not yet consistently been able to prolong allograft rejection, although on occasions prolongation has been observed.

References

1. Brunner, K. T., Mauel, J., Cerottini, J-C., and Chapuis, B. "Quantitative Assay of the Lytic Action of Immune Lymphoid Cells on Cr^{51}-labeled Allogeneic Target Cells," *Immunol.* 14:181 (1968).
2. Freedman, L. R., Cerottini, J-C., and Brunner, K. T. "In Vivo Studies of the Role of Cytotoxic T Cells in Tumor Allograft Immunity," *J. Immunol.* 109:1371 (1972).
3. Rosenau, W. In *Cell Bound Antibodies*, (B. Amos and H. Koprowski, eds. Philadelphia, Pa.: Wistar Institute Press, p. 75 (1963).
4. Brunner, K. T., Mauel, J., Rudolf, H., and Chapuis, B. "Studies of Allograft Immunity in Mice. I. Induction, Development and in Vitro Assay of Cellular Immunity," *Immunol.* 18:499 (1970).
5. Mauel, J., Rudolf, H., Chapuis, B., and Brunner, K. T. "Studies of Allograft Immunity in Mice. II. Mechanism of Target Cell Inactivation in Vitro by Sensitized Lymphocytes," *Immunol.* 18:517 (1970).
6. Henney, C. S., and Bubbers, J. E. "Studies on the Mechanism of Lymphocyte-Mediated Cytolysis. I. The Role of Divalent Cations in Cytolysis by T Lymphocytes," *J. Immunol.* 110:63 (1973).
7. Henney, C. S., Israel, A., and Bloom, B. R. "Studies on the Mechanism of Lymphocyte-Mediated Cytolysis. IV. Comparison of Some Requirements for B and T Cell-Mediated Cytolysis," *J. Immunol.* (in press—1973).
8. Bloom, B. R., and Bennett, B. "Mechanism of a Reaction in Vitro Association with Delayed Type Hypersentivitity," *Science* 153:80 (1966).
9. Cerottini, J-C., and Brunner, K. T. "Reversible Inhibition of Lymphocyte-Mediated Cytotoxicity by Cytochalasin B," *Nature (New Biol.)* 237:272 (1972).
10. Plaut, M., Lichtenstein, L. M., and Henney, C. S. "Studies on the Mechanism of Lymphocyte-Mediated Cytolysis. III. The Role of Microfilaments and Microtubules," *J. Immunol.* 110:771 (1973).
11. Wessels, N. K., Spooner, B. S., Ash, J. F., Bradley, M. O., Luduena, M. A., Taylor, E. L., Wrenn, J. T., and Yamada, K. M. "Microfilaments in Cellular and Developmental Processes. Contractile Microfilament Machinery of Many Cell Types is Reversibly Inhibited by Cytochalasin B," *Science* 171:135 (1971).

12. Lacy, P. E., Howell, S. L., Young, D. A., and Fink, C. J. "New Hypothesis of Insulin Secretion," *Nature* (London) **219**:1177 (1968).

13. Bourne, H. R., Lichtenstein, L. M., and Melmon, K. L. "Pharmacologic Control of Allergic Histamine Release in Vitro: Evidence for an Inhibitory Role of 3',5'-Adenosine Monophosphate in Human Leucocytes," *J. Immunol.* **108**:695 (1972).

14. Weissman, G., Dukor, P., and Zurier, R. B. "Effect of Cyclic AMP on Release of Lysosomal Enzymes from Phagocytes," *Nature (New Biol.)* **231**:131 (1971).

15. Henney, C. S., and Lichtenstein, L. M. "The Role of Cyclic AMP in the Cytolytic Activity of Lymphocytes," *J. Immunol.* **107**:610 (1971).

16. Henney, C. S., Bourne, H. R., and Lichtenstein, L. M. "The Role of Cyclic 3',5'-Adenosine Monophosphate in the Cytolytic Activity of Lymphocytes," *J. Immunol.* **108**:1526 (1972).

17. Henney, C. S. "Quantitation of Cell-Mediated Immune Response. I. The Number of Cytolytically Active Mouse Lymphoid Cells Induced by Immunization with Allogeneic Mastocytoma Cells," *J. Immunol.* **107**:1558 (1971).

18. Bourne, H. R., and Melmon, K. L. "Adenyl Cyclase in Human Leukocytes: Evidence for Activation by Separate Beta Adrenergic and Prostaglandin Receptors," *J. Pharmacol. Exp. Ther.* **178**:1 (1971).

19. Gilman, A. G. "A Protein Binding Assay for Adenosine 3',5'-Cyclic Monophosphate," *Proc. Nat. Acad. Sci.* **67**:305 (1970).

20. Steiner, A. L., Parker, C. W., and Kipnis, D. M. "Radioimmunoassay for Cyclic Nucleotides. I. Preparation of Antibodies and Iodinated Cyclic Nucleotides," *J. Biol. Chem.* **247**:1106 (1972).

21. Lichtenstein, L. M., Gillespie, E., Bourne, H. R., and Henney, C. S. "The Effects of a Series of Prostaglandins on in Vitro Models of Allergic Response and Cellular Immunity," *Prostaglandins* **2**:519 (1972).

22. Pierce, N. F., Greenough, W. B., and Carpenter, C. C. J., "*Vibrio Cholerae* Enterotoxin and Its Mode of Action," *Bact. Rev.* **35**:1 (1971).

23. Lichtenstein, L. M., Henney, C. S., Gillespie, E., and Greenough, W. B. "Effects of Cholera Toxin on in Vitro Models of Immediate and Delayed Hypersensitivity: Further Evidence for the Role of cAMP," *J. Clin. Invest.* **52**:691 (1973).

24. Henney, C. S., Lichtenstein, L. M., Gillespie, E., and Rolley, R. T. "In Vivo Suppression of the Immune Response to Alloantigens by Cholera Enterotoxin" *J. Clin. Invest.* **52**:2853 (1973).

25. Henney, C. S. and Kurata, M. (manuscript in preparation).

26. Strom, T., Deisseroth, A., Morganroth, J., Carpenter, C. B., and Merrill, J. P. "Alteration of the Cytotoxic Action of Sensitized Lymphocytes by Cholinergic Agents and Activators of Adenyl Cyclase," *Proc. Nat. Acad. Sci.* **69**:2295 (1972).

27. Cerottini, J-C., Nordin, A. A., and Brunner, K. T. "Specific in Vitro Cytotoxicity of Thymus-Derived Lymphocytes Sensitized to Alloantigens," *Nature* **228**:1308 (1970).

28. Smith, J. W., Steiner, A. L., Newberry, W. M., and Parker, C. W. "Cyclic Adenosine 3',5'-Monophosphate in Human Lymphocytes. Alterations After Phytohemagglutinin Stimulation," *J. Clin. Invest.* **50**:432 (1971).

MODULATION OF CYTOTOXIC
T LYMPHOCYTE FUNCTION
BY CYCLIC 3',5'-MONONUCLEOTIDES

T. B. STROM, A. DEISSEROTH, J. MORGANROTH, C. B. CARPENTER,
and J. P. MERRILL

Following transplantation with a histoincompatible graft, a small proportion of host thymus-derived (T) lymphocytes contact donor cell structures bearing histocompatibility antigens (Stober and Gowans, 1965). This process is made possible by specific recognition units on lymphocytic surfaces (David et al., 1964). Clonal proliferation of those lymphocytes that bear receptors for donor histocompatibility antigens ensues (Gowans and McGregor, 1965; Turk, 1967). A subpopulation of antigen-sensitized T lymphocytes, which have undergone proliferation, are believed to constitute the effector lymphocyte population, which is cytopathic to donor cells (lymphocyte-mediated cytotoxicity) (Rosenau and Moon, 1961; Cerottini et al., 1970a, b). Effector T lymphocytes elaborate diffusible, biologically active agents collectively termed lymphokines (Lawrence and Landy, 1969). It has not been ascertained whether the cytotoxic lymphocytes are the same effector cells as those responsible for lymphokine secretion. The nature of the cytotoxic process is enigmatic; however, it may be mediated in part via synthesis and secretion of cytotoxic lymphokines (i.e., lymphotoxin) (Ruddle and Waksman, 1967).

Recent data (Henney and Lichtenstein, 1971; Henney et al., 1972) indicate that agents that activate lymphocyte adenylate cyclase also suppress the ability of cytotoxic T lymphocytes to injure histoincompatible target cells. Because cyclic adenosine 3',5'-monophosphate (cyclic AMP) is known to modulate the secretory processes of many biologic systems, they reasoned that cytotoxic T lymphocytes utilize a secretory process to injure target cells.

Other recent studies (Kaliner et al., 1972) have shown that cholinergic agents and 8-bromo cyclic guanosine 3',5'-monophosphate (8-bromo cyclic GMP) enhance IgE-mediated release of histamine and slow-reacting substance of anaphylaxis (SRS-A) from sensitized lung tissue. These data are consistent with the hypothesis that the physiological responses to cholinergic stimulation are mediated through increased intracellular concentrations of cyclic GMP (George et al., 1970; Ferrendelli et al., 1970). We have investigated the influence of cholinergic agents, activators of adenylate cyclase and imidazole (Strom et al., 1972; Strom et al., 1973) upon cytotoxic T lymphocytes using a 4-hr assay that quantitates the cytotoxic effect of allograft-sensitized rat splenocytes on ^{51}Cr-labeled donor thymocytes. Our studies have shown for the first time that cholinergic agents and imidazole augment lymphocyte-mediated cytotoxicity. In addition, we have confirmed and extended the observation (Henney et al., 1972) that lymphocyte adenylate cyclase stimulation results in inhibition of lymphocyte-mediated cytotoxicity.

I. Materials and Methods

Prostaglandin E_1 was kindly donated by Dr. John Pike of the Upjohn Company. We obtained purified cholera toxin (dried lot 1071) prepared under contract for the National Institute of Allergy and Infectious Diseases (NIAID) by R. A. Finkelstein, Ph.D., The University of Texas Southwestern Medical School, Dallas, Texas (Finkelstein and Lo Spalluto, 1970). Acetylcholine, carbamylcholine, atropine, imidazole, and dibutyryl cyclic AMP were purchased from the Sigma Chemical Company. Chromium-51 was purchased from Nuclear Chicago Corporation.

Male Lewis (Lewis × Brown Norway) F_1 (LBN), Brown Norway (BN), and Buffalo rats were obtained from Microbiological Associates (Bethesda, Maryland). Lewis rats were given full thickness skin grafts of standardized size from LBN rats. Recipient Lewis splenocytes and normal BN or Buffalo thymus cells were placed in RPMI-1640 medium with Hepes buffer and heat-inactivated fetal calf serum. The organs were gently expressed through a fine steel mesh, and the cells passed through a tube containing cotton. The suspensions were purified by Ficoll-Hypaque gradient separation (Böyum, 1968), resulting in suspensions consisting almost exclusively of viable lymphocytes.

Lymphocyte-mediated cytotoxicity was quantitated by a modification of Brunner's method (Brunner et al., 1970; Strom et al., 1972). Attacking cell populations consisted of sensitized (by LBN skin grafting) or unsensitized Lewis splenocytes, and target cells were ^{51}Cr-labeled thymocytes. All incubations were done in triplicate. Incubations containing 5×10^4 target cells were used as controls to determine spontaneous ^{51}Cr release. 5×10^6 sensitized or unsensitized Lewis cells were added to 5×10^4 target cells in a final medium volume of 2 ml. At the conclusion of the incubation period, the suspensions were centrifuged. The supernatants were decanted and counted for ^{51}Cr release. In studies with pharmacological agents, 0.1 ml of the agent was added to triplicate control samples containing target cells alone and to triplicate samples of mixtures of attacking and target cells. None of the pharmacological agents in the concentrations reported provoked an appreciable increase in spontaneous ^{51}Cr release from control target cells. In addition, none of the pharmacological agents caused cytotoxicity of the attacking lymphocyte populations as determined by ^{51}Cr release studies. Suspensions of 5 × 10^4 target cells were frozen and thawed in a dry ice ethanol bath three times to determine maximum releasable cell-bound chromium.

II. Results

Control incubations containing only 5 × 10^4 target cells and incubations also containing 5 × 10^6 unsensitized Lewis attacking cells released identical low amounts of ^{51}Cr, averaging 235 cpm (Table 1). The immunological specificity of this assay is shown by the fact that attacking cells injure target cells bearing the sensitizing alloantigens (BN) only and not syngeneic (Lewis) cells; furthermore, Lewis splenocytes sensitized to Brown Norway alloantigens cause only slight detectable injury to Buffalo rat thymocytes (Table 1). All sensitized spleen-attacking cells used in this study were harvested seven or eight days after LBN to Lewis grafting, and produced 32 to 54% specific lysis of Brown Norway target cells.

Table 1. Immunospecific Cytotoxic Effect of Lewis Splenic Lymphocytes Sensitized to (Lewis x Brown Norway) F$_1$ Alloantigens[a]

	% cells lysed		
	Brown Norway target cells	Lewis target cells	Buffalo target cells
Sensitized Lewis attacking cells	43 (41–46) (477–508 cpm)	0 (−2 – +2) (228–282 cpm)	2.7 (1.5–3.5) (1415–1532 cpm)
Unsensitized Lewis attacking cells	0 (−1 – +1) (220–240 cpm)	0 (−1 – +1) (236–278 cpm)	0 (0–0) (1326–1332 cpm)
Target cells alone (cpm$_c$)	(224–243 cpm)	(250–258 cpm)	(1317–1400 cpm)
Freeze-thawed target cells (cpm$_{ft}$)	(812–858 cpm)	(1460–1508 cpm)	(6933–7260 cpm)

Note: The percent of specific cell lysis of ^{51}Cr-labeled thymocytes was ascertained in mixtures containing Lewis cells sensitized to LBN or unsensitized Lewis splenocyte-attacking cells. The numbers listed in parentheses are the range of triplicate samples obtained from a single sensitized Lewis rat. On five other occasions, nonsensitized Lewis rats have failed to injure Brown Norway or Lewis target cells.

CPM$_c$ and CPM$_{ft}$ represent ^{51}Cr from target cells alone and freeze-thawed target cells. Similar symbols are used in Tables 2, 5, 6, and 7. The percent specific lysis is calculated by subtraction of the cpm for cells alone from the freeze-thaw and experimental values (see text).

[a]Table reproduced with the permission of the *Proc. Nat. Acad. Sci.* (U.S.A.).

   Macrophages do not contribute substantially to the cytotoxicity observed in this assay system. Depletion of adherent cells from the splenocyte population results in an enhanced cytotoxic effect when compared to populations in which adherent and nonadherent cells have been reconstituted (Strom et al., unpublished data).

   A decrease in the cytotoxic effect was observed when agents known to elevate intracellular cyclic AMP were added to sensitized attacking and target cell mixtures. Theophylline, an inhibitor of cyclic AMP phosphodiesterase, the intracellular enzyme active in hydrolysis of cyclic AMP to 5′-AMP, generated a mild reduction in the observed cytotoxic effect, as did prostaglandin E_1, an activator of adenylate cyclase in the lymphocyte (Table 2). When a mixture of these two agents was used, a much greater reduction in the cytotoxic effect was observed than in mixtures containing only one of these agents. Dibutyryl cyclic AMP also depressed cytotoxic activity. Prior incubation of the attacking cells with cholera toxin, another adenylate cyclase activator, caused 50% reduction in the cytotoxic effect. The cells treated with cholera toxin were washed before mixing with target cells; hence, the principal effect of this agent was on the attacking and not the target cells.

   Imidazole, an agent which depletes many cell systems of intracellular cyclic AMP (Jost and Rickenberg, 1971), did not result in altered cytotoxicity when added to mixtures of unsensitized splenocytes and target cells (Table 3). Imidazole treatment of sensitized attacking cell and target mixtures did result in augmentation of lymphocyte-mediated cytotoxicity (Table 4). The augmentation of cytotoxic effect by imidazole was proportional to the concentration present.

   With the use of the cholinergic agents, carbamylcholine, and acetylcholine, a 32% augmentation of lymphocyte-mediated cytotoxicity was obtained when the agent was added to mixtures of sensitized attacking and target cells in the concentration range of 1 to 100 pM (Fig. 1). With an increase in the concentration of carbamylcholine or acetylcholine from 0.1 nM to 0.1 mM, augmentation of cytotoxicity decreased. Similarly, if the concentration of acetylcholine or carbamylcholine was decreased from 1 pM to 0.1 fM, the augmentation of cytotoxicity seen at 1 pM disappeared. The augmentation of lymphocyte-mediated cytotoxicity induced by acetylcholine and carbamylcholine was completely abolished by atropine (Table 5). The attacking cells used for the experiments in Fig. 1 and Table 5 were pooled from two Lewis rat spleens obtained seven days after LBN skin grafting. The results of additional experiments using attacking cells obtained from six sensitized Lewis rats (one rat spleen per experiment) are shown in Tables 6 and 7. Cholinergic augmentation of cytotoxicity was consistently observed. The optimal concentration of carbamylcholine was consistently less than that of acetylcholine in producing augmentation of cytotoxicity, but the peak effect of each agent occurred at the same concentrations in each of the

Table 2. Effect of Agents That Elevate Intracellular cAMP upon Lymphocyte-Mediated Cytotoxicity[d]

Agent	% suppression of cytotoxicity[a]	Agent	% suppression of cytotoxicity
Theophylline (10 μM)	18 (17–19) (533–541 cpm) cpm_c (203–217) cpm_{ft} (1335–1362) cpm_u (601–617)	Dibutyryl cAMP (0.1 mM)	12 (10–14) (824–849 cpm) cpm_c (228–235) cpm_{ft} (1928–2002) cpm_u (917–923)
Prostaglandin E_1 (0.25 mM)	38 (35–41) (478–501 cpm) cpm_c (229–256) cpm_{ft} (1290–1356) cpm_u (637–645)	Cholera toxin[b] (4-hr incubation)	52 (50–54) (424–450 cpm) cpm_c (264–268) cpm_{ft} (1101–1198) cpm_u (590–643)
Prostaglandin E_1 (0.25 mM) plus theophylline (10 μM)	54 (51–57) (500–549 cpm) cpm_c (222–241) cpm_{ft} (1632–1714) cpm_u (851–862)	Cholera toxin (2-hr incubation)	44 (42–46) (226–255 cpm) cpm_c (176–188) cpm_{ft} (1101–1198) cpm_u (271–285)
Dibutyryl cAMP MP (1 mM)	88 (85–91) (301–342 cpm) cpm_c (233–247) cpm_{ft} (1928–2002) cpm_u (917–923)	Cholera toxin[c] (4-hr incubation)	59 (57–61) (344–355 cpm) cpm_c (220–232) cpm_{ft} (1022–1088) cpm_u (516–537)

[a]Percent suppression of cytotoxicity as described in text; numbers listed in parentheses are the range of triplicate samples. The attacking cells were a pool obtained from four sensitized animals.

[b]Attacking cells only were incubated for 3 hr with cholera toxin before these washed attacking cells were incubated with target cells for 2 and 4 hr. And 4 μg of cholera toxin was added to 3.6 ml of medium containing 48 × 10⁶ sensitized lymphocytes. Phase microscopic examination and chromium release studies showed no evidence of cytotoxicity in the treated attacking cells.

[c]Attacking cells only were incubated for 3 hr with 0.6 ml (10 μg/ml) cholera toxin added to 3.4 ml of medium containing 48 × 10⁶ sensitized lymphocytes; CPM_c and CPM_ft and defined in Table 1. CPM_u is ⁵¹Cr release in mixtures of sensitized spleen and target cells in the absence of pharmalogic agents. *Note:* Table reproduced with the permission of the *Proc. Nat. Acad. Sci.* (U.S.A.).

Table 3. Imidazole Treatment of Unsensitized Splenic
Lymphocytes

Source of attacking cells	BN target cell lysis[a]
LBN → L 7 days posttransplant	45% (43–47)
Unsensitized L cells	0% (−1 – +1)
Unsensitized L cells plus imidazole, 0.1 μM	0% (0–0)

[a]The percent of specific cell lysis of BN ^{51}Cr-labeled thymocytes was
ascertained using mixtures containing L splenocytes sensitized to
LBN or unsensitized L splenocyte cells. The numbers in parentheses
are the range of triplicate samples.

Table 4. Imidazole-Induced Augmentation of
Lymphocyte-Mediated Cytotoxicity[a]

Source of sensitized cells	Percent augmentation with imidazole concentration given			
	0.1 μM	10 nM	1 nM	0.1 nM
LBN → L 7 days posttransplant[b]	41% (39–43)	19% (18–20)	2% (2–2)	0% (0–0)
LBN → L 7 days posttransplant	37% (36–38)	18% (17–19)	0% (−1 – +1)	0% (−1 – +1)
LBN → L 7 days posttransplant	53% (50–56)	13% (20–26)	5% (4–6)	0% (−2 – +2)

[a]Percent increase in specific lysis is shown. The numbers in parentheses are the range of
triplicate samples.
[b]Pooled splenocytes from two rats were used to determine each dose-response curve.

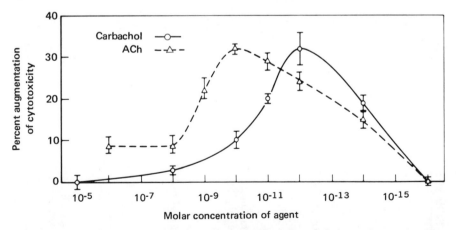

Fig. 1. Cholinergic augmentation of cytotoxicity. The ability of cholinergic agents to augment cytotoxicity when added to mixtures of attacking and target cells in optimal concentrations is depicted. The range of triplicate samples is indicated by brackets. All attacking cells were pooled from two Lewis rats sensitized with LBN skin grafts. This is the same
pool of cells used for the experiments summarized in Table 2. Carbamylcholine (−○−);
acetylcholine (−△−) [figure reproduced with the permission of the *Proc. Nat. Acad. Sci.*
(U.S.A.)].

dose-response curves. Addition of cholinergic agents to unsensitized mixtures of attacking and target cells did not result in specific cell lysis.

III. Discussion

The lymphoid tissue of a histoincompatibility grafted animal responds in part by clonal proliferation of a subpopulation of lymphocytes bearing surface recognition units for the donor transplantation antigens (Gowans and McGregor, 1965; Turk, 1967). From the proliferating cell pool is derived an effector population which includes cytotoxic ("killer") T lymphocytes, which are cytotoxic to donor cells (Cerottini et al., 1970a, b). A 4-hr assay that quantitates the cytotoxic effect of allograft-sensitized rat splenocytes on ^{51}Cr-labeled donor thymocytes has been used to study the possible role of cyclic AMP and cyclic GMP in modula-

Table 5. Effect of Cholinergic Agents on Lymphocyte-Mediated Cytotoxicity When Added Alone or with a Cholinergic Blocking Agent[a]

Agents	Concentration (pM)	% augmentation of cytotoxicity	% augmentation with agent + atropine (1 nM)
Acetylcholine	1000	22 (20–25) (812–834 cpm)	0 (0–1) (721–733 cpm) cpm_c (254–266) cpm_{ft} (1523–1599) cpm_u (717–723)
Acetylcholine	100	32 (31–33) (866–875 cpm)	0 (−2 – +2) (689–741 cpm) cpm_c (248–255) cpm_{ft} (1523–1599) cpm_u (717–723)
Carbamylcholine	10	20 (19–21) (910–935 cpm)	0 (0–1) (813–824 cpm) cpm_c (222–231) cpm_{ft} (1708–1739) cpm_u (810–816)
Carbamylcholine	1	32 (28–36) (976–1022 cpm)	3 (1–5) (829–879 cpm) cpm_c (226–233) cpm_{ft} (1708–1739) cpm_u (810–816)

Note: The range of triplicate samples is shown in parentheses. All attacking cells used in this table were pooled from two Lewis rats sensitized with LBN skin grafts. cpm_c = spontaneous ^{51}Cr release from target cells alone; cpm_{ft} = ^{51}Cr release from freeze-and-thaw-treated target cells; cpm_u = ^{51}Cr release from sensitized mixtures of spleen and target cells that are not pharmacologically treated. Table reproduced with the permission of the *Proc. Nat. Acad. Sci.* (U.S.A.).

Table 6. Cholinergic Augmentation of Cytotoxicity

Source of sensitized cells	Concentration of carbamycholine (pM)				
	1000	100	10	1	0.01
LBN → Lewis 8 days after transplant	7% (5–9) (817–835 cpm)	7% (6–8) (821–831 cpm)	19% (17–22) (880–908 cpm)	15% (14–16) (868–875 cpm)	5% (2–8) (805–832 cpm) cpm_c (231–253) cpm_{ft} (1720–1805) cpm_u (775–801)
LBN → Lewis 7 days after transplant	10% (7–13) (1002–1042 cpm)	—	—	76% (66–68) (1418–1558 cpm)	12% (11–13) (1029–1043 cpm) cpm_c (246–260) cpm_{ft} (1954–2046) cpm_u (941–981)
LBN → Lewis 8 days after transplant	—	6% (4–8) (1066–1000 cpm)	—	(28% (26–30) (1243–1275 cpm)	3% (2–4) (1050–1065 cpm) cpm_c (228–235) cpm_{ft} (1640–1688) cpm_u (1029–1037)

Note: The numbers in parentheses are the range of triplicate samples. A single rat spleen cell population was used to determine each dose-response curve. Table reproduced with the permission of the *Proc. Nat. Acad. Sci.* (U.S.A.).

Table 7. Cholinergic Augmentation of Cytoxicity

Source of sensitized cells	Concentration of acetylcholine (nM)					
	1000	10	1	0.1	0.01	0.001
LBN → Lewis 8 days after transplant	7% (4–10) (876–908 cpm)	10% (8–12) (892–920 cpm)	—	17% (15–19) (939–964 cpm)	—	8% (6–10) (884–909 cpm) cpm_c (228–241) cpm_{ft} (1703–1751) cpm_u (847–857)
LBN → Lewis 7 days after transplant	—	—	9% (8–10) (1028–1061 cpm)	22% (19–25) (1125–1167 cpm)	6% (4–8) (992–1030 cpm) cpm_c (261–267) cpm_{ft} (2103–2145) cpm_u (986–990)	—
LBN → Lewis 7 days after transplant	5% (4.5–6.5) (761–771 cpm)	—	23% (20–26) (840–870 cpm)	41% (38–44) (931–962 cpm)	17% (16–18) (819–830 cpm) cpm_c (218–235) cpm_{ft} (1482–1517) cpm_u (730–746)	—

Note: The numbers in parentheses are the range of triplicate samples. A single rat spleen cell population was used to determine each dose-response curve. Table reproduced with the permission of the *Proc. Nat. Acad. Sci.* (U.S.A.).

tion of the cytotoxic process. Other investigators, using ^{51}Cr release from histoincompatible mouse mastocytoma cells as a measure of lymphocyte-mediated cytotoxicity, have demonstrated that thymus-derived cells are the effector cells in mounting cytotoxicity (Cerottini et al., 1970a, b). B cells are not required (Goldstein et al., 1972). The role of macrophages as amplifiers of tissue injury is not certain; however, macrophages are not required in this or other (Berke et al., 1972) in vitro systems using purified lymphocyte populations. Thus, this assay is a model of effector T lymphocyte or "killer" T lymphocyte function.

Our results show clearly that agents known to elevate intracellular levels of cyclic AMP (dibutyryl cyclic AMP, prostaglandin E_1, theophylline, and cholera toxin) inhibit the cytotoxic effect of lymphocytes on cells to which they are sensitized. These data are in concert with the earlier reported observations of other investigators (Henney and Lichtenstein, 1971; Henney et al., 1972). We first reported the inhibition of lymphocyte-mediated cytotoxicity in a system in which only the sensitized lymphocytes, and not target cells, were exposed to cholera toxin, a potent and prolonged activator of adenylate cyclase (Strom et al., 1972). It is suggested, therefore, that elevation of intracellular cyclic AMP in the cytotoxic lymphocyte-containing population reduces their ability to exert damaging effects on target cells, as postulated by Henney et al. (1972).

The addition of imidazole to mixtures of attaching and target cells results in a dose-dependent augmentation of lymphocyte-mediated cytotoxicity. Imidazole has been shown to stimulate most mammalian cyclic AMP phosphodiesterase systems studied (Jost and Rickenberg, 1971). Furthermore, one study has shown that imidazole treatment inhibited cyclic GMP phosphodiesterase (O'Dea et al., 1970). Our experiment suggests that depletion of intracellular cyclic AMP results in augmented lymphocyte-mediated cytotoxicity, although an effect resulting from accumulation of cyclic GMP cannot be discounted.

Cholinergic agents produced augmentation of cytotoxicity at low optimal concentrations, with a return to base line at higher doses. Augmentation of cytotoxicity produced by cholinergic agents was abolished by atropine, a cholinergic (muscarinic) blocker. These data indicated for the first time that functional cholinergic receptors are present on lymphoid cells. In unpublished studies done in collaboration with M. Kaliner and K. F. Austen, we have observed augmentation of cytotoxicity when the 8-bromo or dibutyryl derivatives of cyclic GMP are added to attacking and target cell mixtures. They (Kaliner et al., 1972) have recently shown that human lung tissue involved in the release of immediate hypersensitivity mediators have cAMP-independent cholinergic receptors; cholinergic agonists, when given in optimal dosages remarkably similar to those used in the present study, caused increased secretion of histamine and SRS-A; treatment of lung tissue with 8-bromo cyclic GMP also resulted in increased mediator release. Cholinergic stimulation of heart (George et al., 1970; Kuo et al., 1972), brain (Ferrendelli et al., 1970; Kuo et al., 1972), and ductus deferens (Schultz et al., 1972) has been reported to result in increased intracellular concentrations of cyclic GMP. Furthermore, cholinergic agents have been noted in an unpublished study to result in accumulation of cyclic GMP in the lymphocyte (Goldberg et al., 1973). These data support the hypothesis that cyclic GMP is implicated in the mediation of events resulting from stimulation by cholinergic agonists (Lee et al., 1972) and certain events antagonistic to those promoted by cyclic AMP (Goldberg et al., 1973).

We propose that in immediate (Kaliner et al., 1973) and delayed hypersensitivity (Strom et al., 1972) systems studied in vitro, antigen-sensitized effector cell function is modulated by hormonal and neurotransmitter agents, which act by altering the intracellular levels of the cyclic 3',5'-mononucleotides, the nucleotides acting as classical second messengers. In vivo proof of killer T lymphocyte modulation by cyclic nucleotides has not yet been obtained, but this is under active investigation at present.

The subcellular and molecular events, which result in destruction of target cells by killer T lymphocytes, are incompletely understood, although contact of attacking cells with target cells is a prerequisite for the cytotoxic process (Ax et al., 1968). The activated T lymphocyte secretes a variety of biologically active substances including cytotoxic factors (Ruddle and Waksman, 1967). Elevations of intracellular cyclic AMP inhibit secretory processes of other biologic systems: platelets (Jost and Rickenberg, 1971), leukocytes (Lichtenstein and Margolis, 1968), and lung (Kaliner et al., 1972). Positive modulation (i.e., enhancement) of IgE-mediated lung secretory processes by cholinergic agonists is likely to occur. Our data strongly suggest that lymphocyte-mediated cytotoxicity, an event profoundly modulated by cyclic nucleotides, is also a secretory process.

Discussion of Dr. Strom et al.'s Paper

DR. HENNEY: If you preincubate target cells with cholinergic agents before the addition of lymphocytes and then wash them, do you still get the same effect?

DR. STROM: No, there is no cytotoxicity. In my system, the target cell is also a lymphocyte (a thymocyte). I have a built-in control for toxicity in this system. If thymocytes exposed to cholinergic drugs release excess chromium (over what control cells incubated without cholinergic agents release), the experiment is discarded. In the experiments that we have reported here and elsewhere, cholinergic agents were used at concentrations that are not cytotoxic to target cells.

I think I should comment on the importance of preincubation time on cholinergic activation of effector cells. Since our initial publication showing that the ability of T lymphocytes to mediate cytotoxicity can be altered by cholinergic agents, we have found that it is necessary to preincubate the attacking cells for 3 to 10 min prior to exposure to target cells. If the target and attacking cells are mixed prior to the addition of 8-bromo cyclic GMP or cholinergic agents, augmentation of cytotoxicity is not seen. And if the preincubation period is too long, the augmentation is not seen.

DR. HENNEY: There are recent data indicating that B lymphocytes plus antitarget cell antibody can lyse target cells. Do you know what the effector cell is in your rat target cell system? Obviously it is not as easy to delineate T and B cells in rats as it is in mice.

DR. STROM: We have unpublished data indicating that the effector cells in our system are T cells and not B cells or macrophages.

DR. HENNEY: The grafted animals used as a source of effector cells do not have agglutinating antibody?

DR. STROM: The only protein source in the medium is heat-inactivated fetal calf serum.

DR. HENNEY: But doesn't the serum from animals bearing the allograft agglutinate incompatible cells?

DR. STROM: Yes.

DR. HENNEY: So the sensitized spleen cell population will certainly contain plasma cells, which might liberate antibodies during the in vitro incubation, resulting in B cell activation.

DR. STROM: It is true that we are working with rat cells, but mouse experiments proving that the cytotoxic cells in sensitized spleens are T derived lymphocytes and that B cells do not contribute to cytotoxicity have been done many times and in many ways.

DR. HENNEY: I don't want to push too hard, but it is a general finding that antitarget cell antibody in the presence of B lymphocytes is a potent cytolytic system.

DR. STROM: I agree, However, there is no evidence that such a mechanism is operative in the Brunner-type system.

DR. HENNEY: Does anyone have data on effects of cyclic GMP on antibody secretion?

DR. STROM: No.

References

Ax, W., Malchow, H., Zeiss, I., and Fisher, H. "The Behavior of Lymphocytes in the Process of Target Cell Destruction," *Exp. Cell. Res.* **53**:108–116 (1968).

Berke, G., Sullivan, K. A., and Amos, B. "Rejection of Ascites Tumor Allografts. I. Isolation, Characterization, and in Vitro Reactivity of Peritoneal Lymphoid Effector Cells from BALB/c Mice Immune to EL4 Leukosis," *J. Exp. Med.* **135**:1334–1350 (1972).

Böyum, A. "Isolation of Leucocytes from Human Blood," *Scand. J. Clin. Lab. Invest., Suppl.* **21**:1–109 (1968).

Brunner, K. T., Manuel, J., Rudolf, H. L., and Chapuis, B. "Studies in Allograft Immunity in Mice. I. Induction, Development and in Vitro Assay of Cellular Immunity," *Immunol.* **18**:501–515 (1970).

Cerottini, J. C., Nordin, A. A., and Brunner, K. T. "In Vitro Cytotoxic Activity of Thymus Cells Sensitized to Alloantigens," *Nature* **227**:72–73 (1970a).

Cerottini, J. C., Nordin, A. A., and Brunner, K. T. "In Vitro Cytotoxic Activity of of Thymus-Derived Lymphocytes Sensitized to Alloantigens," *Nature* **228**:1308–1309 (1970b).

David, J. R., Lawrence, H. S., and Thomas, L. "The in Vitro Desensitization of Sensitive Cells by Trypsin," *J. Exp. Med.* **120**:1189–1200 (1964).

Ferrendelli, J. A., Steiner, A. L., McDougal, D. R., and Kipnis, D. M. "The Effect of Oxotremorine and Atropine on cGMP and cAMP Levels in Mouse Cerebral Cortex and Cerebellum," *Biochem. Biophys. Res. Commun.* **42**:844–849 (1970).

Finkelstein, R. J., and Lo Spalluto, J. "Production of Highly Purified Choleragen and Choleragenoid," *J. Infec. Dis.* **121**:563–572 (1970).

George, W. J., Polson, J. B., O'Toole, A. G., and Goldberg, N. D. "Elevation of Guanosine 3',5'-Cyclic Phosphate in Rat Heart After Perfusion with Acetylcholine," *Proc. Nat. Acad. Sci.* (U.S.A.) **66**:398–403 (1970).

Goldberg, N. D., Haddox, M. K., Hartle, D. K., and Hadden, J. W. In *Proc. V. International Congress Pharmacology*. Basel:Karger (1973).

Goldstein, P., Wigzell, H., Blomgren, H., and Svedmyr, E. A. J. "Cells Mediating Specific in Vitro Cytotoxicity. II. Probable Autonomy of Thymus Processed Lymphocytes (T Cells) for the Killing of Allogeneic Target Cells," *J. Exp. Med.* **135**:890–906 (1972).

Gowans, J. L., and McGregor, D. D. "The Immunological Activites of Lymphocytes," *Progr. Allerg.* **9**:1–78 (1965).

Henney, C. S., and Lichtenstein, L. M. "The Role of Cyclic AMP in the Cytolytic Activity of Lymphocytes," *J. Immunol.* **107**:610–612 (1971).

Henney, C. S., Bourne, H. E., and Lichtenstein, L. M. "The Role of Cyclic 3',5'-Adenosine Monophosphate in the Specific Cytolytic Activity of Lymphocytes," *J. Immunol.* **108**:1526–1534 (1972).

Jost, J. P., and Rickenberg, H. V. "Cyclic AMP," *Ann. Rev. Biochem.* **40**:741–744 (1971).

Kaliner, M., Orange, R. P., and Austen, K. F. "Immunological Release of Histamine and Slow Reacting Substance of Anaphylaxis from Human Lung. IV. Enhancement by Cholinergic and Alpha Adrenergic Stimulation," *J. Exp. Med.* **136**:556–567 (1972).

Kuo, J., Lee, T., Reyes, P. L., Walton, K. G., Donnelly, Jr., T. E., and Greengard, P. "Cyclic Nucleotide Dependent Protein Kinases X. An Assay Method of the Measurement of Guanosine 3',5'-Monophosphate in Various Biological Materials and a Study of Agents Regulating Its Level in Heart and Brain," *J. Biol. Chem.* **247**:16–22 (1972).

Lee, T. P., Kuo, J. F., and Greengard, P. "Role of Muscarinic Cholinergic Receptors in Regulation of Guanosine 3',5'-Cyclic Monophosphate Content in Mammalian Brain, Heart Muscle, and Intestinal Smooth Muscle," *Proc. Nat. Acad. Sci.* (U.S.A.) **69**:3287–3291 (1972).

Lawrence, H. S., and Landy, M. *Mediators of Cellular Immunity.* New York, London: Academic Press (1969).

Lichtenstein, L. M., and Margolis, K. "Histamine Release in Vitro: Inhibition by Catecholamines and Methylxanthines," *Science* 161:902–903 (1968).

O'Dea, R. F., Haddox, K. M., and Goldberg, N. D. "Kinetic Analysis of a Soluble Rat Brain Cyclic Nucleotide Phosphodiesterase," *Fed. Proc.* 29:473 (Abst.) (1970).

Rosenau, W., and Moon, H. D. "Lysis of Homologous Cells by Sensitized Lymphocytes in Tissue Culture," *J. Nat. Cancer Inst.* 27:471–477 (1961).

Ruddle, N. H., and Waksman, B. H. "Cytotoxic Effect of Lymphocyte-Antigen Interaction in Delayed Hypersensitivity," *Science* 157:1060–1062 (1967).

Schultz, G., Hardman, J. G., Davis, J. W., Schultz, K., and Sutherland, E. W. "Determination of Cyclic GMP by a New Enzymatic Method," *Fed. Proc.* 31:440 (Abst.) (1972).

Strober, S., and Gowans, J. L. "The Role of Lymphocytes in the Sensitization of Rats to Renal Homografts," *J. Exp. Med.* 122:347–360 (1965).

Strom, T. B., Deisseroth, A., Morganroth, J., Carpenter, C. B., and Merrill, J. P. "Alteration of the Cytotoxic Action of Sensitized Lymphocytes by Cholinergic Agents and Activators of Adenylate Cyclase," *Proc. Nat. Acad. Sci.* (U.S.A.) 69:2995–2999 (1972).

Strom, T. B., Deisseroth, A., Morganroth, J., Carpenter, C. B., and Merrill, J. P. "Regulatory Role of the Cyclic Nucleotides in Alloimmune Lymphocyte-Mediated Cytotoxicity. Effect of Imidazole," *Transplant. Proc.* (in press — 1973).

Turk, J. L. "Action of Lymphocytes in Transplantation," *J. Clin. Pathol., Suppl.* 20:423–429 (1967).

CYCLIC GMP: VESTIGE OR ANOTHER INTRACELLULAR MESSENGER?

JOEL G. HARDMAN, GÜNTER SCHULTZ, and EARL W. SUTHERLAND

I. Introduction

For some time, the possibility has been entertained that cyclic AMP might not be unique as an intracellular mediator of effects of hormones and other alterations in the extracellular environment. The discovery of cyclic GMP in rat urine by Ashman et al. (1963) has stimulated a substantial amount of investigation of this nucleotide as a potential regulatory agent. Progress in elucidating the role(s) of cyclic GMP has been slow. This has been due in part to the difficulty of methodology available for studying the endogenous nucleotide and in part to the inherently tedious task of finding an unknown function of a substance that is present in very small amounts.

There should be tolerance of individuals who have entertained the possibility that cyclic GMP might be a vestige in mammalian systems. Because of the high endogenous levels of cyclic GMP in tissues of lower phyla (Ishikawa et al., 1969), the extraordinarily high activity of guanylate cyclase in the sperm of an invertebrate (Gray et al., 1970), the relative abundance of cyclic GMP-stimulable protein kinase in arthropoda (Kuo and Greengard, 1970; Kuo et al., 1971), and the elusiveness of firm evidence for a role for cyclic GMP in higher organisms, such a possibility has not always seemed farfetched. Today, however, the serious consideration of cyclic GMP only as a vestige probably would be motivated more from despair than from keen intuition. A rapidly growing list of agents that have been observed to elevate cyclic GMP levels in intact mammalian cells may be providing important leads to understanding the function of this nucleotide (see articles by Dr. Hadden and Dr. Goldberg in this book). At least such observations show that the metabolism of cyclic GMP is dynamic and can be rapidly controlled by certain alterations in the extracellular environment.

This research is supported in part by NIH grants GM-16811, HL-08332, HL-13996, and AM 07462, and by the Deutsche Forschungsgemeinschaft.

223

What follows is a brief review of some of the main features of the biosynthetic and degradative systems for cyclic GMP, some comments about effects of exogenous cyclic GMP, and a discussion of a possible role for Ca^{2+} in the control of cyclic GMP levels in intact cells.

II. Biosynthesis of Cyclic GMP

The formation of cyclic GMP is analogous to that of cyclic AMP in that cyclization involves the alpha phosphate group of the corresponding nucleoside triphosphate precursor (Rall and Sutherland, 1962; Hardman, et al., 1971). The possibility that some adenylate cyclase systems may use GTP as an alternate substrate cannot be entirely ruled out, but it is clear that the formation of cyclic GMP in most cell-free systems is catalyzed by a separate enzyme system, guanylate cyclase.

In homogenates of most mammalian tissues examined so far, guanylate cyclase appears in both the soluble and particulate fractions (Hardman and Sutherland, 1969; White and Aurbach, 1969; Schultz et al., 1969; Böhme, 1970) in contrast to adenylate cyclase, which is confined to particulate fractions. A soluble guanylate cyclase from bovine lung has been purified several hundred-fold (White et al., 1972).

Although guanylate cyclase appears largely in soluble fractions of several tissues including lung, liver, and kidney, the use of detergents has revealed that substantial amounts of the enzyme are present in particulate fractions of these tissues (Hardman et al., 1973). Furthermore, virtually all of the detectable guanylate cyclase activity in the small intestine (Ishikawa et al., 1969) and isolated fat cells (Neumann et al., 1972) of the rat is associated with particulate fractions, and the richest source (by 2 or 3 orders of magnitude) of guanylate cyclase yet reported, the sperm of the sea urchin, *Strogylocentrotus purpuratus*, also contains a totally particulate activity (Gray et al., 1970). What structures this particulate activity is associated with, and how its properties compare with those of the soluble enzyme are unknown at present. Some guanylate cyclase activity has been detected in partially purified plasma membrane preparations from rat kidney (Schultz et al., 1972c).

Guanylate cyclase exhibits a strong dependence of Mn^{2+} for maximum activity (Hardman and Sutherland, 1969; White and Aurbach, 1969; Schultz et al., 1969). Mg^{2+} will substitute only very ineffectively for Mn^{2+}, and Ca^{2+} is even less effective than Mg^{2+} when used alone. Ca^{2+} can increase guanylate cyclase substantially in the presence of low concentrations of Mn^{2+}, an effect that is partially reproduced by Ba^{2+} but not by Mg^{2+} (Hardman et al., 1971, 1973).

Several hormones that increase the activity of adenylate cyclase in broken cell systems have had no effect on guanylate cyclase activity. These include epinephrine, glucagon (Hardman and Sutherland, 1969), ACTH (McMillan et al., 1971), neurohypophyseal hormones, parathyroid hormone, and calcitonin (Schultz et al., 1972c). Whether these negative results reflect differences in the abilities of the two cyclase systems to interact directly with hormones or differences in the structural or functional association of the cyclases with hormone receptors is unknown.

III. Degradation of Cyclic GMP

The only established pathway for the degradation of cyclic GMP is, as for cyclic AMP, conversion to the corresponding 5'-monophosphate by phosphodiesterases that are found in both soluble and particulate fractions of tissues. In general, both cyclic nucleotides can serve as substrates for a single enzyme, but phosphodiesterases from different sources can vary markedly in their relative affinities for the two nucleotides. The soluble enzyme prepared from bovine heart by Butcher and Sutherland (1962) has a much higher apparent affinity for cyclic GMP than for cyclic AMP, but a particulate fraction from bovine heart contains a phosphodiesterase with a higher apparent affinity for cyclic AMP than for cyclic GMP (Beavo et al., 1970). In contrast, most, if not all, of the soluble phosphodiesterase present in rat erythrocytes has a much higher apparent affinity for cyclic AMP than for cyclic GMP (Patterson et al., 1971; Hardman et al., 1971). In a very important series of studies, Thompson and Appleman (1971 a, b) have shown that two peaks of phosphodiesterase activity can be resolved by gel filtration of extracts of several rat tissues. A higher-molecular-weight peak (about 400,000) resembles the preparation of Butcher and Sutherland in having a higher affinity for cyclic GMP than for cyclic AMP; a lower-molecular-weight peak (about 200,000) has, like the enzyme in rat erythrocytes, a much higher affinity for cyclic AMP than for cyclic GMP and exhibits negative cooperative behavior (Russell et al., 1972). Russell et al. (1973) have recently reported the existence in liver of a small amount of phosphodiesterase with an apparently absolute specificity for cyclic GMP.

Which form of phosphodiesterase plays the most important role in controlling cyclic nucleotide levels in intact cells is not known. It might be assumed that the form of the enzyme with the highest relative affinity for one of the cyclic nucleotide would be the form predominantly involved in degrading that nucleotide. There is no real evidence to indicate that this is the case, however.

As would be expected, cyclic AMP and cyclic GMP can competitively inhibit the hydrolysis of each other where they serve as substrates for the same enzyme (Rosen, 1970; Beavo et al., 1970; Murad et al., 1970; O'Dea et al., 1971; Thompson and Appleman, 1971a). Under some conditions, however, cyclic GMP may inhibit cyclic AMP hydrolysis noncompetitively (Murad et al., 1970; Thompson and Appleman, 1971 a). Furthermore, depending on the concentrations of the two nucleotides, cyclic GMP can actually stimulate the hydrolysis of cyclic AMP by crude phosphodiesterase preparations from several tissues (Beavo et al., 1970, 1971; Franks and MacManus, 1971; Klotz and Stock, 1972; Russell et al., 1973). Cyclic AMP does not appear to be able to stimulate cyclic GMP hydrolysis under any conditions so far tested. Whether or not the two cyclic nucleotides interact under physiological circumstances to modify the hydrolysis of each other in either a positive or negative direction is unknown.

IV. Effects of Cyclic GMP

A. *Cell-Free Systems*

It has been assumed that cyclic GMP would be found to play a regulatory role perhaps analogous to that of cyclic AMP and perhaps involving responses of cells to alterations in their extracellular environement. However, to date little can be said in more than a speculative way about the function of cyclic GMP.

Cyclic GMP was found very early to be less than 1% as effective as cyclic AMP in activating the liver phosphorylase system (Rall and Sutherland, 1962). More recently, cyclic GMP has been found to be similarly ineffective in activating cAMP-dependent protein kinases from a variety of tissues (e.g., Schlender et al., 1969; Corbin and Krebs, 1969; Kuo and Greengard, 1969; Gill and Garren, 1970; Reimann et al., 1971). Because of these findings and the observations that cyclic GMP levels in tissues are usually only about 10% or less of the cyclic AMP levels (Goldberg et al., 1969; Ishikawa et al, 1969; Steiner et al., 1972), it seems very unlikely that cyclic GMP serves physiologically as a substitute for cyclic AMP to regulate cAMP-dependent protein kinases. Protein kinases with specificity for cyclic GMP have been found in arthropoda (Kuo and Greengard, 1970; Kuo et al., 1971) and in mammalian cerebellum (Hofmann and Sold, 1972), but the natural protein substrates and functions of these enzymes are unknown.

Cyclic GMP competes with cyclic AMP for binding sites on a receptor protein from *E. coli* and, in doing so, prevents the stimulation by cyclic AMP of messenger RNA synthesis in cell-free systems (Zubay et al., 1970; Emmer et al., 1970). Goldberg (1972) has reported the ability of low concentrations of cyclic GMP to inhibit the binding of cyclic AMP to protein kinase from skeletal muscle. Other evidence suggesting antagonistic or dualistic roles for cyclic GMP and cyclic AMP will be discussed by Dr. Goldberg in another chapter in this book.

B. *Intact Cell Systems*

Space does not permit a review of all reported effects of exogenous cyclic GMP or its derivatives on intact cells. In some cases, such effects have appeared superficially to be functionally similar to or, on the other hand, functionally opposite to effects of cyclic AMP; in other instances, cyclic GMP or its derivatives have been without effect on cells that were sensitive to cyclic AMP or its derivatives. It is probably fair to say that attempts to elucidate the biological role of cyclic GMP by applying the nucleotide to intact cells have not been invariably successful.

Part of the problem with this approach is the difficulty of distinguishing effects of exogenous cyclic GMP that could reflect physiological functions of the nucleotide form effects that are perhaps pharmacologically interesting, but physiologically meaningless. Effects of the latter type may occur for a number of reasons. The unexpected potency of exogenous cyclic GMP in reproducing effects of cyclic AMP in the perfused rat liver appears to be a result of the ability of cyclic GMP to accumulate in the liver in concentrations high enough to activate cAMP-dependent protein kinases (Exton et al., 1971). Furthermore, since cyclic GMP may either stimulate or inhibit the hydrolysis of cyclic AMP, it may produce effects in some intact cells that are secondary to alterations in endogenous cyclic AMP levels. For example, high concentrations of exogenous cyclic GMP have been shown to elevate the cyclic AMP levels and stimulate lipolysis in isolated fat cells (Murad et al., 1970), and it has been suggested that effects of low concentrations of exogenous cyclic GMP to inhibit the mitogenic action of cyclic AMP on thymocytes may result from a stimulation by cyclic GMP of cyclic AMP hydrolysis (Franks and MacManus, 1971).

Cyclic GMP could conceivably produce nonspecific effects in intact cells by altering the activity of an enzyme that is physiologically regulated by other guanine nucleotides. Such an effect might be easily recognized as nonspecific with the use of related nucleotides as controls (a minimum precaution that is omitted with disturbing frequency). However, such an effect might give the appearance of being quite specific if, as is likely, cyclic GMP has a greater ability to enter intact cells than do 5′-GMP, GDP, and GTP.

While the subject of related nucleotides is at hand, a word of caution should be offered concerning the interpretation of results obtained with acylated and brominated derivatives of cyclic GMP. By analogy with acylated derivatives of cyclic AMP, acylated derivatives of cyclic GMP may be more effective than the parent compound in producing some physiological responses. However, until some physiologically meaningful responses of cyclic GMP can be defined, there will be no way of knowing whether or not derivatives of cyclic GMP are in fact accurate models of the parent compound in biochemical systems.

Finally, the possibility of indirect effects of cyclic GMP in systems containing multiple cell types should be recognized. Exogenous cyclic GMP might bring about the release of a substance from one cell type that would in turn produce the detected response in another cell type. Observations of Puglisi et al. (1971), for example, raise the possibility that cyclic GMP can stimulate smooth muscle contraction in some cases by releasing endogenous acetylcholine.

These points have been raised, not to create a negative attitude toward the use of exogenous nucleotides, but to try to encourage careful planning of such studies and thoughtful interpretation of results obtained with them.

V. Importance of Ca^{2+} for the Control of Cyclic GMP Levels

The observation by George et al. (1970) that acetylcholine elevated cyclic GMP levels in the perfused rat heart appears to have been a very significant discovery. Subsequently, several other observations have directed attention to the possible involvement of cyclic GMP in responses to acetylcholine. Cyclic GMP in rat brain rose in response to the in vivo administration of oxotremorine (Ferrendelli et al., 1971), an agent thought to act by increasing the level of free acetylcholine, and cyclic GMP in heart and brain slices rose in response to the in vitro application of acetylcholine (Kuo et al., 1972). The ability of cholinergic agents to elevate cyclic GMP in smooth muscle has been demonstrated in the ductus deferens of the rat (Schultz et al., 1972 a, b, 1973 a, b) and in guinea pig intestinal smooth muscle (Lee et al., 1972; Schultz et al., 1973a). Cyclic GMP in dog thyroid and rat liver slices was elevated by acetylcholine (Yamashita and Field, 1972), and the level of the nucleotide was elevated in slices of rat submaxillary glands by methacholine (Schultz et al., 1973a).

Repeated unsuccessful attempts in our laboratory to demonstrate effects of cholinergic agents on guanylate cyclase activity in cell-free systems of several tissues led us to consider the possibility that effects of cholinergic agents to increase cyclic GMP levels in tissues might occur secondarily, perhaps to alterations in membrane permeability to ions (Schultz et al., 1973a,b). The acetylcholine-induced increase in cyclic GMP levels in segments of the ductus deferens of the rat in vitro was unchanged when NaCl was replaced in the incubation medium by an equiosmolar amount of sucrose. However, the omission of Ca^{2+} from the medium produced a striking effect (Schultz et al., 1973a, b). Not only did acetylcholine fail to elevate cyclic GMP, but the basal level of cyclic GMP was reduced to about 20% of normal when tissues were incubated in Ca^{2+}-free medium for 30 min. Furthermore, the readdition of Ca^{2+}-free medium caused a return toward normal of both the basal cyclic GMP levels and the response to acetylcholine. Cyclic AMP levels were unaltered by Ca^{2+} omission or readdition.

Similar observations were made with slices of a noncontractile tissue, the rat submaxillary gland (Schultz et al., 1973a), indicating that the

Ca^{2+}-dependent elevation of cyclic GMP levels by cholinergic agents does not require a contractile response of the involved tissue.

The apparent requirement for external Ca^{2+} in the effect of acetylcholine on cyclic GMP levels suggested that other substances that could increase the cytoplasmic Ca^{2+} concentration might also elevate cyclic GMP levels. To test this possibility, a depolarizing contraction-producing concentration of K^+ was tested for its effect on cyclic GMP levels in the ductus deferens (Schultz et al., 1973 a, b). Exposure of the tissue to 125 mM KCl for 3 min resulted in a twofold increase in cyclic GMP levels. Moreover, as was seen with acetylcholine, KCl was incapable of elevating cyclic GMP in Ca^{2+}-free medium. The high K^+ concentration, like acetylcholine, had no effect on cyclic AMP levels in either the presence or the absence of Ca^{2+}.

In collaboration with Dr. Leon Hurwitz, we have also studied responses of the longitudinal smooth muscle of the guinea pig small intestine. Cholinergic agents and high K^+ concentrations increased cyclic GMP levels in the intestinal smooth muscle and required Ca^{2+} to produce their effects (Schultz et al., 1973a). Furthermore, histamine, another agent that produces contraction of the intestinal muscle, also increased cyclic GMP levels in that tissue and required external Ca^{2+} to do so. An effect of histamine to elevate cyclic GMP levels has also been observed in rat brain (Kuo et al., 1972) and in pig coronary arteries (Sutherland et al., 1973).

Thus, cholinergic agents, histamine, and depolarizing concentrations of K^+ appear to require Ca^{2+} to elevate cyclic GMP in smooth muscle and probably in other cell types. This requirement for Ca^{2+} and the striking alteration in basal cyclic GMP levels in the ductus deferens caused by the omission and readdition of Ca^{2+} strongly suggest that the elevation in cyclic GMP caused by these and perhaps other agents is secondary to increased cytoplasmic Ca^{2+} concentrations. Indeed, most, if not all, agents reported to increase tissue or plasma levels of cyclic GMP are thought to elevate cytoplasmic Ca^{2+} concentrations. These agents include, in addition to the substances already discussed, alpha adrenergic agents (Ball et al., 1972), NaF (Yamashita and Field, 1972; Whitney and Sutherland, 1972), and phytohemagglutinin (Hadden et al., 1972; Whitney and Sutherland, 1972).

The possibility that Ca^{2+} could be involved in the regulation of cyclic GMP metabolism was suggested by earlier observations that urine levels of cyclic GMP in man rose in response to Ca^{2+} infusions (Kaminsky et al., 1970). More recently, effects of Ca^{2+} on the activity of guanylate cyclase in cell-free systems have been observed (Böhme, 1970; Hardman et al., 1971, 1973). Although guanylate cyclase from rat lung is virtually inactive in the presence of Ca^{2+} alone, it undergoes a marked increase in activity when exposed to Ca^{2+} *and* small amounts of Mn^{2+}. Dr. David Garbers in our laboratory has obtained evidence that Ca^{2+}:GTP binds to sea urchin sperm guanylate cyclase, but it is unclear whether the binding site is a catalytic or an allosteric one (Garbers and Hardman, 1973). Whether or not these observations will eventually lead to an explanation of the apparent effects of Ca^{2+} on cyclic GMP levels in intact tissues remains to be seen.

VI. Concluding Remarks

There has been no intention to suggest here that cyclic GMP functions, in general, as a mediator or co-mediator of some effects of agents such as acetylcholine and alpha adrenergic agents that probably increase cytoplasmic Ca^{2+} concentrations. This could turn out to be the case, but present evidence makes such a suggestion seem premature. Cyclic GMP could participate in the overall responses of certain cells to agents that increase cytoplasmic Ca^{2+} without necessarily acting as an intracellular mediator of effects of those agents. It is conceivable, for example, that cyclic GMP could participate in a negative feedback system to restore cytoplasmic Ca^{2+} to prestimulus levels. Cyclic AMP, which may be able to accomplish the same end result (see, for example, Andersson and Nilsson, 1972), could be unable to participate in such a system if Ca^{2+} can inhibit adenylate cylase in intact cells as it can in some cell-free systems (e.g., Drummond and Duncan, 1970; Jakobs et al., 1972).

If one adopts the view that cyclic GMP does participate in the mediation of effects of agents such as acetylcholine and alpha adrenergic agents, then several observations make it necessary to recognize the likelihood that cyclic GMP and cyclic AMP can, in at least some situations, elicit the same cellular response, although perhaps by acting on different systems. For example, acetylcholine can produce effects similar to those of TSH and cyclic AMP in the thyroid (Pastan et al., 1961; Altman et al., 1966). In some species both alpha and beta adrenergic agents can induce similar hepatic metabolic effects (Sherline et al., 1972; Haylett and Jenkinson, 1972) and can relax gastrointestinal smooth muscle (Andersson and Mohme-Lundholm, 1970; Bowman and Hall, 1970). Furthermore, both acetylcholine and beta adrenergic agents appear to enhance, whereas alpha adrenergic agents inhibit insulin secretion (Mayhew et al., 1969; Burr et al., 1971), and both acetylcholine and beta adrenergic agents produce apparently similar saluretic effects in the dog kidney (Pearson and Williams, 1968). Although phosphodiesterase inhibitors produce in smooth muscle effects that are generally ascribed to cyclic AMP, these agents elevate the levels of both cyclic AMP and cyclic GMP in affected tissues (Schultz et al., 1972a, b, 1973 a, b; Sutherland et al., 1973). All of these observations are difficult to fit within the framework of a concept that requires cyclic GMP and cyclic AMP to play invariably antagonistic regulatory roles.

A number of biochemical processes have evolved that have specific requirements for either ATP or GTP. The systems specifically requiring one of the triphosphates are not necessarily, or even in general, functionally antagonistic or complimentary to those requiring the other triphosphate. By analogy, there may have evolved biochemical systems that are regulated specifically by either cyclic AMP or cyclic GMP. Systems regulated by one cyclic nucleotide could be functionally antagonistic or functionally complimentary to systems regulated by the other one. On the other hand, functional relationships between systems regulated by cyclic AMP or cyclic GMP may be no more systematic than are relationships between systems that specifically require ATP or GTP.

Discussion of Dr. Hardman et al.'s Paper

DR. AUSTEN: You showed that calcium was required in the environment to achieve changes in depolarization and cyclic GMP concentrations with acetylcholine. Can you achieve the same result if you use manganese in place of the calcium?

DR. HARDMAN: We have not done experiments with manganese as yet.

DR. STROM: Do you have any comment on the observations of Greengard's laboratory suggesting that the muscarinic, but not nicotinic, effects of the cholinergic agents are mediated through the GMP system?

DR. HARDMAN: I think that fits perfectly well with the data in other laboratories: Dr. Goldberg's observations in the perfused heart and the effects my colleagues and I have found in smooth muscle are blocked by atopine and are presumably muscarinic.

References

Altman, M., Oka, H., and Field, J. B. "Effect of TSH, Acetylcholine, Epinephrine, Serotonin and Synkavite on ^{32}P Incorporation into Phospholipids in Dog Thyroid Slices," *Biochim. Biophys. Acta* **116**:586 (1966).

Andersson, R., and Mohme-Lundholm, E. "Metabolic Actions in Intestinal Smooth Muscle Associated with Relaxation Mediated by Adrenergic Alpha and Beta Receptors," *Acta Physiol. Scand.* **79**:244 (1970).

Andersson, R., and Nilsson, K. "Cyclic AMP and Calcium in Relaxation in Intestinal Smooth Muscle," *Nature (New Biol.)* **238**:119 (1972).

Ashman, D. F., Lipton, R., Melicow, M. M., and Price, T. D. "Isolation of Adenosine 3',5'-Monophosphate and Guanosine 3',5'-Monophosphate from Rat Urine," *Biochem. Biophys. Res. Commun.* **11**:330 (1963).

Ball, J. H., Kaminsky, N. I., Hardman, J. G., Broadus, A. E., Sutherland, E. W., and Liddle, G. W. "Effects of Catecholamines and Adrenergic-Blocking Agents on Plasma and Urinary Cyclic Nucleotides in Man," *J. Clin. Invest.* **51**:2124 (1972).

Beavo, J. A., Hardman, J. G., and Sutherland, E. W. "Hydrolysis of Cyclic Guanosine and Adenosine 3',5'-Monophosphates by Rat and Bovine Tissues," *J. Biol. Chem.* **245**:5649 (1970).

Beavo, J. A., Hardman, J. G., and Sutherland, E. W., "Stimulation of Adenosine 3′,5′-Monophosphate Hydrolysis by Guanosine 3′,5′-Monophosphate," *J. Biol. Chem.* **245**:5649 (1971).

Böhme, E. "Gyanyl cyclase. Bildung von Guanosin-3′,5′-monophosphat in Niere und anderen Geweben der Ratte," *Eur. J. Biochem.* **14**:422 (1970).

Bowman, W. C., and Hall, M. T. "Inhibition of Rabbit Intestine Mediated by Alpha and Beta Receptors," *Brit. J. Pharmacol.* **38**:399 (1970).

Burr, I. M., Taft, H. P., Stauffacher, W., and Renold, A. E. "On the Role of Cyclic AMP in Insulin Release. II. Dynamic Aspects and Relations to Adrenergic Receptors in the Perifused Pancreas of Adult Rats," *Ann. N.Y. Acad. Sci.* **185**:245 (1971).

Butcher, R. W., and Sutherland, E. W. "Adenosine 3′,5′-Phosphate in Biological Materials. I. Purification and Properties of Cyclic 3′,5′-Nucleotide Phosphodiesterase and Use of This Enzyme to Characterize Adenosine 3′,5′-Phosphate in Human Urine," *J. Biol. Chem.* **237**:1244 (1962).

Corbin, J. D., and Krebs, E. G. "A Cyclic AMP-Stimulated Protein Kinase in Adipose Tissue," *Biochem. Biophys. Res. Commun.* **36**:328 (1969).

Drummond, G. I., and Duncan, L. "Adenyl Cyclase in Cardiac Tissue," *J. Biol. Chem.* **245**:976 (1970).

Emmer, M., de Crombrugghe, B., Pastan, I., and Perlman, R. "Cyclic AMP Receptor Protein of E. coli: Its Role in the Synthesis of Inducible Enzymes," *Proc. Nat. Acad. Sci.* **66**:480 (1970).

Exton, J. H., Hardman, J. G., Williams, T. F., Sutherland, E. W., and Park, C. R. "Effects of Guanosine 3′,5′-Monophosphate on the Perfused Rat Liver," *J. Biol. Chem.* **246**:2658 (1971).

Ferrendelli, J. A., Steiner, A. L., McDougal, D. B., and Kipnis, D. M. "The Effect of Oxotremorine and Atropine on cGMP and cAMP Levels in Mouse Cerebral Cortex and Cerebellum," *Biochem. Biophys. Res. Commun.* **41**:1061 (1970).

Franks, D. J., and MacManus, J. P. "Cyclic GMP Stimulation and Inhibition of Cyclic AMP Phosphodiesterase from Thymic Lymphocytes," *Biochem. Biophys. Res. Commun.* **42**:844 (1971).

Garbers, D. L., and Hardman, J. G. "Protective Effects of Metals and Nucleotides on Guanylate Cyclase from Sea Urchin Sperm," *Fed. Proc.* (in press—1973).

George, W. J., Polson, J. B., O'Toole, A. G., and Goldberg, N. D. "Elevation of Guanosine 3′,5′-Cyclic Phosphate in Rat Heart After Perfusion with Acetylcholine," *Proc. Nat. Acad. Sci.* **66**:398 (1970).

Gill, G. N., and Garren, L. D. "A Cyclic 3′,5′-Adenosine Monophosphate-Dependent Protein Kinase from the Adrenal Cortex: Comparison with a Cyclic AMP Binding Protein," *Biochem. Biophys. Res. Commun.* **39**:335 (1970).

Goldberg, N. D. "Possible Biological Role(s) of Cyclic 3′,5′-Guanosine Monophosphate (Cyclic GMP)." Abstracts of Invited Presentations. Fifth International Congress on Pharmacology, San Francisco, 1972, p. 229.

Goldberg, N. D., Dietz, S. B., and O'Toole, A. G. "Cyclic Guanosine 3′,5′-Monophosphate in Mammalian Tissues and Urine," *J. Biol. Chem.* **244**:4458 (1969).

Gray, J. P., Hardman, J. G., Bibring, T., and Sutherland, E. W. "High Guanyl

Cyclase Activity in Sea Urchin Spermatozoa," *Fed. Proc.* **29**:608 (abst.) (1970).

Hadden, J. W., Hadden, E. M., Haddox, M. K., and Goldberg, N. D. "Guanosine 3',5'-Cyclic Monophosphate: A Possible Intracellular Mediator of Mitogenic Influences in Lymphocytes," *Proc. Nat. Acad. Sci.* **69**:3024 (1972).

Hardman, J. G., Beavo, J. A., Gray, J. P., Chrisman, T. D., Patterson, W. D., and Sutherland, E. W. "The Formation and Metabolism of Cyclic GMP," *Ann. N.Y. Acad. Sci.* **185**:27 (1971).

Hardman, J. G., Chrisman, T. D., Gray, J. P., Suddath, J. L., and Sutherland, E. W. "Guanylate Cyclase: Alteration of Apparent Subcellular Distribution and Activity by Detergents and Cations," *Proc. Fifth International Congress on Pharmacology* (San Francisco), Vol. 5. Basel: S. Karger (in press—1973).

Hardman, J. G., and Sutherland, E. W. "Guanyl Cyclase: An Enzyme Catalyzing the Formation of Guanosine 3',5' Monophosphate from Guanosine Triphosphate," *J. Biol. Chem.* **244**:6363 (1969).

Haylett, D. G., and Jenkinson, D. H. "The Receptors Concerned in the Actions of Catecholamines of Glucose Release, Membrane Potential and Ion Movements in Guinea Pig Liver," *J. Physiol.* **225**:751 (1972).

Hofmann, F., and Sold, G. "A Protein Kinase Activity from Rat Cerebellum Stimulated by Guanosine 3',5'-Monophosphate," *Biochem. Biophys. Res. Commun.* **49**:1100 (1972).

Ishikawa, E., Ishikawa, S., Davis, J. W., and Sutherland, E. W. "Determination of Cyclic GMP in Tissues and of Guanyl Cyclase in Rat Intestine," *J. Biol. Chem.* **244**:6371 (1969).

Jakobs, K. H., Schultz, K., and Schultz, G. "Hemmung von Adenyl-Cyclase-Präparationen aus der Rattenniere durch Calciumionen und verschiedene Diuretica," *Naunyn-Schmiedebergs Arch. Pharmakol.* **273**:248 (1972).

Kaminsky, N. I., Broadus, A. E., Hardman, J. G., Jones, D. J., Ball, J. H., Sutherland, E. W., and Liddle G. W. "Effects of Parathyroid Hormone on Plasma and Urinary Adenosine 3',5'-Monophosphate in Man," *J. Clin. Invest.* **49**2387 (1970).

Klotz, U., and Stock, K. "Influence of Cyclic Guanosine 3',5'-Monophosphate on the Enzymatic Hydrolysis of Adenosine 3',5'-Monosphate," *Naunyn-Schmiedebergs Arch. Pharmakol.* **274**:54 (1972).

Kuo, J. F., and Greengard, P. "Cyclic Nucleotide-Dependent Protein Kinases. IV. Widespread Occurence of Adenosine 3',5'-Monophosphate-Dependent Protein Kinase in Various Tissues and Phyla of the Animal Kingdom," *Proc. Nat. Acad. Sci.* **64**:1349 (1969).

Kuo, J. F., and Greengard, P. "Cyclic Nucleotide Dependent Protein Kinases. VI. Isolation and Partial Purification of Protein Kinase Activated by Guanosine 3',5'-Monophosphate," *J. Biol. Chem.* **245**:2493 (1970).

Kuo, J. F., Lee, T. P., Reyes, P. L., Walton, K. G., Donnelly, Jr., T. E., and Greengard, P. "Cyclic Nucleotide-Dependent Protein Kinases. X. An Assay Method for the Measurement of Guanosine 3',5'-Monophosphate in Various Biological Materials and a Study of Agents Regulating Its Levels in Heart and Brain," *J. Biol. Chem.* **247**:16 (1972).

Kuo, J. F., Wyatt, G. R., and Greengard, P. "Cyclic Nucleotide-Dependent Protein Kinases. IX. Partial Purification and Some Properties of Guanosine 3',5'-Monophosphate-Dependent and Adenosine 3',5'-Monophosphate-Dependent Protein Kinases from Various Tissues and Species of Arthropoda," *J. Biol. Chem.* **246**:7159 (1971).

Lee, T. P., Kuo, J. F., and Greengard, P. "Role of Muscarinic Cholinergic Receptors in Regulation of Guanosine 3',5'-Cyclic Monophosphate Content in Mammalian Brain, Heart Muscle, and Intestinal Smooth Muscle," *Proc. Nat. Acad. Sci.* **69**:3287 (1972).

Mayhew, D. A., Wright, P. H., and Ashmore, J. "Regulation of Insulin Secretion," *Pharm. Rev.* **21**:183 (1969).

McMillan, B. H., Ney, R. L., and Schorr, I. "Guanyl Cyclase Activity in Normal Adrenals and a Corticosterone Producing Adrenal Cancer of the Rat," *Endocrinology* **89**:281 (1971).

Murad, F., Manganiello, V., and Vaughan, M. "Effects of Guanosine 3',5'-Monophosphate on Glycerol Production and Accumulation of Adenosine 3',5'-Monophosphate by Fat Cells," *J. Biol. Chem.* **245**:3352 (1970).

Neumann, E. P., Mocikat, S., and Böhme, E. "Formation of Guanosine 3',5'-Monophosphate in Isolated Fat Cells of Rats," *Naunyn-Schmiedebergs Arch. Pharmakol., Suppl.* **274**:R82 (1972).

O'Dea, R. F., Haddox, M. K., and Goldberg, N. D. "Interaction with Phosphodiesterase of Free and Kinase-Complexed Cyclic Adenosine 3',5'-Monophosphate," *J. Biol. Chem.* **246**:6183 (1971).

Pastan, I., Herring, B., Johnson, P., and Field, J. B. "Stimulation in Vitro of Glucose Oxidation in Thyroid by Acetylcholine," *J. Biol. Chem.* **236**:340 (1961).

Patterson, W. D., Hardman, J. G., and Sutherland, E. W. "Metabolism of Cyclic Nucleotides in Rat Blood," *Fed. Proc.* **30**:220 (abst.) (1971).

Pearson, J. E., and Williams, R. L. "Analysis of Direct Renal Actions of Alpha and Beta Adrenergic Stimulation upon Sodium Excretion Compared to Aceytlcholine," *Brit. J. Pharmacol.* **33**:223 (1968).

Puglisi, L., Berti, F., and Paoletti, R. "Antagonism of Dibutyryl-Guo-3',5'-P and Atropine on Stomach Smooth Muscle Contraction," *Experientia* **27**:1187 (1971).

Rall, T. W., and Sutherland, E. W. "Adenyl Cyclase. II. The Enzymatically Catalyzed Formation of Adenosine 3',5'-Phosphate and Inorganic Pyrophosphate from Adenosine Triphosphate," *J. Biol. Chem.* **237**:1228 (1962).

Reimann, E. M., Walsh, D. A., and Krebs, E. G. "Purification and Properties of Rabbit: Skeletal Muscle Adenosine 3',5'-Monophosphate-Dependent Protein Kinases," *J. Biol. Chem.* **246**:1986 (1971).

Rosen, O. M. "Preparation and Properties of a Cyclic 3',5'-Nucleotide Phosphodiesterase Isolated from Frog Erythrocytes," *Arch. Biochem. Biophys.* **137**:435 (1970).

Russell, T. R., Terasaki, W. L., and Appleman, M. M. *J. Biol. Chem.* (in press— 1973).

Russell, T. R., Thompson, W. J., Schneider, F. W., and Appleman, M. M." 3',5'-Cyclic Adenosine Monophosphate Phosphodiesterase: Negative Cooperativity," *Proc. Nat. Acad. Sci.* **69**:1791 (1972).

Schlender, K. K. Wei, S. H., and Villar-Palasi, C. "UDP-Glucose: Glycogen alpha-4-Glucosyl Transferase I Kinase Activity of Purified Muscle Protein Kinase. Cyclic Nucleotide Specificity," *Biochim. Biophys. Acta* 191:272 (1969).

Schultz, G., Böhme, E., and Munske, K. "Guanyl Cyclase. Determination of Enzyme Activity," *Life Sci.* 8:1323 (1969).

Schultz, G., Hardman, J. G., Davis, J. W., Schultz, K., and Sutherland, E. W. "Determination of Cyclic GMP by a New Enzymatic Method," *Fed. Proc.* 31440 (abst.) (1972a).

Schultz, G., Hardman, J. G., Hurwitz, L., and Sutherland, E. W. "Importance of Calcium for the Control of Cyclic GMP Levels," *Fed. Proc.* (in press—1973a).

Schultz, G., Hardman, J. G., Schultz, K., Baird, C. E., Parks, M. A., Davis, J. W., and Sutherland, E. W. "Cyclic GMP and Cyclic AMP in Ductus Deferens and Submaxillary Gland on the Rat." Abstracts of Volunteer Papers. Fifth International Congress on Pharmacology, p. 206 (1972b).

Schultz, G., Hardman, J. G., and Sutherland, E. W. "Cyclic Nucleotides and Smooth Muscle Function." In *Asthma: Immunopharmacology and Treatment,* F. Austen and L. Lichtenstein, eds. New York: Academic Press (in press—1973b).

Schultz, G., Jakobs, K. H., Böhme, E., and Schultz, K. "Einfluss verschiedener Hormone auf die Bildung von Adenosin-3':5'-monophosphat und Guanosin-3':5'-monophosphat durch particuläre Präparationen aus der Rattenniere," *Eur. J. Biochem.* 24:520 (1972c).

Sherline, P., Lynch, A., and Glinsmann, W. H. "Cyclic AMP and Adrenergic Receptor Control of Rat Liver Glycogen Metabolism," *Endocrinology* 91:680 (1972).

Steiner, A. L., Pagliara, A. S., Chase, L. R., and Kipnis, D. M. "Radioimmunoassay for Cyclic Nucleotides. II. Adenosine 3',5'-Monophosphate and Guanosine 3',5'-Monophosphate in Mammalian Tissues and Body Fluids," *J. Biol. Chem.* 247:1114 (1972).

Sutherland, C. A., Schultz, G., Hardman, J. G., and Sutherland, E. W. "Effects of Vasoactive Agents on Cyclic Nucleotide Levels in Pig Coronary Arteries," *Fed. Proc.* (in press—1973).

Thompson, W. J., and Appleman, M. M. "Multiple Cyclic Nucleotide Phosphodiesterase Activities from Rat Brain," *Biochemistry* 10:311 (1971a).

Thompson, W. J., and Appleman, M. M. "Characterization of Cyclic Nucleotide Phosphodiesterases of Rat Tissues," *J. Biol. Chem.* 246:3145 (1971b).

White, A. A., and Aurbach, G. D. "Detection of Guanyl Cyclase in Mammalian Tissues," *Biochim. Biophys. Acta* 191:686 (1969).

White, A. A., Northup, S. J., Zenser, T. V. "Guanyl Cyclase: Partial Purification and Assay." In *Methods in Cyclic Nucleotide Research (Methods in Molecular Biology,)* Vol. 3, M. Chasin, ed. New York: Marcel Dekker, p. 125 (1972).

Whitney, R. B., and Sutherland, R. M. "Enhanced Uptake of Calcium by Transforming Lymphocytes," *Cell. Immunol.* 5:137 (1972).

Yamashita, K., and Field, J. B. "Elevation of Cyclic Guanosine 3',5'-

Monophosphate Levels in Dog Thyroid Slices Caused by Acetylcholine and Sodium Fluoride," *J. Biol. Chem.* **247**:7062 (1972).

Zubay, G., Schwartz, D., and Beckwith, J. "Mechanism of Activation of Catabolite-Sensitive Genes: a Positive Control System," *Proc. Nat. Acad. Sci.* **66**:104 (1970).

CYCLIC GMP AND CYCLIC AMP IN LYMPHOCYTE METABOLISM AND PROLIFERATION

JOHN W. HADDEN, ELGA HADDEN, and NELSON D. GOLDBERG

Two major questions concerning the role of cyclic adenosine 3′,5′-monophosphate (cyclic AMP) in lymphocyte metabolism and function have been focused upon in recent years:

(1) What role does cyclic AMP play in mediating hormonal influences to modify lymphocyte proliferation and function?

(2) Do mitogens initiate proliferation by using a second-messenger system like cyclic AMP?

The many contributions to this book attest to the variety of answers obtained to these questions. It is obviously beyond the scope of this chapter to review these many observations; so we would like to focus on a few observations we have made concerning cyclic AMP in the lymphocyte and add some recent observations concerning another cyclic nucleotide, guanosine cyclic 3′,5′-monophosphate (cyclic GMP), in these cells. The evidence, in sum, points to important roles for both nucleotides in the lymphocyte and implicates cyclic GMP as an intracellular messenger mediating both cholinergic modulation and mitogenic stimulation.

Our original experiments in this area were designed to investigate the effects of adrenergic agents on phytohemagglutinin (PHA)-induced lymphocyte proliferation. Although early experiments with relatively high concentrations of these agents showed them to be uniformly inhibitory (Hadden et al., 1970), subsequent pulse exposure experiments with nano- to micromolar concentrations demonstrated selective effects

This work was made possible through the contributions of many, including Gerald Meetz, Mari Haddox, and Robert A. Good, and through the support of the American Heart Association (in cooperation with the Minnesota Heart Association), The National Cystic Fibrosis Research Foundation, The Minnesota Medical Foundation, and the U. S. Public Health Service (AI-00292, AI-00798, NB-05979, HE-07939).

of alpha and beta adrenergic agents. Table 1 shows a summary of these data and indicates that while alpha adrenergic stimulation (an event presumed to diminish cellular cyclic AMP) enhances DNA synthesis in PHA-treated lymphocytes, beta adrenergic stimulation (an event shown to increase cellular cyclic AMP) inhibits. These data indicate that when cellular cyclic AMP is raised concomitant with PHA stimulation, inhibition of mitogen-induced DNA synthesis results. In an effort to understand how the stimulation of cellular cyclic AMP might be antagonistic to the action of PHA, we examined the influence of adrenergic stimulation on lymphocyte utilization of glucose (Hadden et al., 1971a). During the first 24 hr in culture, lymphocytes stimulated by PHA actively take up glucose from the medium and accumulate glycogen, an event usually suppressed by cyclic AMP in other tissues. During this period, alpha adrenergic stimulation was found to promote, and beta adrenergic stimulation to inhibit glucose uptake and glycogen accumulation (Table 1). The latter action is consistent with a well-recognized effect of cellular cyclic AMP to promote glycogenolysis through its action on the protein kinase-phosphorylase activation cascade. The interpretation that subsequent DNA synthesis might be impaired in part through the activation of glycogenolysis during a period of glycogen accumulation has been suggested; however, many other intracellular actions of cyclic AMP might be implicated to account for this antagonism of PHA action by cyclic AMP. Perhaps the more pertinent information derived from these observations is that metabolic events initiated by PHA are consistent with the *absence* of cyclic AMP action and that events promoted by its presence are inconsistent with mitogenic action.

When studying the effects of adrenergic stimulation over a more restricted period in actively dividing, PHA-stimulated lymphocytes, we found that during the peak period of DNA synthesis—68 to 72 hr of culture (Hadden et al., 1971b)—alpha adrenergic stimulation produced an effect similar to that seen with stimulation at the onset of culture. On the other hand, beta adrenergic stimulation during this period results in *enhanced* DNA synthesis and enhanced glycogenolysis. It is during this

Table 1. Effects of Adrenergic Stimulation on the Metabolism of PHA-Stimulated Lymphocytes[a]

Stimulation	Onset of transformation			Peak of transformation		
	DNA synthesis	Glucose uptake	Glycogen content	DNA synthesis	Glucose uptake	Glycogen content
Alpha adrenergic	↑22%	↑15%	↑22%	↑18%	↑31%	↑28%
Beta adrenergic	↓13%	↓10%	↓22%	↓15%	↓42%	↓23%

[a]Data expressed in percent relative to the control. Alpha versus beta effects where divergent are all significant at $p < 0.01$.

period of the cell cycle that glycogenolysis has been shown to be actively in process, an event consistent with the presence of cyclic AMP action. Stimulation of glycogenolysis during this period is now an event in concert with the progression to mitosis.

These data indicated to us that alteration of cellular cyclic AMP by hormonal stimulation may indeed play a role in modulating metabolic events associated with lymphocyte transformation and that the effect of cyclic AMP changes from inhibitory to stimulatory as lymphocytes progress from "G_0" or early "G_1" to late "G_1" and "S" phase of the cell cycle. This latter interpretation is similar to that made in the fibroblast proliferation system (Willingham et al., 1972) and is consistent with the observations that in rapidly dividing stem cells of the cortex of the thymus, cyclic AMP may stimulate the progression of cells from late "G_1" through to mitosis (MacManus and Whitfield, 1973).

Because cholinergic agents like alpha adrenergic agents produce "cAMP-antagonistic" effects in a number of tissues, the influence of acetylcholine on lymphocyte metabolism was studied (Hadden et al., 1973). Cells were continuously incubated with the isotopic precursors during a period of 24 hr, a time calculated to include a single cycle of replication in cells stimulated to divide by PHA. Table 2 shows a summary of the results of these experiments. Acetylcholine (10^{-6} M) induces in resting lymphocytes significant increases in both RNA and protein synthesis and augments PHA-induced increases in RNA, protein, and DNA synthesis. These effects of acetylcholine are mimicked by Carbachol and methylcholine at the same and somewhat lower concentrations (10^{-9} M range) and are blocked by the muscarinic blocking agent atropine (10^{-5} M). Since it had been shown by Goldberg and his collaborators (George et al., 1970) that cholinergic action in rat myocardium is associated with cellular accumulation of cyclic GMP, the influence of acetylcholine on the levels of this nucleotide in lymphocytes was also examined. We found that acetylcholine (10^{-6} M) does, indeed, stimulate cyclic GMP accumulation in lymphocytes. Increases of about 300% were detected at 10 min after cholinergic stimulation, and by 20 min, the levels returned to those of the control (data not shown).

The T lymphocyte is known to synthesize and to secrete a collection of nonglobulin proteins (mol. wt. 20,000 to 40,000 range) termed

Table 2. Effect of Acetylcholine on the Metabolism of Resting and PHA-Stimulated Lymphocytes

Agent	RNA synthesis (%)	Protein synthesis (%)	DNA synthesis (%)
Acetylcholine[a]	↑15	↑34	↑0
Acetylcholine[a] + PHA	↑38	↑32	↑12

[a]The concentration used was 10^{-6} M.

lymphokines. The most thoroughly studied of these factors is the macrophage migration inhibitory factor (MIF). The importance of this factor, above others, lies in its presumed effects to produce macrophage activation and thereby promote an amplification of the expression of the efferent limb of cellular immunity.

In our studies of the effects of acetylcholine to influence MIF production, we employed lymph node lymphocytes from guinea pigs. The animals were immunized with bovine gamma globulin (BGG) in complete Freunds 14 to 17 days previously; the lymphocytes obtained were incubated (20×10^6/ml) for 24 hr with and without antigen and in the presence and absence of acetylcholine (10^{-6} M) in media devoid of serum. Following incubation, the supernatant fractions were collected and the addition of 0.1 unit of acetylcholinesterase (Sigma Chemical Co., St. Louis, Missouri) insured the elimination of acetylcholine from the supernatant fractions. BGG was added back to the controls not stimulated with antigen. The supernatant fractions including appropriate controls were assayed on migration of induced peritoneal exudate cells as described (Leu et al., 1972).

Table 3 shows that acetylcholine significantly augments the production of MIF by antigen-stimulated lymph node lymphocytes. Based on our observations concerning the influence of MIF dose and the migration response (Leu et al., 1972), we can estimate that the difference in MIF concentration (units) in the supernatants represents a fourfold increase due to acetylcholine. Acetylcholine, acetylcholinesterase, or their combination had no observable effect on the migration response in the absence of antigen. We conclude from these observations that the T lymphocyte is sensitive to cholinergic stimulation, and such stimulation modulates mitogen-induced proliferation and antigen-induced lymphokine production, as well as the modulation of lymphocytotoxicity described by Strom et al. (1972). The fact that acetylcholine was found to stimulate cyclic GMP accumulation in lymphocytes indicates that all of these cholinergic effects may be linked to an intracellular action of cyclic GMP.

Our observations taken collectively indicate to us that the peripheral blood, PHA-sensitive T lymphocyte, is sensitive to both cholinergic and adrenergic modulation and that cyclic GMP can be related to the

Table. 3 Effect of Acetylcholine on MIF Production by Sensitized Guinea Pig Lymph Node Cells Stimulated by Antigen (BGG)

Agent	% migration compared with control	MIF in units
Reconstituted control	100	0
Antigen stimulated	82	1.5
Antigen stimulated + acetylcholine (10^{-6} M)	59	6.0

cholinergic, while cyclic AMP can be related to the beta adrenergic actions.

The second question raised earlier concerning the mechanism of mitogen action was of interest to us because many of the features of PHA action suggest that this mitogen, like a number of hormones, appears to find its final site of action at the plasma membrane and that the signal to divide is of the trigger type such that once the lymphocyte is induced to divide, multiple divisions follow without apparent requirement for further stimulation of PHA (Hadden et al., 1972). Two observations relevant to this point of view are that pulse exposure of cells to PHA for 30 min followed by 30 min of trypsinization (5%) or incubation with *N*-acetylgalactosamine (to reverse binding) and then by multiple washings does not interfere with the PHA-induced proliferation signal and produces proliferation equal or greater in magnitude to incubation with PHA continuously (Hadden—unpublished data). Furthermore, PHA bound to sepharose beads and therefore unable to enter the cell has been shown to have an action similar to that of the native mitogen (Greaves and Bauminger, 1972).

In contrast to the foregoing view of cyclic AMP as simply a mediator of hormonal influences in the lymphocyte are the observations of Mac-Manus and Whitfield (1973) that implicate cyclic AMP as mitogenic in cells of the thymus and those of Smith et al. (1971) that indicate that under certain circumstances, PHA can induce increases in lymphocyte cyclic AMP levels. Because many observations in a variety of tissues have indicated an antimitogenic role for cyclic AMP, we initiated studies to help clarify the role that cyclic nucleotides might play in expressing the intracellular action of the mitogenic signal sequence initiated by PHA. Recent evidence implicating cyclic GMP in the mediation of events antagonistic to those promoted by cyclic AMP (Goldberg et al., 1973a) prompted us to consider this nucleotide, as well, as a potential candidate for participation in this sequence.

The initial studies (Hadden et al., 1972) were conducted with three mitogenic preparations:

(1) *PHA-M* (Difco Laboratories, Detroit, Michigan), a highly agglutinating form of phytomitogen derived from the red kidney bean (*Phaseolus vulgaris*).

(2) *PHA-MR69* (Wallace Diagnostic Reagents, Research Triangle Park, N. C.), a form of PHA with agglutinating potential 1/100th that of PHA-M.

(3) *Concanavalin A* (Con A), a form of phytomitogen with mimimal agglutinating activity derived from the jack bean (*Concanavalin ensiformis*).

Each of the mitogens was comparable in its mitogenic potential at maximal mitogenic cocentration. Optimal doses were employed in both

the nucleotide assays and in the agglutination assay. The results summarized in Table 4 show that PHA-M, the only preparation that was
highly agglutinating, was also the only preparation found to increase
cyclic AMP levels (< 1-fold). We, therefore, attribute the effect of PHA-
M to increase lymphocyte cyclic AMP levels to nonmitogenic, perhaps
agglutinating, characteristics of the preparation.

The important observation indicated in Table 4 is that each of the
mitogenic preparations produced by 20 min a 10- to 50-fold increase in
cellular cyclic GMP concentration. The time course of this increase for
each was similar in that the effect was apparent as early as 5 min and rose
progressively to 20 min. The magnitude of these increases is several
times greater than that seen with acetylcholine.

We initially questioned if the difference in the magnitude of the
cyclic GMP rise in concentration was sufficient to separate a mitogenic
influence from the nonmitogenic cholinergic influence. Since calcium is
well recognized as a requisite for both proliferation and cholinergic
stimulation in several tissues, we investigated the influence of PHA and
acetylcholine on calcium[45] (Ca^{45}) flux in lymphocytes. Consistent with
the observations of Allwood et al. (1971), we observed that PHA induces
Ca^{45} entry into lymphocytes as early as 20 min following exposure. In addition, we observed that acetylcholine produces a similar effect.
Although these observations are consistent with the view that the
generation and action of cyclic GMP may be intrinsically related to
calcium (Schultz et al., 1972; Goldberg, 1973b; Whitfield and MacManus,
1972), they do not serve to separate the mitogenic from the cholinergic
influence.

Our observations outlined in the ensuing chapter (Goldberg et al.,
1973c) do offer insight into this distinction. The mitogenic influences of
phorbol myristate acetate and insulin in mouse fibroblasts are associated
with discreet increases in cellular cyclic GMP levels of 10- to 40-fold,
and the levels of cyclic GMP achieved in these cells correlate closely
with the degree of proliferation induced. These observations lend support to the view that the magnitude of the increase in cellular cyclic
GMP concentration is a major determinant in the type of cellular

Table 4. Effects of the Phytomitogens on the Concentration of Cyclic
Nucleotides and the Agglutination of Lymphocytes

Agent	Cyclic nucleotide content[a]		Agglutination index
	Cyclic AMP	Cyclic GMP	
Control	0.57	0.12	0
PHA-M (250 μg/ml)	0.97	1.52	++++
PHA-MR69 (3 μg/ml)	0.57	2.10	±
Con A (25 μg/ml)	0.59	5.80	±

[a]Given in pmoles per 10^6 cells at 20 min.

response produced (i.e., mitogenic or hormonal). In addition, that a decline of cellular cyclic AMP has been observed in the fibroblast in relation to mitogenic action, but not in the lymphocyte, implies that although a decline in cellular cyclic AMP may be facilatory in the induction of division, it is not a requisite.

In an effort to implicate cyclic GMP in a more direct way in the initiation of events leading to cell replication, we have studied the effects of cyclic GMP on the metabolism of isolated lymphocyte nuclei. Pogo et al. (1966) have observed that within minutes of stimulation of intact lymphocytes by PHA, acetylation of arginine-rich histones occurs followed within 30 min by nuclear RNA synthesis.

We (Hadden et al., 1973) have observed that cyclic GMP in the presence of calcium stimulates the incorporation of H^3 uridine into RNA of isolated lymphocyte nuclei (Fig. 1). This stimulation of RNA synthesis occurs in the absence of theophylline, however, the effect is magnified (> 2-fold) by its presence. The effect of cGMP to increase RNA synthesis is linear for 30 min and is optimal with concentrations of cyclic GMP between 10^{-11} and 10^{-10} M. Cyclic AMP has a similar effect, which occurs at concentrations between 10^{-7} and 10^{-6} M. Neither cyclic nucleotide in these concentrations has an effect on RNA synthesis of intact peripheral blood lymphocytes. We interpret these data to indicate that extremely low concentrations of cyclic GMP in a discreet concentration range have an action to stimulate nuclear RNA synthesis, an event

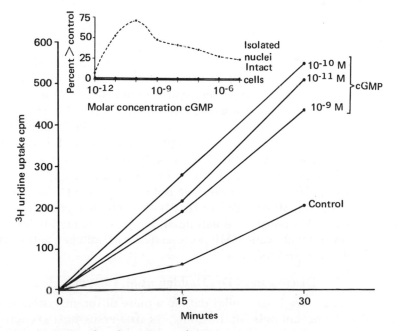

Fig. 1. Cyclic GMP and nuclear RNA synthesis.

known to occur when mitogenesis is induced in the intact cell. The observation that cyclic AMP has an effect to stimulate nuclear RNA synthesis is consistent with the observations of Jost and Sahib (1971) in liver and may relate to the effect of cyclic AMP in lysine-rich histone modification described by Langan (1970). How cyclic GMP may be involved in promoting this metabolic event is under study.

The present evidence indicates that the action of mitogens to induce clonal proliferation in lymphocytes is associated with striking increases in lymphocyte levels of cyclic GMP without effect on cellular cyclic AMP. The magnitude of these increases in association with the observation that corresponding concentrations of cyclic GMP stimulate RNA synthesis in isolated lymphocyte nuclei makes compelling the view that cyclic GMP is involved as an intracellular mediator in triggering cell division.

Summary

The following observations concerning the roles that cyclic nucleotides play in modifying lymphocyte metabolism and proliferation are presented: *cyclic AMP—*beta adrenergically induced increases in cellular cyclic AMP lead to an inhibition of PHA-induced DNA synthesis in lymphocytes when the adrenergic stimulation occurs concomitant with mitogenic stimulation; however, such increases in cyclic AMP induced in PHA-stimulated lymphocytes actively undergoing DNA synthesis enhance the proliferative process. The beta adrenergically induced alterations are associated with decreases in glucose uptake and glycogen content of lymphocytes. Purified PHA and Con A have no effect on lymphocyte cyclic AMP levels. Cyclic AMP (10^{-7} to 10^{-6} M) has an effect to stimulate RNA synthesis in isolated lymphocyte nuclei. *Cyclic GMP—* Cholinergically induced increases in cellular cyclic GMP are associated with stimulation of RNA and protein synthesis in resting lymphocytes, of RNA, protein, and DNA synthesis in PHA-stimulated lymphocytes, and of secretion of MIF by antigen-stimulated lymphocytes. Purified PHA and Con A produce striking increases in lymphocyte cyclic GMP levels. Both mitogenic and cholinergic stimulations of lymphocytes are associated with early changes in Ca^{45} flux. In the presence of calcium, cGMP (10^{-11} to 10^{-10} M) stimulates RNA synthesis in isolated lymphocyte nuclei. The accumulated observations indicate that both cyclic nucleotides are important in the hormonal modulation of lymphocyte metabolism and proliferation and that cyclic GMP is associated with mitogen action to induce lymphocyte proliferation.

Discussion of Dr. Hadden et al.'s Paper

DR. STROM: We have similar data in a quite different system utilizing cells from animals undergoing graft-versus-host reactions. Lymphocytes from these animals have a high blastogenic rate and also release leukemia viruses. The blastogenic response is modulated by prostaglandins, dibutyryl cyclic AMP, and cholinergic agents. We think both the proliferative and effector cell responses of lymphocytes are modulated by cholinergic agents.

DR. WEISSMANN: In referring to increased RNA synthesis by isolated nuclei, will you tell us what the experimental system was?

DR. HADDEN: We measured the incorporation of tritiated uridine into acid-insoluble material, sensitive to the action of RNAase. We have not examined the class of RNA stimulated.

DR. MACMANUS: If you stimulate cells with PHA in the presence of EGTA, do you still see rises in cyclic GMP?

DR. HADDEN: That is an important question. Unfortunately, we have not done the experiment. Of course, if you add EGTA to phytohemagglutinin-stimulated lymphocytes at the onset of culture, it prevents lymphocyte proliferation.

References

Allwood, G., Asherson, G. L., Davey, M. Jean, and Goodford, P. J. "The Early Uptake of Radioactive Calcium by Human Lymphocytes Treated with Phytohemagglutinin." *Immunology*, 21:500-516 (1971).

George, W. J., Polson, J. B., O'Toole, A. G., and Goldberg, N. D. "Elevation of Guanosine 3′,5′-Cyclic Phosphate in Rat Heart After Perfusion with Acetylcholine." *Proc. Nat. Acad. Sci.*, 66:398-403 (1970).

Goldberg, N., Haddox, M., Hartle, D., and Hadden, J. "The Biological Role of Cyclic 3′,5′ Guanosine Monophosphate. *Proc. V International Congress Pharmacology.*" Basel: Karger (in press) (1973a).

Goldberg, N. D., Haddox, M. R., and O'Dea, R. F. *Advances in Cyclic Nucleotide Research: Cyclic GMP*, vol. 3. New York: Raven Press (in press) (1973b).

Goldberg, N. D., Haddox, M. K., Lopez, C., Estensen, R., and Hadden, J. W. "Biological Dualism Between Cyclic GMP and Cyclic AMP in Cell Proliferation and Other Cellular Processes." (In this volume.)

Hadden, J. W., Hadden, E. M., and Middleton, E., Jr. "Lymphocyte Blast Transformation. I. Demonstration of Adrenergic Receptors in Human Peripheral Lymphocytes." *Cellular Immunology* 1:583-595 (1970).

Hadden, J. W., Hadden, E. M., Middleton, E., Jr., and Good, R. A. "Lymphocyte Blast Transformation. II. The Mechanism of Action of Alpha Adrenergic Receptor Effects." *International Archives of Allergy and Applied Immunology* 40:526-539 (1971a).

Hadden, J. W., Hadden, E. M., and Good, R. A. "Adrenergic Mechanisms in Human Lymphocyte Metabolism." *Biochem. Biophys. Acta* 237:339-347 (1971b).

Hadden, J. W., Hadden, E. M., Haddox, M. K., and Goldberg, N. D. "Guanosine 3′:5′ Cyclic Monophosphate: A Possible Intracellular Mediator of Mitogenic Influences in Lymphocytes." *Proc. Nat. Acad. Sci.* 69:3024-3027 (1972).

Hadden, J. W., Hadden, E. M., Meetz, G., Good, R. A., Haddox, M. K., and Goldberg, N. D. "Guanosine Cyclic 3′5′ Monophosphate: A Cholinergic

Mediator in Lymphocytes Modulating Cellular and Nuclear Metabolism." *Proc. Nat. Acad. Sci.* (In preparation.)

Jost, J. P., and Sahib, M. K. "Role of Cyclic Adenosine 3′,5′-Monophosphate in the Induction of Hepatic Enzymes." *J. Biol. Chem.* **246**:1623-1629 (1971).

Langan, T. A. "Phosphorylation of Histones in Vivo Under the Control of Cyclic AMP and Hormones." In *Role of Cyclic AMP in Cell Function.* New York: Raven Press (1970).

Leu, R. W., Eddleston, A. L. W. F., Hadden, J. W., and Good, R. A. "Mechanism of Action of Migration Inhibitory Factor (MIF)." *J. Exp. Med.* **136**:589-603 (1972).

MacManus, J. P., and Whitfield, J. F. "Control of Normal Cell Proliferation (in Vivo and Vitro) by cAMP and Several Agents That Use This Cyclic Nucleotide as Their Mediator." (In this volume.)

Pogo, B. G. T., Allfrey, V. G., and Mirsky, A. E. "RNA Synthesis and Histone Acetylation During the Course of Gene Activation in Lymphocytes." *Proc. Nat. Acad. Sci.* **55**:805-812 (1966).

Schultz, G., Hardman, J. G., Davis, J. W., Schultz, K., and Sutherland, E. W. "Cyclic GMP and Cyclic AMP in Ductus Deferens and Submaxillary Gland of the Rat." *Fifth International Congress of Pharmacology: Abstracts of Volunteer Papers,* p. 206 (1972).

Smith, J. W., Steiner, A. L., Newberry, W. M., Jr., and Parker, C. W. "Cyclic Adenosine 3′, 5′-Monophosphate in Human Lymphocytes. Alterations After Phytohemagglutinin Stimulation." *J. Clin. Invest.* **50**:432-448 (1971).

Strom, T. B., Deisseroth, A., Morganroth, J., Carpenter, C. B., and Merrill, J. P. "Alteration of the Cytotoxic Action of Sensitized Lymphocytes by Cholinergic Agents and Activators of Adenylate Cyclase." *Proc. Nat. Acad. Sci.* **69**:2995-2999 (1972).

Whitfield, J. F., and MacManus, J. P. "Calcium-Mediated Effects of Cyclic GMP on the Stimulation of Thymocyte Proliferation by Prostaglandin E." *Proc. Soc. Exp. Biol. and Med.* **139**:818-824 (1972).

Willingham, M. C., Johnson, G. S., and Pastan, I. "Control of DNA Synthesis and Mitosis in 3T3 Cells by Cyclic AMP." *Biochem. Biophys. Res. Comm.* **48**:743-748 (1972).

EVIDENCE OF A DUALISM BETWEEN CYCLIC GMP AND CYCLIC AMP IN THE REGULATION OF CELL PROLIFERATION AND OTHER CELLULAR PROCESSES

N. D. Goldberg, M. K. Haddox, R. Estensen, J. G. White, C. Lopez, and J. W. Hadden

This book and numerous others with similar titles emphasize the fact that cyclic 3′,5′-adenosine monophosphate (cyclic AMP) has been accorded center stage in the field of biological regulation during the past 10 years. Almost a decade ago, a second cyclic nucleotide, cyclic 3′,5′-guanosine monophosphate (cyclic GMP), was found to occur in mammalian urine (Ashman et al., 1963) and subsequently to be a naturally occurring component in mammalian tissues (Goldberg et al., 1969; Ishikawa et al., 1969) and in lower phylogenetic forms including bacteria (Goldberg et al., 1973). The biological importance of cyclic GMP has remained relatively obscure since its discovery primarily because of the technical difficulties associated with its detection and what appear now to be a number of subtle properties that may distinguish its biological characteristics from those of cyclic AMP.

The first insight into the possible biological importance of cyclic GMP was provided by the observation that cholinergically induced suppression of cardiac contractility is associated with a rapid cellular accumulation of cyclic GMP (George et al., 1970). From the direction provided by the latter and the investigations that have followed, a clearer picture of the role of cyclic GMP in biological regulation has begun to emerge.

The focus of attention on cyclic AMP over the past decade led to the general acceptance of a concept of biological regulation in which bidirectional changes in cellular concentrations of cyclic AMP alone were thought to provide for both the stimulatory and inhibitory regulatory influences imposed. According to this concept, which we term

This work was supported by U.S. Public Health Service Grants NB-05979, HE-07939, AI-00292, and AI-00798.

the *unitary concept* of regulation through cyclic AMP, two types of biological systems exist: those that are facilitated by an elevation of cellular cyclic AMP levels, the A-type, and those suppressed by an increase in its concentration, the B-type. In the A-type system, the opposing or suppressive regulatory signal is generally understood to be represented by a lowering of tissue cyclic AMP levels, and, by the same token, in the B-type system, a lowering of cyclic AMP concentration is viewed as the positive or facilitory intracellular signal.

One implication of this unitary concept is that a critical steady-state level of cellular cyclic AMP is required for maintaining the "normal" activity of a number of biological systems. However, such a role for cyclic AMP would not be entirely consistent with its recognized function as an intracellular mediator of hormone actions, since hormones are generally viewed as biological agents that adjust rather than maintain the ongoing activity of different cellular functions in the organism. It would seem more reasonable, therefore, to expect that control of a biological system, susceptible to facilitory and suppressive regulatory influences, would be imposed through different, rather than the same, regulatory effector and that "normal" activity would be determined by certain self-sustaining characteristics inherent to the system itself.

The results of a number of experiments carried out in our laboratory indicate that cyclic GMP may be involved in promoting a variety of cellular events that are considered to be antagonistic or opposite to those mediated by cyclic AMP. These observations have provided the basis for proposing the *dualism* or *yin-yang hypothesis* of biological control (Goldberg et al., 1973). This hypothesis describes a form of regulation imposed through dual, opposing actions of the two cyclic nucleotides.

According to the dualism concept (Fig. 1), the facilitory influence in an A-type system is mediated through the elevation of cellular cyclic AMP, while the opposing cellular event or suppressive influence is promoted by an elevation in the concentration of cyclic GMP. In a B-type system, it is proposed, the positive or facilitory influence to a particular cellular event is represented by an elevation of cellular cyclic GMP, rather than a passive lowering of the basal cyclic AMP concentration, and an elevation in the concentration of cyclic AMP suppresses the event stimulable by cyclic GMP and/or promotes an opposing cellular process.

According to this hypothesis, the expression of a cyclic AMP-linked signal would be maximal when the level of cyclic GMP in the reactive cellular pool is minimal, and the expression of a cyclic GMP-linked signal maximal when the reactive pool of cyclic AMP within the cell is reduced to basal levels.

It should also be pointed out that the dual opposing actions of cyclic AMP and cyclic GMP are thought to be expressed only in systems considered to be controlled *bidirectionally.* Such a system could be defined as one comprised of definable antagonistic or opposing metabolic or functional cellular processes, for example, glycogenolysis versus glycolysis, lipolysis versus lipogenesis, contraction versus relaxation, and so on.

In contrast to the dualism type of control proposed for bidirectionally controlled cellular functions, it is also suggested that a somewhat different type of regulation would be required for a biological process that may not be opposed by another cellular event. Biological processes that respond only to a stimulatory signal can be designated as *monodirectionally* controlled systems. These systems are envisaged as proceeding at a basal rate of activity, which can only be accelerated in response to an appropriate signal but which return (passively) to the basal level of activity when the stimulus disappears. It is conceivable that a monodirectionally controlled system may respond in its characteristic manner to different physiological stimuli, and that either cyclic GMP or cyclic AMP may serve as the intracellular mediator for the different extracellular signals. Both cyclic nucleotides, in this case, would promote the same rather than opposing cellular events.

I. Bidirectionally regulated systems
Antagonistic influences on opposing cellular events
(yin-yang or dualism hypothesis)

A-type

B-type

II. Monodirectionally regulated systems
Facilitory influences on unopposed cellular events

Signal X $\longrightarrow$ $\uparrow$ [cyclic AMP] (+)
Signal Y $\longrightarrow$ $\uparrow$ [cyclic GMP] (+)

Fig. 1. Proposed regulatory influences of cyclic GMP and cyclic AMP in various biological systems. Bidirectionally controlled systems are those comprised of opposing cellular processes. Depending upon the type of bidirectional system (i.e., A- or B-type), an increase in the concentration of cyclic AMP or cyclic GMP could represent the facilitory influence, while the suppressive influence and/or opposing cellular event would be promoted by an increase in the cellular level of the other cyclic nucleotide. A monodirectionally controlled system is an unopposed cellular process, which only responds to a stimulatory signal. Cyclic GMP or cyclic AMP may represent intracellular mediators of different extracellular signals, but both cyclic nucleotides, in this case, would promote the same cellular event.

In this chapter, only evidence supporting the concept that cyclic GMP and cyclic AMP may act antagonistically in bidirectionally regulated systems will be presented. The evidence derives from the results of a number of experiments, in which it can be demonstrated that hormonal and other biologically active agents that promote a relatively rapid accumulation of cellular cyclic GMP also promote cellular events that oppose those now believed to be mediated through an elevation of cellular cyclic AMP levels.

Correlating changes in steady-state levels of endogenous cellular cyclic GMP with alterations in cell function produced by different agents, although less than definitive proof of cause and effect, is, we believe, a necessary first step in helping to define the biological role(s) for this cyclic nucleotide. This approach, although similar to one taken by Sutherland and Rall in elucidating the biological role of cyclic AMP, is considerably more difficult in the case of cyclic GMP, because the tissue levels of this cyclic nucleotide are 1/10 to 1/50 those of cyclic AMP in most biological material. Furthermore, unlike the situation with regard to cyclic AMP and the early recognition of its role in promoting activation of the phosphorylase system, no enzymic system exquisitely sensitive to activation by cyclic GMP, which could be used for its detection, was uncovered until relatively recently (Kuo and Greengard, 1970). The first procedure to meet the demands of sensitivity and specificity for measuring tissue cyclic GMP concentrations was an enzymic conversion and amplification procedure comprised of two guanine nucleotide specific enzymes, which were used to convert cyclic GMP, first purified free of interfering tissue metabolities, to GTP. Millimolar concentrations of pyruvate, proportional to the GTP present, were generated by a specific enzymic cycling system, and the pyruvate then measured by a routine enzymic fluorometric procedure (Goldberg et al., 1969; Goldberg and O'Toole, 1971). The latter procedure was employed in most of the studies to be reported here. More recently, however, a radioimmune method similar to that developed by Steiner et al. (1972), employing antibody and isotopically labeled, derivatized cyclic GMP antigen (supplied by Collaborative Research, Waltham, Mass.), has been used for the measurement of this cyclic nucleotide in tissue.

Modifications of the published radioimmune procedure include (1) partial purification of the cyclic GMP in tissue extracts by the use of Dowex-1 formate anion exchange columns (Murad et al., 1971) and (2) collection of the antibody-antigen complex on millipore filters after a 20-hr incubation (0°) of antibody, isotopically labeled antigen, and partially purified tissue extract. Phosphodiesterase-treated extracts, in which a known amount of ^{3}H-cyclic GMP is completely hydrolyzable under the conditions employed, serve as the control samples (blank) and the samples to which internal standards are added. The procedure has proven accurate and reliable when corrections are made for blanks (i.e., phosphodiesterase-treated control extracts), and for recovery of ^{3}H-cyclic GMP carried through the extraction and purification procedures.

The search for the biological importance of cyclic GMP began by first establishing that agents, such as epinephrine and glucagon, which promote the elevation of tissue cyclic AMP levels, have no effect on those of cyclic GMP (Goldberg et al., 1969). Concluding that the cellular levels of the two cyclic nucleotides are probably under separate hormonal and/or metabolic control and that the biological role of cyclic GMP is therefore probably different from that of cyclic AMP, a series of investigations were conducted to uncover a biologically active agent that could stimulate the cellular accumulation of cyclic GMP. The first definable hormone-induced alteration in cell function related to an elevation of tissue cyclic GMP levels was established (George et al., 1970) when it was discovered that acetylcholine produced a prompt (10 sec) elevation of almost threefold in the level of myocardial cyclic GMP, which was coincident with the very first signs of cholinergically induced depression of cardiac contractility and rate. Myocardial cyclic AMP levels were unaffected initially, but at 20 sec were found to decrease about 40%.

It was noted then and subsequently that atropine could prevent the cholinergic effect on both heart function and cyclic GMP accumulation (Goldberg et al., 1972). Furthermore, isoproterenol, which stimulates the rate and force of cardiac contractility as well as cyclic AMP generation in this tissue, was found to lower the levels of cyclic GMP by approximately 50% (George et al., 1970).

Since the relationship between cholinergic action and cyclic GMP accumulation was uncovered in rat heart, the levels of tissue cyclic GMP have been shown to be elevated in other tissues and cells by cholinergic agents (Goldberg et al., 1973; Kuo et al., 1972; Schultz et al., 1972a, 1972b; Lee et al., 1972; Yamashita and Field, 1972). In each of the examples shown in Table 1, the increase in cyclic GMP concentration was detected at the earliest time tested after introduction of the active agent, and in each case no change or a small decrease was found in the tissue level of cyclic AMP. It should be noted that electrical stimulation of the sciatic nerve has no effect on promoting cyclic GMP accumulation in the gastrocnemius muscle of the rat. This observation, combined with the demonstration that atropine can block the cholinergically induced accumulation of cyclic GMP in rat heart, led to the conclusion that only the muscarinic and not the nicotinic actions of acetylcholine are linked to an enhanced generation of cellular cyclic GMP (Goldberg et al., 1973). The latter has recently been confirmed by Lee et al. (1972).

One characteristic common to cholinergic actions of the muscarinic type is that the alterations produced in cell function are usually opposite to those elicited by agents that stimulate cyclic AMP generation. The question that arose, therefore, was whether any other agents that pro-

Table 1. Alterations in Tissue Cyclic GMP Levels Observed with Cholinergic Stimulation

Tissue	Agent	Conditions	Earliest observed increase in cyclic GMP	Cyclic GMP	Cyclic AMP
				% of control[a]	
Rat heart	Acetylcholine	Perfused, in vitro	10 sec	270	60
Rabbit lung	Methacholine	Slices, in vitro	60 sec	290	—
Rat uterus	Methacholine	in vitro	30 sec	450	90
Mouse cerebellum	Maaloxone[b]	in vivo	30 sec	185	98
Human peripheral blood lymphocytes	Acetylcholine	in vitro	10 min	270	85
Rat skeletal muscle	Sciatic nerve stimulation	in vivo	5 to 15 sec	95	98

[a]Maximal change with respect to control at any time tested.
[b]An inhibitor of acetylcholinesterase.

mote cyclic AMP-antagonistic cellular events also stimulate cellular cyclic GMP accumulation. Every agent of this type tested thus far has been found to do so.

For example, in the rat uterus, oxytocin, serotonin, and prostaglandin $F_{2\alpha}$ ($PGF_{2\alpha}$) like methacholine all stimulate contractility of uterine smooth muscle from diethylstilbesterol-treated rats, and each of these agents also promotes 2½- to 5-fold increases in tissue cyclic GMP levels within seconds (Fig. 2). There is virtually no change in uterine cyclic AMP levels associated with the action of these hormones. The effect of beta-adrenergic agents to relax uterine smooth muscle is, of course, associated with an elevation of cyclic AMP levels in this tissue (Triner et al., 1971).

$PGF_{2\alpha}$ can also be shown (Fig. 3) to stimulate bovine vein smooth muscle contractility in association with a two- to threefold rise in the concentration of cyclic GMP in this tissue with no significant change in the level of cAMP.°

A cellular process well known to be inhibited by cyclic AMP or agents that stimulate its generation is the aggregation of human blood platelets. In Fig. 4 it is shown that epinephrine, through what appears to be an alpha-adrenergic mechanism (Cole et al., 1971), stimulates both platelet aggregation and almost a fourfold increase in platelet cyclic GMP concentration at a time (30 sec) when the very first indication of the hormone effect is detectable. Collagen, another agent that stimulates

°These studies were conducted in collaboration with Dr. Earl Dunham of the Department of Pharmacology at the University of Minnesota.

platelet aggregation, was also found to promote a rise in cellular cyclic GMP concentration (Fig. 4). The concentration of collagen used in these experiments induced a slower rate of aggregation than the concentration of epinephrine employed and a correspondingly smaller (80%) increase in platelet cyclic GMP levels. Neither aggregating agent caused a significant change in the levels of cyclic AMP in these cells.

Phorbol myristate acetate (PMA), a biologically active component of croton oil, has also been shown to act as a potent platelet-aggregating agent (Zucker et al., 1972). Exposure of platelets to PMA at a concentration of 0.1 μg/ml promotes a doubling at 15 sec and, by 30 sec, a fourfold increase in platelet cyclic GMP concentration (Fig. 5). The increase in the cyclic GMP levels of these cells is coincident with the action of PMA to stimulate aggregation and cell degranulation (not shown).

Evidence from several laboratories (cf. Sheppard in this book and Willingham, et al., 1972) has appeared indicating that cyclic AMP (or its

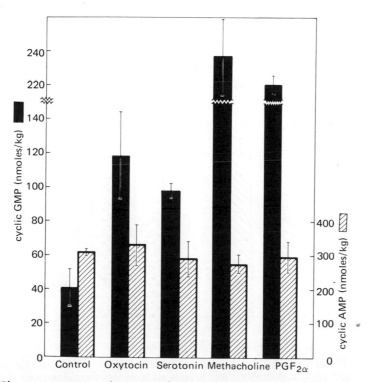

Fig. 2. Changes in uterine cyclic GMP and cyclic AMP concentrations induced by agents that stimulate contractility. Uteri from diethylstilbesterol-treated rats were equilibrated (2 min) in Munsick's media, then exposed to oxytocin (0.1 units/ml), serotonin (10^{-6} M), methacholine (4×10^{-5} M), or prostaglandin $F_{2\alpha}$ ($PGF_{2\alpha}$) 10^{-6} M for 45 sec (when contraction was marked) before they were quick-frozen in Freon-12 ($-150°$C). Cyclic GMP was determined by the method of Goldberg and O'Toole (1971) and cyclic AMP by the procedure of Gilman (1970).

dibutyryl deviatives) or agents that elevate the cellular concentration of cyclic AMP inhibit the proliferation of cells exhibiting "malignant" growth characteristics, i.e., loss of density-dependent contact-inhibited growth. Furthermore, the transformation of cells with "normal" growth characteristics into cells exhibiting uncontrolled growth has been shown to be related to a lowering of cellular cyclic AMP concentration. The general concept emerging from these investigations is that an increase in cellular cyclic AMP levels inhibits cell proliferation, and a lowering promotes or permits cell division. My colleague, Dr. John Hadden, has presented evidence at this meeting in support of our view (Hadden et al., 1972; Goldberg et al., 1973) that the active intracellular signal for initiating cell division of human peripheral blood lymphocytes may be represented by an elevation of cellular cyclic GMP concentration rather than by a passive lowering of the cyclic AMP concentration. Further support for this concept was found when the effects of other mitogenic agents were tested on mouse 3T3 fibroblasts grown in culture.

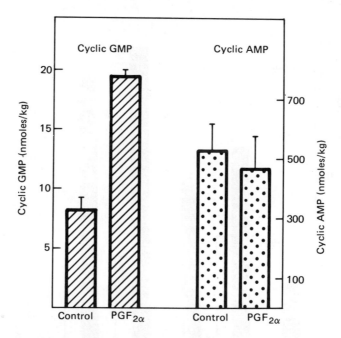

Fig. 3. Cyclic nucleotide levels in bovine vein in control animals and 3 min after $PGF_{2\alpha}$. The veins were dissected shortly after the animals were slaughtered and paired helical strips cut from biforcations of the common dorsal digital vein. The strips were suspended (0.5 to 1.0 g tension) and superfused (10 ml/min) with Krebs physiological salt solution (37°C). The veins were frozen between stainless steel blocks (−190°) 30 to 180 sec after perfusion with 7×10^{-6} M $PGF_{2\alpha}$, which produced a response averaging 60% of a "standard contraction" (near maximal) elicited by 105mM KC1. The values are the averages from 4 paired helical strips. The modified radioimmune method of Steiner et al. (1972) was used for the analysis of cyclic GMP.

PMA, which was shown earlier to promote platelet aggregation and an accompanying elevation of platelet cyclic GMP concentration, is more commonly recognized for its potent mitogenic activity in a variety of tissues and cultured cells including mouse 3T3 fibroblasts (Sivak, 1972). Exposure of 3T3 cells in culture to PMA (0.1 μg/ml) was found to produce a very rapid and striking elevation of cellular cyclic GMP concentration (Fig. 6). Treatment with PMA produced a doubling of cyclic GMP levels at 15 sec, a 10-fold increase at 30 sec, and a peak increase of 50-fold by 45 sec before declining detectably toward control levels at 60 sec. This temporally discrete increase confined to a period of only seconds is consistent with the proposal of a "trigger type" mechanism for initiating the process of proliferation (Hadden et al., 1972). It may be worthwhile noting that the increases in cellular cyclic GMP concentrations associated with the mitogenic action of PMA in 3T3 cells are approximately an order of magnitude greater than those associated with the nonmitogenic action of this same agent in platelets. The increases in cellular cyclic GMP associated with mitogenic action appear, in general, to be greater than the increases found in association with hormonal modulation of other cell functions.

For a number of years, insulin has been viewed as a hormone that produces a number of its effects through a lowering of cellular cyclic

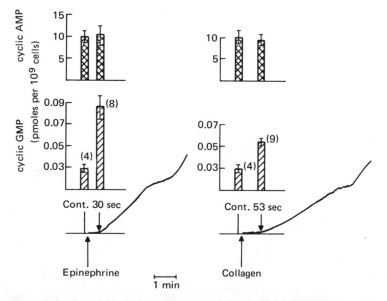

Fig. 4. Elevation of platelet cyclic GMP concentration by epinephrine and collagen. The concentration of cyclic GMP (Goldberg and O'Toole, 1971) and cyclic AMP (Gilman, 1970) was determined in platelets from human citrated platelet-rich plasma (PRP) at the time that the very first sign of aggregation was detectable in an identical PRP sample used to monitor the progress of the aggregation process in a nephelometer (see tracings). Epinephrine was used at a concentration of 5×10^{-6} M and collagen at 2 mg/ml.

AMP concentration (Jefferson et al., 1967), although evidence to the contrary has also been reported (Goldberg et al., 1967; Nichols and Goldberg, 1972). In addition to the well-recognized influences of insulin on glucose transport and intermediary metabolism, this hormone has also been shown to induce cell division (Timen, 1967). Figure 7(*a*) shows the accumulation of cyclic GMP in cells stimulated by insulin at a concentration of 10 munits/ml. In this experiment, there was a time-dependent increase in the level of cyclic GMP of about 10-fold over a 20-min period. It was also demonstrated, as shown in Fig. 7(*b*) and (*c*), that the extent of the increase in cellular cyclic GMP depends on the concentration of insulin employed over a concentration range which, although high, is representative of the concentrations generally used to stimulate cell division. The mechanism underlying the action of insulin as a "growth factor" may be unrelated to the mechanism through which its metabolic effects are brought about. However, if the metabolic actions of this hormone are ultimately found to be associated with cellular-cyclic GMP accumulation, the magnitude of such an increase would probably be much smaller than that related to its mitogenic action. Such is certainly the case with regard to the changes in cyclic GMP levels found to be associated with the mitogenic and nonmitogenic actions of PMA (see above).

In summary, the results contained in this report demonstrate that cellular cyclic GMP accumulation can be induced in a number of different tissues and cells by a variety of chemically diverse, biologically

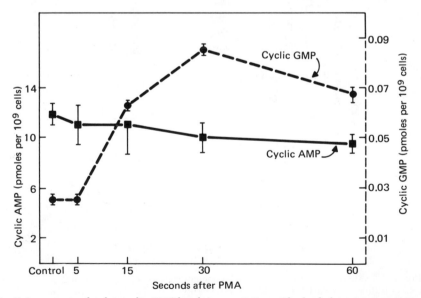

Fig. 5. Increase in platelet cyclic GMP levels in association with phorbol myristate acetate-induced aggregation. Albumin-washed human platelets were used, and the cyclic GMP measured by the modified radioimmune procedure. PMA was employed at a concentration of 100 ng/ml.

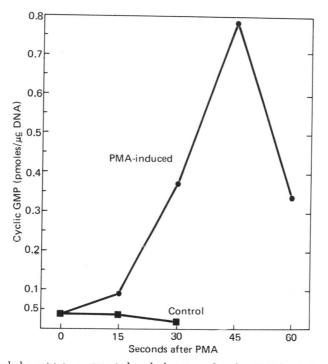

Fig. 6. Phorbol myristate acetate-induced elevation of cyclic GMP levels in contact-in-hibited mouse 3T3 fibroblasts grown in culture. The cells were grown to confluency (3 days) in 10% calf serum. Cells were washed twice with a balanced salt solution, then trichloroacetic acid (10%) was added at the times indicated after addition of 100 μg/ml PMA. Cyclic GMP levels were determined by the modified radioimmune procedure. In 48 hr, cell numbers (in triplicate samples) increased 2.1 times when compared with controls.

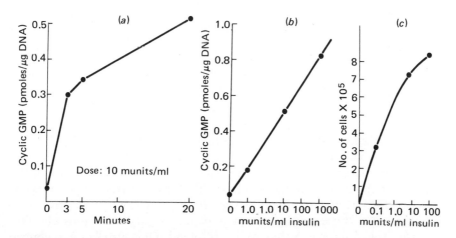

Fig. 7. Cyclic GMP levels in mouse 3T3 fibroblasts in association with insulin-induced cell proliferation. The cells were harvested with the procedures described in Fig. 6 after the addition of the amounts of insulin indicated. Cells were incubated for 10 min with insulin in the experiments illustrated in panels (*b*) and (*c*).

active agents that modulate a broad spectrum of cellular functions. The cellular effects produced by agents that promote cyclic GMP accumulation are opposite to those promoted by agents known to stimulate the generation of cyclic AMP in these same tissues and cells. It is hoped that the relationships uncovered will help provide a more fruitful course for future investigations dealing with cyclic GMP and ultimately a more complete definition of the elements basic to biological regulation.

Discussion of Dr. Goldberg et al's Paper

DR. STECHSCHULTE: Do you think the lowering effect of acetylcholine on cyclic AMP is a direct or an indirect effect?

DR. GOLDBERG: I have no idea. I think Dr. Hardman's finding that cyclic GMP can stimulate phosphodiesterase fits very nicely into that scheme. If the first event is an elevation of cellular cyclic GMP, one of its effects would be the activation of phosphodiesterase and that would produce a lowering of cyclic AMP. I think a discharge of cyclic AMP from the protein kinase, making it susceptible to hydrolysis, is another possible mechanism.

DR. STECHSCHULTE: The fact that you see an elevation of cyclic GMP before cyclic AMP would also be in agreement with that?

DR. GOLDBERG: We have to do more experiments to show that this is truly the case.

DR. AUSTEN: Would you comment further on the effect of epinephrine on cyclic GMP accumulation in platelets?

DR. GOLDBERG: In platelets, epinephrine is predominantly an alpha adrenergic stimulator, and this may be the reason that it is able to raise cyclic GMP.

DR. JOHN HADDEN: Dr. Schultz and Dr. Hardman have shown that norepinephrine stimulation in rat vas deferens is associated with cyclic GMP accumulation. In the lymphocyte, we can only cite preliminary observations to support the idea that alpha adrenergic stimulation is related to cellular accumulation of cyclic GMP.

DR. AUSTEN: Was there a fall in cyclic AMP or only an elevation in cyclic GMP?

DR. HADDEN: We have not discerned a fall in cyclic AMP.

DR. PARKER: In approximately 12 experiments, some of which were done two years ago and some subsequent to your recent paper in the *Proc. Nat. Acad. Sci.*, we have failed to find substantial changes in cyclic GMP in PHA-stimulated lymphocytes. In the recent work, we have tried to reproduce the conditions in your paper as exactly as possible. Obviously, the discrepancy has to be resolved. One important technical difference is that we are measuring cyclic GMP by radioimmunoassay, whereas in your published work you are using the assay that you have developed.

DR. HADDEN: Our observations on the effects on mitogens in lymphocytes involved two assays: Dr. Goldberg's enzyme cycling assay and the radioimmunoassay. And as Dr. Goldberg described, a variety of mitogens has a similar effect on cellular cyclic GMP levels.

DR. GOLDBERG: Do you purify your lymphocyte extracts or carry out your measurements on unpurified homogenates?

DR. PARKER: We have been working with crude extracts. There is no column purification or concentration step. We have a trichloroacetic acid precipitation step to inactivate phosphodiesterase, which is similar to what you use.

DR. GOLDBERG: We are unable to make cyclic GMP measurements of any validity in unpurified extracts. In our original method, purification was an integral part of the procedure, but even with the radioimmunoassay, we still have to partially purify. We cannot get stoichiometry in the assay unless we purify the extract.

DR. PARKER: What purification procedure do you use?

DR. GOLDBERG: The purification scheme used in most of our experiments involves column and thin-layer chromatography steps. We also purify extracts prior to radioimmunoassay, in part because most of the antibodies we have used cross-react to some extent with other nucleotides.

DR. PARKER: All I can say is that we get very good recovery of added cyclic GMP in our system, and the background levels of cGMP-like activity we are measuring in unstimulated cells are low. In view of the magnitude of the changes you have described, I think we should be able to demonstrate a sizable increase, even if our assay is not linear. I have two other comments: One, I take it that there is still no direct evidence that there is guanylate cyclase in the plasma membrane, or that it can be directly stimulated by hormonal agents. Two, am I correct in assuming that Sutherland's postulates for cyclic AMP have not yet been fulfilled for cyclic GMP in lymphocytes and other cells? Do you have any evidence that when cyclic GMP levels are raised in the cell, as with acetylcholine or cyclic GMP itself, that a potent mitogenic response is obtained? By that I mean a 50- to 100-fold increase in DNA synthesis analogous to what would occur with PHA. The twofold increase in thymidine incorporation Dr. Hadden described with acetylcholine is very similar to what is observed with cyclic AMP itself or dibutyryl cyclic AMP. Our own observations indicate that cyclic GMP inhibits mitogenesis.

DR. GOLDBERG: We have not fulfilled the Sutherland criteria for a cyclic nucleotide as an intracellular mediator, but the Sutherland criteria were set up for cyclic AMP. I don't think the idea that cyclic GMP is an important mediator for the events under discussion can be disregarded because we have not met the Sutherland criteria. The exact nature of guanylate cyclase remains to be established. Actually, we don't know a great deal more about the adenylate cyclase system. You can stimulate it, but the relationship of the hormonal receptor to the catalytic region and the mechanism of the stimulation remains to be seen. Dr. Hardman mentioned that there is preliminary evidence for a particulate guanylate cyclase, which is unmasked after treatment with detergent, and that may be a clue that agonists bring about a similar unmasking of guanylate cyclase activity. The membrane location of this activity is encouraging and is architecturally consistent with our present notions of the involvement of cyclases in the mediation of certain hormone actions. But it should be kept in mind that guanylate or even adenylate cyclase could be involved in the mediation of hormone action in some cases even if they were cytoplasmic enzymes. It would be necessary to postulate that a component is generated or released by the membrane, which, in turn, interacts with the soluble cyclase or a cyclase associated with the membrane of another cell organelle such as the nucleus.

References

Ashman, D. F., Lipton, R., Melicow, M. M., and Price, T. D. "Isolation of Adenosine 3',5'-Monophosphate and Guanosine 3',5'-Monophosphate from Rat Urine," *Biochem. Biophys. Res. Commun.* 11:330–334 (1963).

Cole, B., Robison, G. A., and Hartmann, R. C. "Studies on the Role of Cyclic AMP in Platelet Function." In *Cyclic AMP and Cell Function. Ann. N.Y. Acad. Sci.*, G. A. Robison, G. G. Nahas, and L. Triner, eds. 185:477–487 (1971).

George, W. J., Polson, J. B., O'Toole, A. B., and Goldberg, N. D. *Proc. Nat. Acad. Sci.* (U.S.A.) 66:398 (1970).

Gilman, A. G. "A Protein Binding Assay for Adenosine 3',5'-Cyclic Monophosphate," *Proc. Nat. Acad. Sci.* (U.S.A.) 67:305–312 (1970).

Goldberg, N. D., Haddox, M. K., Hartle, D. K., and Hadden, J. W. "The Biological Role of Cyclic 3',5'-Guanosine Monophosphate." *Fifth International Congress on Pharmacology* (in press—1973).

Goldberg, N. D., and O'Toole, A. G. "Analysis of Cyclic 3',5'-Adenosine Monophosphate and Cyclic 3',5'-Guanosine Monophosphate." *Methods of Biochemical Analysis*, Vol. 10. New York: Wiley, pp. 1-39 (1971).

Goldberg, N. D., Dietz, S. B., and O'Toole, A. G. "Cyclic Guanosine 3',5'-Monophosphate in Mammalian Tissues and Urine," *J. Biol. Chem.* 244:4458–4466 (1969).

Goldberg, N. D., Villar-Palasi, C., Sasko, H., and Larner, J. "Effects of Insulin-Treatment on Muscle 3',5'-Cyclic Adenylate Levels in Vivo and in Vitro," *Biochim. Biophys. Acta* 148:665–672 (1967).

Hadden, J. W., Hadden, E. M., Haddox, M. K., and Goldberg, N. D. "Guanosine 3',5'-Cyclic Monophosphate: A Possible Intracellular Mediator of Mitogenic Influences in Lymphocytes," *Proc. Nat. Acad. Sci. (U.S.A.)* 69:3024–3027 (1972).

Ishikawa, E., Ishikawa, S., Davis, J. W., and Sutherland, E. W. "Determination of

Guanosine 3',5'-Monophosphate in Tissues and Guanyl Cyclase in Rat Intestine," *J. Biol. Chem.* **244**:6371–6376 (1969).

Jefferson, L. S., Exton, J. H., Butcher, R. W., Sutherland, E. W., and Park, C. R. "Role of Adenosine 3',5'-Monophosphate in the Effects of Insulin and Antiinsulin Serum on Liver Metabolism," *J. Biol. Chem.* **243**:1031–1038 (1967).

Kuo, J. F., Ling, T. P., Reyes, P. L., Walton, K. G., Donnelly, T. E., Jr., and Greengard, P. "Cyclic Nucleotide-Dependent Protein Kinases. X. An Assay Method for the Measurement of Guanosine 3',5'-Monophosphate in Various Biological Materials and a Study of Agents Regulating Its Levels in Heart and Brain," *J. Biol.'Chem.* **247**:16–22 (1972).

Kuo, J. F., and Greengard, P. "Cyclic Nucleotide-Dependent Protein Kinases. VI. Isolation and Partial Purification of a Protein Kinase Activated by Guanosine 3',5'-Monophosphate, *J. Biol. Chem.* **245**:2493–2498 (1970).

Lee, T. P., Kuo, J. F., and Greengard, P. "Role of Muscarinic Cholinergic Receptors in Regulation of Guanosine 3',5'-Cyclic Monophosphate Content in Mammalian Brain, Heart Muscle, and Intestinal Smooth Muscle," *Proc. Nat. Acad. Sci. (U.S.A.)* **69**:3287–3291 (1972).

Munsick, R. A. "Effect of Magnesium Ion on the Response of the Rat Uterus to Neurohypophyseal Hormones and Analogues," *Endocrinology* **66**:451–457 (1960).

Murad, F., Manganiello, V., and Vaughan, M. "A Simple, Sensitive Protein-Binding Assay for Guanosine 3',5'-Monophosphate," *Proc. Nat. Acad. Sci. (U.S.A.)* **68**:736–739 (1971).

Nichols, W. K., and Goldberg, N. D. "The Relationship Between Insulin and Apparent Glucocorticoid Promoted Activation of Hepatic Glycogen Synthetase," *Biochim. Biophys. Acta* **279**:245–258 (1972).

Schultz, G., Hardman, J. G., Davis, J. W., Schultz, K., and Sutherland, E. W. "Determination of Cyclic GMP by a New Enzymatic Method," *Fed. Proc.* **31**:440 (1972a).

Schultz, G., Hardman, J. G., Schultz, K., Baird, C. E., Parks, M. A., Davis, J. W., and Sutherland, E. W. "Cyclic GMP and Cyclic AMP in Ductus Deferens and Submaxillary Gland of the Rat." *Fifth International Congress on Pharmacology*, abstracts of volunteer papers, p. 206 (1972b).

Sivak, A. "Induction of Cell Division; Role of Cell Membrane Sights," *J. Cell Physiol.* **80**:167-173 (1972).

Steiner, A. L., Parker, C. W., and Kipnis, D. M. "Radioimmunoassay for Cyclic Nucleotides. I. Preparation of Antibodies and Iodinated Cyclic Nucleotides," *J. Biol. Chem.* **247**:1106–1113 (1972).

Timen, H. M. "Studies on Carcinogenesis by Avian Sarcoma Viruses. VI. Differential Multiplication of Uninfected and of Converted Cells in Response to Insulin, *J. Cell. Physiol.* **69**:377–384 (1967).

Triner, L., Nahas, G. G., Vulliemoz, Y., Overweg, N. I. A., Verosky, M., Habif, D. V., and Ngai, S. H. "Cyclic AMP and Smooth Muscle Function." In *Cyclic AMP and Cell Function. Ann. N.Y. Acad. Sci. (U.S.A.)*, G. A. Robison, G. G. Nahas, and L. Triner, eds. **185**:458-476 (1971).

Willingham, M. C., Johnson, G. S., and Pastan, I. "Control of ANA Synthesis and Mitosis in 3T3 Cells by Cyclic AMP." *Biochem. Biophys. Res. Commun.* **48**:743–748 (1972).

Yamashita, K., and Field, J. B. "Elevation of Cyclic Guanosine 3',5'-Monophosphate Levels in Dog Thyroid Slices Caused by Acetylcholine and Sodium Fluoride", *J. Biol. Chem.* **247**:7062–7066 (1972).

Zucker, M. B., Kim, S. J., Balman, S., Troll, W. "Phorbol Ester-A Potent New Platelet Aggregating Agent." Abstracts of the 3rd Congress, International Society of Thrombosis and Haeomostasus, Washington, D. C., p. 266 (1972).

BORDETELLA PERTUSSIS AND BETA ADRENERGIC BLOCKADE

STEPHEN I. MORSE

A variety of effects are produced in mice and rats following the injection of intact killed phase I *Bordetella pertussis,* extracts of the cells, or culture supernatant fluids. Among these effects are

(1) Leucocytosis with a predominating lymphocytosis (Morse, 1965; Morse and Bray, 1969).

(2) Increased sensitivity to the lethal properties of histamine, serotonin, acetylcholine, and bradykinin as well as increased sensitivity to nonspecific stimuli such as endotoxin and cold stress (reviewed by Munoz and Bergman, 1968).

(3) Enhanced antibody formation (Dresser et al., 1970) including a greater production of homocytotropic antibody in response to a given antigen (Mota and Peixoto, 1966).

(4) Heightened susceptibility to both active (Malkiel and Hargis, 1952) and passive systemic anaphylaxis (Munoz and Anacker, 1959).

(5) Hypoglycemia (Stronk and Pittman, 1955) and a decreased hyperglycemic response to epinephrine (Fishel and Szentivanyi, 1963), effects that are presumed to be due to beta adrenergic blockade.

The bacterial products responsible for these effects have not been isolated or characterized, nor is it at all clear how many effects may be mediated by a single factor. In our laboratory, we have been primarily interested in isolating the lymphocytosis-promoting factor (LPF) from culture supernatants and have found that histamine sensitization occurs in fractions rich in LPF (Morse and Morse, 1970). In collaboration with Dr. Charles W. Parker, we have now looked at the ability of the fractions to induce alterations in adrenergic responsiveness in vitro, as measured by changes in the cyclic AMP metabolism of human lymphocytes. (Details of these experiments are presented in a paper in press, Parker and Morse, 1973.)

Bordetella pertussis strain 3779B (NIH-114) was cultured under conditions previously described (Morse and Bray, 1969). The culture supernatant fluids were fractionated according to the scheme given in Table 1. All fractions were in 0.1 M tris-0.5 M NaCl at pH 10 (tris-NaCl), as it had been shown previously that the lymphocytosis-promoting factor aggregated and lost activity unless conditions of both high pH and hypertonic salt were maintained. Leucocyte counts were performed three days after the intravenous injection of material into young adult Albany strain female mice. And 1 mg of histamine dihydrochloride was then injected intraperitoneally into the same mice, and the number of deaths that occurred within 2 hr was noted. The effects of the pertussis fractions on cyclic AMP metabolism were assayed in human lymphocytes prepared from dextran sedimented blood. Polymorphonuclear leucocytes and monocytes were removed by passing the cells through nylon fiber colums, and low-speed centrifugation was used to remove platelets (Smith and Parker, 1971). The cell preparations contained over 98% lymphocytes. The lymphocytes were suspended in Gey's solution at cencentrations of 1 to 5 × 10⁶ cells per milliliter. Changes in the cyclic AMP content of the lymphocytes were determined by radioimmunoassay by methods previously described in detail (Steiner et al., 1969).

As can be seen in Table 1, the lymphocytosis-promoting factor was precipitated by 90% $(NH_4)_2SO_4$. The activity resided in the fraction that was insoluble in water but soluble in tris-NaCl. When this material was put on a cesium chloride gradient, the activity was restricted to the load

Table 1. Induction of Leucocytosis and Histamine Sensitization by Fractions from Culture Supernatants of *B. pertussis*

		Response to 1 μg	
Fraction	Preparation	WBC × 10³/mm³	Histamine (D/T)
1	Supernatant fluid	48.6	1/4
	90% $(NH_2)_4SO_4$ ppt		
2	H₂O soluble	12.0	0/4
3	H₂O insoluble	109.0	1/4
4	CsCl load	135.8	3/4
	G-150		
4A	Excluded peak	23.7	0/4
4C	1st included peak	108.1	2/4
5	CsCl remainder	48.2	2/4

Note: Fractions were all in 0.1 M tris-0.5 M NaCl, pH 10.0, and were injected intravenously. Leukocyte counts (WBC) were determined three days later; then the same mice received 1 mg of histamine diHCl. Results are expressed as no. dead per total no. (D/T).

volume (Fx 4). The gradient consisted of 6 ml volumes of CsCl of densities 1.2, 1.25, 1.3, and 1.5, and the load was 6 ml of Fx 3. Centrifugation of the SW-25 rotor was at $50,000 \times g$ for 3½ hr. When Fx 4 was subjected to Sephadex G-150 chromatography, the active material was contained in the first included peak (Fx 4C); there was no activity in the large, excluded peak (Fx 4A). There was less consistency with respect to the histamine-sensitizing activity, which was present in Fx 5 as well as Fx 3, suggesting that more than one factor, perhaps endotoxin, could be involved.

Table 2. The Effect of Fraction 1 on the Cyclic AMP Response of Lymphocytes to Isoproterenol

Isoproterenol	Fx 1	Picomoles cAMP per 10^7 lymphocytes
1 mM	None	44.0
	1:100	5.0
	1:10,000	21.0
10 mM	None	54.0
	1:100	10.2
	1:1000	44.0
	1:10,000	44.0
None	1:100	5.8
	None	5.6

Note: Lymphocytes were preincubated with Fx 1 or buffer for 90 min at 37°C prior to the addition of isoproterenol. Suspensions were then incubated for another 10 min.

Initial experiments on the effects of *B. pertussis* products on cyclic AMP metabolism were performed with Fraction 1 (crude dialyzed supernatant fluid in tris-NaCl). Suspensions of human lymphocytes were preincubated with various dilutions of Fx 1 at 37°C for 90 min. Then isoproterenol, 1 mM or 10 mM, was added, and incubation continued for another 10 min.

As seen in Table 2, preincubation with a 1:100 dilution of Fx 1 resulted in almost complete inhibition of the cyclic AMP response to 1 mM isoproterenol and nearly complete inhibition of the response to 10 mM isoproterenol. At the lower concentration of isoproterenol, there was inhibition when the fraction was diluted to 1:10,000. It was of note that when the preincubation period was shortened to 20 min, there was little or no inhibition of the cyclic AMP response. It was also found that when preincubation was at 0° instead of 37°, the response to isoproterenol was only slightly decreased. Therefore, an incubation time of 90 min and a temperature of 37° were chosen for all subsequent experiments.

Dilutions of the other fractions were then tested for their ability to inhibit the isoproterenol response. The dilutions represented equivalent dilutions with respect to the volume of the original cultured supernatant fluid, and are therefore directly comparable. Comparing Table 1 with Table 3, it can be seen that the same fractions that had leucocytosis-promoting activity and histamine-sensitizing activity, namely, Fx 3 and Fx 4, were most potent in inhibiting the isoproterenol effect on cyclic AMP. The same fractions also had marked inhibitory effects on the cyclic AMP response to prostaglandin E_1.

Table 3. The Effect of Various *B. Pertussis* Fractions on the Cyclic AMP Response to 1 mM Isoproterenol

	Dilution and percent inhibition		
Fraction	1:30	1:100	1:1000
1	—	73	32
2	12	3	—
3	98	84	36
4	92	100	54
5	67	62	28

Note: 1.2×10^6 lymphocytes were preincubated with the diluted fractions for 90 min at 37°C prior to the addition of isoproterenol (final concentration, 10 mM) for an additional 10 min. In control tubes, 10 mM isoproterenol increased the cyclic AMP concentration from 20.8 to 124 pmoles per 10^7 lymphocytes.

Known concentrations of Fraction 4 were then studied (Table 4). The cyclic AMP responses of human lymphocytes to 0.1, 1.0, and 10 mM isoproterenol, 3 μM and 30 μM PGE_1, and 3 mM methacholine were all inhibited by concentrations of 500 ng/ml of Fraction 4. It was also found that a concentration of as little as 5 ng/ml (or 3.1 ng per 10^6 lymphocytes), Fx 4 strikingly inhibited the cyclic AMP response to both 1 mM isoproterenol and 30 μM PGE_1 (Table 4).

When Fraction 4 was chromatographed on Sephadex G-150, it was found that the inhibition of the cyclic AMP response was present in the first included peak (Fx 4C), a finding that paralleled the occurrence of the leucocytosis-promoting factor and histamine-sensitizing factor (Table 1).

It is clear from the studies reported here that a substance is present in culture supernatant fluids of *Bordetella pertussis* that inhibits the increment of intracellular cyclic AMP produced by isoproterenol, PGE_1, and methacholine. (At the same time, it should be noted that the resting level of cyclic AMP was not lowered following preincubation of human lymphocytes with the bacterial fractions.) Since the effects of adrenergic drugs are mediated through cyclic AMP, it is likely that the in vivo production of beta adrenergic blockade by *Bordetella pertussis* may be due to this same factor.

It was of great interest that the highest specific activity with respect to inhibition of cyclic AMP response was found in the same fractions that caused lymphocytosis. The lymphocytosis due to *Bordetella pertussis* has been shown to be the result of a redistribution of circulating lymphocytes from the tissues into the blood. Cells in the blood are no longer able to traverse the postcapillary venules in lymph nodes at a normal rate (Morse and Barron, 1970; Iwasa et al., 1970; Taub et al., 1972). Hence, the major reason for the accumulation of blood lymphocytes appears to be a decrease in the rate of egress of the cells from the blood, while the rate of entry is normal. Although evidence presented to date suggests that the primary effect is on the lymphocytes, rather than on the endothelium of postcapillary venules, this latter possibility has not yet been completely excluded. The question can then be raised as to whether there is a direct relationship between the aberration of lymphocyte recirculation and the alteration in cyclic AMP metabolism. The parallelism between the two activities is close enough to warrant further examination.

Table 4. Effect of Fraction 4 on the Cyclic AMP Response to Isoproterenol, PGE_1, and Methacholine

		Fx 4 concentration (ng/ml)			
Agent	Concn.	0	5	50	500
None	—	9[a]	—	—	13
Isoproterenol	0.1 mM	35	—	—	15
Isoproterenol	1 mM	30	18	14	18
Isoproterenol	10 mM	128	—	—	14
PGE_1	30 μM	112	63	46	43
PGE_1	3 μM	120	—	—	33
Methacholine	3 mM	15	—	—	9

Note: 1.6×10^6 lymphocytes in 1.0 ml were preincubated with the indicated concentrations of Fx 4 for 90 min at 37°C. Drug or buffer was then added and the incubation continued for another 10 min.
[a]Picomoles cyclic AMP per 10^7 cells.

The mode of action of the pertussis factions on inhibiting the cyclic AMP response of lymphocytes has yet to be determined. Presumably, the effect is not on receptors for the various stimulating substances, since simultaneous changes in prostaglandin, catecholamine, and cholinergic receptors would be required. One would suspect that the most likely candidates for sites of action are (1) adenylate cyclase and (2) phosphodiesterase. Studies to distinguish among the various possibilities are in progress.

Discussion of Dr. Morse's Paper

DR. PLESCIA: You mentioned that the factor had adjuvant activity in stimulating antibody formation, and Dr. Braun has shown that

isoproterenol also has adjuvant activity. Nevertheless, your factor inhibits or blocks the cAMP-stimulating activity of isoproterenol, and that
seems to me to be an apparent contradiction.

DR. MORSE: I think the problem is in communication. What I meant
to say was that the whole organism has conventional adjuvant activity.
We have no evidence for adjuvant activity with the soluble
lymphocytosis-promoting factor.

References

Dresser, D. W., Wortis, H. H., and Anderson, H. R. "The Effect of Pertussis Vac-
    cine on the Immune Response of Mice to Sheep Red Blood Cells," *Clin.
    Exp. Immunol.* 7:817–831 (1970).

Fishel, C. W., and Szentivanyi, A. "The Absence of Adrenaline-Induced Hy-
    perglycemia in Pertussis-Sensitized Mice and Its Relation to Histamine and
    Serotonin Hypersensitivity," *J. Allergy* 34:439–545 (1965).

Iwasa, S., Yoshikawa, T., Fukumara, K., and Kurokawa, M. "Effects of the
    Lymphocytosis-Promoting Factor from *Bordetella pertussis* on the Func-
    tion and Potentiality of Lymphocytes. I. Ability of Lymphocytes to Recir-
    culate in the Body," *Japan J. Med. Sci. Biol.* 23:47–60 (1970).

Malkiel, S., and Hargis, B. J. "Anaphylactic Shock in the Pertussis-Vaccinated
    Mouse," *J. Allergy* 23:352–358 (1952).

Morse, J. H., and Morse, S. I. "Studies on the Ultrastructure of *Bordetella per-
    tussis*. I. Morphology, Origin, and Biological Activity of Structures Present
    in the Extracellular Fluid of Liquid Cultures of *Bordetella pertussis*," *J.
    Exp. Med.* 131:1342–1357 (1970).

Morse, S. I. "Studies on the Lymphocytosis Induced in Mice by *Bordetella per-
    tussis*," *J. Exp. Med.* 121:49–68 (1965).

Morse, S. I., and Barron, B. A. "Studies on the Leukocytosis and Lymphocytosis
    Induced by *Bordetella pertussis*. III. The Distribution of Transfused
    Lymphocytes in Pertussis-Treated and Normal Mice," *J. Exp. Med.*
    132:663–672 (1970).

Morse, S. I., and Bray, K. K. "The Occurrence and Properties of Leukocytosis
    and Lymphocytosis Stimulating Material in the Supernatant Fluids of *Bor-
    detella pertussis* Cultures," *J. Exp. Med.* 129:523–550 (1969).

Mota, I., and Peixoto, J. M. "A Skin-Sensitizing and Thermolabile Antibody in the
    Mouse," *Life Sci.* 5:1723–1728 (1966).

Munoz, J., and Anacker, R. L. "Anaphylaxis in *Bordetella pertussis*-Treated
    Mice. II. Passive Anaphylaxis with Homologous Antibody," *J. Immunol.*
    83:502–506 (1959).

Munoz, J., and Bergman, R. K. "Histamine-Sensitizing Factors from Microbial
    Agents with Special Reference to *Bordetella pertussis*," *Bacter. Rev.*
    32:103–126 (1968).

Parker, C. W., and Morse, S. I. "The Effect of *B. pertussis* on Cyclic AMP
    Metabolism," *J. Exp. Med.* 137: (in press—1973).

Smith, J. W., and Parker, C. W. "Cyclic Adenosine 3′,5′-Monophosphate in
    Human Lymphocytes. Alterations After Phytohemagglutinin Stimulation,"
    J. Clin. Invest. 50:432–441 (1971).

Steiner, A. L., Kipnis, D. M., Utiger, R., and Parker, C. W. "Radioimmunoassay for the Measurement of Adenosine 3',5'-Cyclic Monophosphate," *Proc. Nat. Acad. Sci.* **64**:367–373 (1969).

Stronk, M. G., and Pittman, M. "The Influence of Pertussis Vaccine on Histamine Sensitivity of Rabbits and Guinea Pigs and on the Blood Sugar in Rabbits and Mice," *J. Infec. Dis.* **96**:152–161 (1955).

Taub, R. N., Rosett, W., Adler, A., and Morse, S. I. "Distribution of Labeled Lymph Node Cells in Mice During the Lymphocytosis Induced by *Bordetella pertussis*," *J. Exp. Med.* **136**:1581–1593 (1972).

CYCLIC AMP AND CHEMOTAXIS OF LEUKOCYTES

JOSEF H. WISSLER, VERA J. STECHER, and ERNST SORKIN

I. Introduction

We wish to report here on the possible role of cAMP in chemotaxis of leukocytes (leukotaxis). Chemotaxis of leukoytes is a cell-stimulatory process by which chemical changes in the environment are recognized. As a consequence of this cellular recognition process, the cell changes its migratory behavior from random to directional locomotion along a positive concentration gradient (Wissler et al., 1972a; Sorkin et al., 1970) without changing the speed of migration (Carruthers, 1966, 1967; Ramsey 1972a, b). Chemotactic responses are shown by phagocytes— neutrophils, eosinophils, and macrophages (for review see Sorkin et al., 1970). Recently, it has also been reported that basophils (Kay and Austen, 1972) and lymphocytes (Ward et al., 1971) give a chemotactic response. It is likely that other tissue cells, like mast cells, histiocytes, etc., show chemotaxis, but the lack of suitable assay systems restricts our knowledge of their chemotactic response. In view of the possible fundamental importance of tactic reactions (chemotaxis, osmotaxis, thermotaxis, phototaxis, aerotaxis) for the development, differentiation, organization, and behavior of living cells, we would like to regard leukotaxis as a model system for many kinds of recognition phenomena.

The importance of leukotaxis is not limited only to immunological events, although the immunological defense system can hardly be thought to work efficiently without leukotaxis. This has been clearly established by the detection of the "lazy-leukocyte syndrome" (Miller et

This work was supported by the Schweizerischer Nationalfonds Grant No. 32 46 69, the Deutsche Forschungsgemeinschaft, Grant No. Wi 406 and F. Hoffmann-La Roche AG, Basel.

al., 1971). However, it should be recalled that huge numbers of cells die without relation to immunological events daily in an organism. Altered or dead cells require removal by phagocytes after having been recognized as being different from normal cells. This process has been called "necrotaxis" (Bessis and Burte, 1964, 1965).

Chemotactic attractants have been identified as being mostly proteins or peptides. Recently, we succeeded in the first crystallization of the components of a binary leukotactic serum peptide system (ACL system), consisting of classical anaphylatoxin (CAT) and cocytotaxin (CCT) as peptide components (Wissler, 1971, 1972a, b; Wissler et al., 1972a, b, c, d; Bernauer et al., 1971, 1972; Sorkin et al., 1972). It was only recently that Wilkinson and McKay (1971; Wilkinson, 1971, 1972) succeeded in outlining a first model for environmental recognition by neutrophils. They have shown that denatured proteins, and not the native ones, are the chemotactic attractants, indicating that neutrophils recognize the random (denatured) form of proteins. Our investigations on the mechanism of chemotactic action of the ACL system have demonstrated that unfolding of one peptide (CAT) results in chemotactic attraction of neutrophils by the ACL system (Wissler et al., 1972d, 1973a; Wissler and Sorkin, 1973). Furthermore, we demonstrated that neither the primary, secondary (helices, random coil, beta-sheet), tertiary, nor quarternary structural *state* of a protein per se is responsible for chemotactic attraction of neutrophils. The neutrophil recognizes the Gibbs free energy of conformational transitions of proteins only above the energy quanta limit of about $-\Delta F = 10$ kcal/mole with a possible upper limit of about 45 kcal/mole (molecular quanta range: 1.66×10^{-23} kcal to 7.5×10^{-23} kcal) (Wissler and Sorkin, 1973; Wissler et al., 1973b). These findings have first shown that leukotaxis is a quanta-regulated process, as is the liberation of neural transmitter substances (Katz, 1969, 1971), and thus put chemotaxis of leukocytes on a common basis with other tactic reactions of cells, like phototaxis, thermotaxis, osmotaxis, and recognition mechanisms through the sensory organs of multicellular organisms (quanta range of human visual recognition: 6.85×10^{-23} kcal to 11.3×10^{-23} kcal). It seemed, therefore, not unlikely that cAMP, as general intracellular messenger, plays also a role in cellular recognition phenomena. It is known to be of importance in recognition processes in functionally specialized sensory organs of multicellular organisms, e.g., visual receptor of the eye (Bitensky et al., 1971; Delbrück, 1972). Indeed, concerning chemotaxis of slime molds, evidence has been presented that cAMP is a chemotactic factor (Konijn et al., 1967; Bonner et al., 1969) and that the adenyl cyclase system is intimately involved in the regulation and control of their chemotactic response (Rossomando and Sussmann, 1972; Gerisch, 1968). Unfortunately, the role of cAMP in chemotaxis of leukocytes is less clarified, and there exists considerable controversy.

II. Experimental

Details about the formation of CAT and CCT in serum, their isolation, crystallization, and preparation for use in biological assays are described elsewhere (Wissler, 1972a, b; Wissler et al., 1972b). Assays for chemotactic activity for neutrophils, eosinophils, and macrophages have been performed in the Boyden chamber (Boyden, 1962; Sorkin et al., 1970) as described by us for neutrophil leukocytes (Wissler et al., 1972b). Three-μ pore size filters were used for experiments with neutrophils; 8μ pore size filters for studies of macrophages and eosinophils. Incubation time was 3, 4, and 5 hr for neutrophils, eosinophils, and macrophages, respectively. Details on different staining techniques for eosinophils and macrophages will be reported elsewhere (Wissler et al., 1973b). Neutrophils were obtained from rabbit, guinea pig, and rat peritoneal exudates (Wissler et al., 1972b; Wissler et al., 1973b). Sterile procedure was used throughout. Guinea pig eosinophils and rabbit macrophages were obtained from peritoneal exudates after injection of sterile horse serum and sterile paraffin oil, respectively (Wissler et al., 1973b). Chemotactic activity is presented as cell count per microscopic field using an algebraic average result of four fields counted (Wissler et al., 1972b).

Nucleotide phosphates (sodium salts of free acids) and all other compounds tested were purchased from Boehringer, Mannheim, Germany, and used without further purification. Fresh solutions were used throughout after sterilizing them by filtration. Cyclic AMP and its dibutyryl derivative have also been obtained from Calbiochem, San Diego, California, and Sigma, St. Louis, Missouri.

III. Cyclic AMP and Chemotaxis of Leukocytes

A. Is cAMP a Chemotactic Factor for Leukocytes?

In view of the proven activity of cAMP for slime molds, its possible chemotactic activity for different types of phagocytes has been investigated by several laboratories. Based on statistical analysis, Leahy et al. (1970) found a probability of 58% for cAMP at 10^{-5} M being a chemotactic attractant for polymorphonuclear leucocytes. Becker (1971) has reached the same conclusion. Other laboratories (Symon et al., 1972; Kaley and Weiner, 1971a, b; Wilkinson, 1973) and our own (Wissler et al., 1972c) could not confirm the positive findings concerning chemotactic activity of cAMP. The negative finding was true for neutrophils, macrophages (Symon et al., 1972; Wilkinson, 1973; Wissler et al., 1972c, and the authors' unpublished observations), as well as for eosinophils (Wissler et al., 1972c) and for polymorphonuclear leukocytes (Kaley and Weiner, 1971a, b). The concentration range tested was 10^{-3} to 10^{-7} M in the case of Symon et al.'s and Wilkinson's experiments; 10^{-4} to 10^{-5} M for cAMP, its dibutyryl derivative, and for cGMP in Kaley's and Weiner's experiments; and 10^{-3} to 10^{-13} M of cAMP and its dibutyryl derivative in our experiments. The same negative results in these concentration ranges for the abovementioned types of leukocytes were found by us for cAMP analogues as listed in Table 1. There are also listed other "energy rich" components, for which no chemotactic activity for eosinophils has been found, including ATP and other nucleotides. The components were commercially purchased and were assayed without further physicochemical or chemical manipulations.

Table 1. Examination of Chemotactic Activity for Neutrophils and Eosinophils Under Standard Conditions of Various Nucleotides and "Energy Rich" Components (Results Were Negative Throughout.)

| | | Assay | |
Compound	Concentration (M)	For neutrophils	For eosinophils
cAMP	10^{-13} to 10^{-3}	—	—
Dibutyryl cAMP	10^{-13} to 10^{-4}	—	—
Deoxy cAMP	10^{-9} to 10^{-4}	—	—
ATP	10^{-12} to 10^{-3}	—	—
cGMP	10^{-9} to 10^{-4}	—	ND
GTP	10^{-10} to 10^{-3}	—	ND
Deoxy cGMP	10^{-9} to 10^{-4}	—	ND
ITP	10^{-9} to 10^{-4}	—	ND
NAD$^+$	10^{-9} to 10^{-4}	—	ND
NADH	10^{-10} to 10^{-4}	—	ND
NADP$^+$	10^{-10} to 10^{-5}	—	ND
NADPH	10^{-9} to 10^{-6}	—	ND
Deoxy cTMP	10^{-9} to 10^{-5}	—	ND
cCMP	10^{-8} to 10^{-5}	—	ND
Phosphoenolpyruvate	10^{-9} to 10^{-4}	ND	—
S-adenosyl-L-methionine	10^{-10} to 10^{-4}	ND	—
Creatinphosphate	10^{-10} to 10^{-4}	ND	—
Acetylphosphate	10^{-11} to 10^{-4}	ND	—
Acetylcoenzyme A	10^{-10} to 10^{-4}	ND	—
cUMP (dibutyryl)	10^{-9} to 10^{-5}	—	ND

ND = not done.
— = negative results.

Wilkinson (1973) has also investigated the chemotactic activity of cAMP and its dibutyryl derivative from different sources. The results were uniformly negative, using human blood neutrophils and guinea pig peritoneal macrophages. The preparations of cAMP were purchased from the following companies: Calbiochem, San Diego, U.S.A.; Sigma, St. Louis, U.S.A.; Koch-Light, London, England.

In summary, one gains the impression that cAMP is not at all chemotactic for the different types of phagocytes. Why some exceptions have been reported is unclear at present, although it should be noted that the sources of the chemotactically active cAMP preparations were not given.

B. Extracellular cAMP and the Regulation of Chemotaxis of Granulocytes by the Anaphylatoxin-Related Leukotactic Binary Peptide System

Recently, we reported that the major chemotactic activity for neutrophils of activated serum is due to the action of the binary leukotactic peptide system (ACL system), the peptide components of which (CAT and CCT) we have obtained in the crystalline, monodisperse state. The cooperative action of the two peptides attracts neutrophils and eosinophils, either cell specifically or cell selectively, this being dependent on only the absolute concentration and the molar ratio of the two peptides (Wissler, 1971, 1972a,b; Wissler et al., 1972a,b,c,d; Sorkin et al., 1972). Furthermore, we showed that one of the peptides (CCT) can be replaced by nucleotides, including cAMP. However, this results in loss of the regulatory function of the ACL system, namely, its ability for cell-specific activity for either neutrophils or eosinophils (Wissler et al., 1972a,c).

We have further investigated the ability of various nucleotides to render CAT chemotactically active for neutrophils. Figure 1 shows the dependence of chemotactic activity for neutrophils on the concentration of various nucleotides, including cAMP, when in combination with a constant concentration of CAT. For comparison, the same dependence is shown in this figure for varying concentrations of CCT at the standard concentration of CAT (Wissler et al., 1972a,c). For reasons of the less pronounced concentration dependence of chemotactic activity for neutrophils and lack of a pronounced nonresponsiveness phase of chemotactic activity, the combination of CAT and cAMP or other neucleotides is unable to display regulatory cell-specific or cell-selective chemotaxis, as does the combination of CAT and CCT—ACL system (Wissler et al., 1972a,c). A nonresponsiveness phase for chemotaxis of neutrophils is displayed only by some nucleotides (including cAMP) when in combination with CAT at concentrations of more than 10^{-4} M (in the case of cAMP of more than 7×10^{-5} M). On the other hand, a concentration of CCT of more than 10^{-6} M (dependent on the absolute concentration of CAT) and especially a molar ratio of CCT : CAT of $\geq 8 : 1$ (independent of the absolute concentration of the two peptides) is sufficient to produce nonresponsiveness (Wissler et al., 1972a,c). Furthermore, the dependence of the activity on the concentration of nucleotides in combination with CAT is strikingly different from the bell-shape-like activity curves of the CAT-CCT combination.

Whatever the exact reason for the special shape of activity curves may be, it should be noted that these bell-shape curves first found by us for neutrophils and similar ones for eosinophils (Wissler et al., 1972a,c) are analogous to the activity curves of *E. coli* in their chemotactic response to certain sugar molecules, such as galactose (Hazelbauer and Adler, 1971). This obviously general behavior of the chemotactic response of cells in dependence of the concentration of chemotactic stimulants underlines the failure of extracellular cAMP at physiological concentrations to regulate the chemotactic response of leukocytes as does CCT. This might be significant in those circumstances where CCT is not present or its regulatory function destroyed in a CAT-forming process, whereupon nucleotides can replace the CCT peptide activity without replacing its regulatory function.

Since it is well established that nucleotides are locally released from cells during tissue injury, e.g., by immune processes (Mustard et al., 1969;

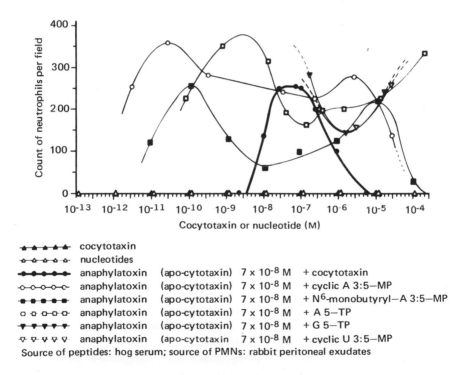

▲-▲-▲-▲-▲	cocytotaxin	
△-△-△-△-△	nucleotides	
●-●-●-●-●	anaphylatoxin (apo-cytotaxin) 7 x 10⁻⁸ M	+ cocytotaxin
○-○-○-○-○	anaphylatoxin (apo-cytotaxin) 7 x 10⁻⁸ M	+ cyclic A 3:5—MP
■-■-■-■-■	anaphylatoxin (apo-cytotaxin) 7 x 10⁻⁸ M	+ N⁶-monobutyryl—A 3:5—MP
□-□-□-□-□	anaphylatoxin (apo-cytotaxin) 7 x 10⁻⁸ M	+ A 5—TP
▼-▼-▼-▼-▼	anaphylatoxin (apo-cytotaxin) 7 x 10⁻⁸ M	+ G 5—TP
▽-▽-▽-▽-▽	anaphylatoxin (apo-cytotaxin) 7 x 10⁻⁸ M	+ cyclic U 3:5—MP

Source of peptides: hog serum; source of PMNs: rabbit peritoneal exudates

Fig. 1. Replacement of cocytotaxin (hog) by different nucleotides and demonstration of the regulatory effect of cocytotaxin on chemotactic activity for neutrophils (rabbit), displayed by the anaphylatoxin-related leukotactic peptide system. The figure shows chemotactic activity (in neutrophil counts per microscopic field) versus concentration either of cocytotaxin or various nucleotides, e.g., ATP, cyclic AMP, GTP, etc., at a constant level of classical anaphylatoxin hog (7.0×10^{-8} M). Note that whereas cocytotaxin regulates chemotactic activity within the system (ACL system), nucleotides, including cyclic AMP, exert only very little regulatory effects on leukotaxis.

Kishuo et al., 1958), extracellular cAMP might play a role in the uncontrolled attraction of neutrophils in some pathological tissue changes.
This could be exemplified by the accumulation of leukocytes in the synovial joint fluid in primary chronic polyarthritis.

Formation of the split product of the complement component C5,
C5a (anaphylatoxin) has been demonstrated in these fluids (Ward and
Zvaifler, 1971). It is quite likely that the role of extracellular cAMP and
its possible intracellular activity can be imitated by cAMP derivatives,
like dibutyryl cAMP, and cAMP analogues such as cGMP or ATP, etc. So
far, various analogues of cAMP show a similar concentration behavior in
terms of rendering CAT chemotactically active for neutrophils (Fig. 1),
although the chemotactic response of neutrophils to CAT-nucleotide
combinations may have different values when compared at a certain concentration of the nucleotide and a distinct concentration of CAT (Fig. 2).

Otherwise, the chemotactically active combination of CAT and
cAMP or of CAT and other nucleotides might lead one to suppose that
cAMP may finally act as a chemotactic attractant in contrast to evalua-
tions described in Section IIA in the presence of CAT. However, that the
contrary is true is shown by the results presented in Table 2. It is clearly
established that only when the concentration gradient of CAT is positive,
neutrophils will migrate, provided that nucleotides, i.e., cAMP, are present with either a positive, a negative, or even no concentration gradient
(indifferent concentration). This clearly shows that the combination of
cAMP and CAT does not act differently in chemotactic attraction of

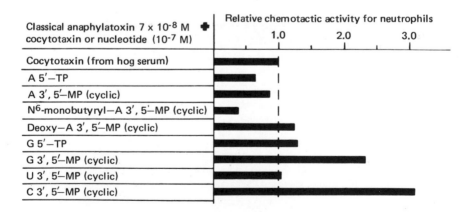

Fig. 2. Replacement of the cocytotaxin peptide by various nucleotides: relative chemotactic
activities for neutrophils (rabbit) by anaphylatoxin-nucleotide combinations under standard
conditions. Reference value (relative chemotactic activity = 1.0) is the ACL system. All
values are based on an anaphylatoxin concentration (hog) of 7.0×10^{-8} M and cocytotaxin
(hog) or nucleotide concentration of 10^{-7} M. According to the results presented in Fig. 1, activity relations are changed if concentrations of reactants are changed.

Table 2. Demonstration of the Absolute Requirement for a
Concentration Gradient of Anaphylatoxin for Neutrophil
Chemotaxis

Anaphylatoxin gradient	A 3',5'-MP, cyclic gradient	Chemotactic activity cells per field
No anaphylatoxin	+	0
+	+	360
+	±0	320
+	no cAMP	0
±0	−	0
−	+	0

Standard conditions: Anaphylatoxin (hog serum) concentration 7 ×
10^{-8} M, cocytotaxin (hog) 1.18 × 10^{-7} M, cyclic- A 3',5'- MP concentration 5 × 10^{-11} M. *Assay system:* Boyden chamber; for details see
Section II, "Experimental."
Note: At the given concentrations of peptide and of nucleotide,
leukocytes migrate along a positive concentration gradient of
anaphylatoxin, irrespective of whether the cAMP gradient is positive,
zero, or negative. This is identical to the behavior of cocytotaxin within
the peptide system (Wissler et al., 1972a).

neutrophils from the binary ACL system with CAT and CCT as components (Wissler et al., 1972a), and that only CAT when modified by an appropriate component (CCT, cAMP, ATP, etc.) is the chemotactic attractant for neutrophils.

On these grounds, we selected the name "cocytotaxin" for the newly
detected, regulatory basic peptide (Wissler, 1972b; Wissler et al.,
1972a,b) in analogy to the regulatory action of many coenzymes and
cofactors. Furthermore, for reasons of the obvious nontoxic, but regulated action of anaphylatoxin within the ACL system, one can also consider the function of CAT to be the "apocytotaxin" of the ACL system.
Both names are in accordance with the proposed classification of activity
principles involved in chemotaxis (Sorkin et al., 1970) as well as with the
classical nomenclature used in enzymology.

Although we can as yet not fully interpret the cooperative action of
CAT and cAMP and other nucleotides, it seems to us unlikely that the action of cAMP is primarily a direct one on the cell. Most likely, CAT is rendered chemotactically active by cAMP by an unfolding process, as occurs
with CCT and other Brønsted acids (e.g., CH_3COOH) (Wissler et al.,
1972d; Wissler et al., 1973a,b). Spectrophotometrical and viscosimetric
studies have to clarify this point. In addition, for the range of higher concentrations of cAMP at a constant low CAT concentration, where
chemotactic activity is decreasing (cAMP 7 × 10^{-5} M), there are two
possible interpretations:

(1) Cyclic AMP at higher concentrations is able to stabilize the helix

conformation of CAT as is known to occur with nucleotides and peptides (Rifkind and Eichhorn, 1970).

(2) The behavior of the CAT-cAMP combination is based on a similar mechanism as is the inhibition of allergic histamine release from basophil leukocytes by similar high concentrations of extracellular cAMP (Lichtenstein et al., 1970; Lichtenstein, 1969).

It should be mentioned that CAT (unfolded or native) is a potent histamine liberator in guinea pigs (Wissler et al., 1973a), and therefore a similar mechanism between allergic histamine release and chemotactic stimulation of different cell types should be considered.

C. Role of Intracellular cAMP in Chemotaxis of Leukocytes

As already mentioned, there are no direct results available on the possible role of intracellular cAMP in chemotaxis of leukocytes. However, its well-established importance in many kinds of cell-stimulatory processes and also in recognition processes by functionally specialized sensory organs of multicellular organisms indicates possibly a role for cAMP in the membrane during cellular recognition phenomena.

How environmental changes can be transmitted to cells by the ACL system is a fundamental problem in determining the mechanism of chemotactic recognition by neutrophil leukocytes. The following points may be considered; cAMP, adenyl cyclase, as well as ATP are components of primary energy transducers of the cell (for review see Green and Baum, 1970). Environmental recognition by neutrophil leucocytes, leading to the chemotactic response, occurs by registration of the Gibbs free energy of conformational transitions of proteins in a distinct quanta range (see Section I, Introduction). Registration and transduction into a cell or organism of such external energy signals occur in nature through poised energy transducers. These neither conserve nor utilize the input of energy for performing work, nor are they actuated and, in contrast to primary energy transducers, do not have the capability to produce ATP. The input of energy through poised energy transformation merely triggers the release of potential energy, which has been already accumulated by an independent mechanism. This is known to occur, for example, in sensory nerve membranes, as well as in the outer segments of retinal rods. The poised systems are energized through other primary energy transformations, which accomplish the preliminary "cocking" of the poised system (Green and Baum, 1970). If a poised system is involved in a transmitting process, a latent period in the recovery of the cell (deactivation of the cell, tachyphylaxis) is an expression of the participation of such an ancillary independent transduction. Chemotactic recognition by leukocytes is characterized by the phenomena of tachyphylaxis or deactivation—latent period of recovery (Wissler et al., 1973b; Becker, 1971).

Furthermore, the energy quanta range of recognition by neutrophil leukocytes is in the order of vibrational transitions of molecular groups and frames—10 to 45 kcal/mole (Wissler et al., 1973b; Wissler and Sorkin, 1973). Therefore, we wish to propose the concept that a conformational change in the membrane by direct transduction of vibrational energy (resonance) is the basis of poised energy transformation in chemotaxis of neutrophils; such conformational changes are typical and form the basis of other types of energy transformations through membranes (e.g., transduction of chemical energy from the inside to the outside of mitochondria). The input of an energy quantum for recognition during chemotactic stimulation is most likely not directly correlated with the activity of the adenyl cyclase system as primary energy transducer. If such a primary energy transducer should be directly involved, no tachyphylactic phenomena at all could be observed. However, we suggest that adenyl cyclase is involved in the regeneration of the stimulated cell to the unstimulated state. If this concept is true, one should observe a sharp quantum-dependent drop in the activity of adenyl cyclase at the moment of chemotactic stimulation and a simultaneous decrease in the rate of intracellular cAMP formation. The increase in the original activity approaches the regenerated state (servoregulation). These proposed mechanism and role of cAMP are deduced from results of chemotactic experiments as well as from findings on light-induced suppression of adenyl cyclase activity in retinal rods during the recognition process (Bitensky et al., 1971). The proposed role for adenyl cyclase holds true most likely also for the regeneration of other cell-stimulatory processes, such as phagocytosis and histamine liberation from basophils and mast cells, which are characterized by the occurrence of tachyphylactic states and supposed to be regulated on a quantum basis (Wissler et al., 1973b).

Summary

The ability of extracellular cAMP to attract leukocytes has been investigated. No activity for the different types of phagocytes was found.

Anaphylatoxin, which is chemotactically inactive per se for neutrophils, can be rendered active by cAMP. The importance of cAMP in the mechanism of leukocyte migration is discussed. It is proposed that cAMP acts by regenerating the stimulated cell back to the unstimulated state and is not involved in the direct transduction of the chemotactic impulse to the cell.

Discussion of Dr. Wissler et al.'s Paper

DR. BOURNE: Mark Twain said, "there is a wonderful thing about science: You get such an enormous return in speculation from such a trifling investment of fact." I wonder to what extent you can claim that cyclic

AMP is really involved here when you have not measured anything having to do with synthesis or breakdown of cyclic AMP under the conditions that you are using, and when cyclic GMP appeared to do better than cyclic AMP, as did several other nucleotides.

DR. WISSLER: I did not claim that cyclic AMP is directly involved in the transmission of the external signal (e.g., in the cellular recognition process itself). And the other exogenously applied nucleotides only work better in terms of the effects obtained in combination with anaphylatoxin. The major point I was trying to make about energy requirements was that, as far as is known, phototaxis and neural transmitter liberation are quanta dependent and quanta regulated, and our experiments indicate that this may also be true for chemotaxis of leukocytes. It has been shown by Bitensky et al. that the input of a photon in the visual receptor will sharply decrease the adenyl cyclase activity in the outer segments of retinal rods.

References

Becker, E. L. "Biochemical Aspects of the Polymorphonuclear Response to Chemotactic Factors." In *Biochemistry of the Acute Allergic Reactions,* K. F. Austen and E. L. Becker, eds. Oxford: Blackwell Scientific Publications, pp. 243–252 (1971).

Bernauer, W., Hahn, F., Wissler, J., Nimptsch, P. and Filipowski, P. "Untersuchungen mit klassischem und kristalliertem Anaphylatoxin an isolierten Herzpräparaten," *Arch. Pharmakol. Exp. Path.* **269**:43 (1971).

Bernauer, W., Hahn, F., Nimptsch, P., and Wissler, J. "Studies on Heart Anaphylaxis. V. Cross-Desentization Between Antigen, Anaphylatoxin, and Compound 48/80 in the Guinea Pig Papillary Muscle," *Int. Arch. Allergy* **42**:136–151 (1972).

Bessis, M., and Burte, B. "Chimiotactisme apres destruction d'une cellule par microfaisceau Laser," *C. R. Soc. Biol.* **158**:1995 (1964).

Bessis, M., and Burte, B. "Positive and Negative Chemotaxis As Observed After the Destruction of a Cell by UV or Laser Microbeams," *Texas Reports Biol. Med.* **23**:204 (1965).

Bitensky, M. W., Gorman, R. E., and Miller, W. H. "Adenyl Cyclase As a Link Between Photon Capture and Changes in Membrane Permeability of Frog Photoreceptors," *Proc. Nat. Acad. Sci.* (U.S.A.) **68**:561–562 (1971).

Bonner, J. T., Barkley, D. S., Hall, E. M., Konijn, T. M., Mason, J. W., O'Keefe, III, G., and Wolfe, P. B. "Acrasin, Acrasinase and Sensitivity to Acrasin in Dictyostelium Discoideum," *Develop. Biol.* **20**:72 (1969).

Boyden, S. V. "The Chemotactic Effect of Mixtures of Antibody and Antigen on Polymorphonuclear Leucocytes," *J. Exp. Med.* **115**:453–466 (1962).

Carruthers, B. M. "Leukocyte Motility. I. Method of Study, Normal Variation, Effect of Physical Alterations in Environment, and Effect of Iodoacetate," *Can. J. Physiol. Pharmacol.* **44**:475–485 (1966).

Carruthers, B. M. "Leukocyte Motility. II. Effect of Absence of Glucose in Medium: Effect of Presence of Deoxyglucose, Dinitrophenol, Puromycin,

Actionmycin D, and Trypsin on the Response to Chemotactic Substance; Effect of Segregation of Cells from Chemotactic Substance," *Can. J. Physiol. Pharmacol.* **45**:269–280 (1967).

Delbrück, M. "Signalwandler: terra incognita der Molekularbiologie," *Angew. Chem.* **84**:1–7 (1972).

Gerisch, G. *Current Topics in Developmental Biology*, A. Monroy and A. A. Moscona, eds., Vol. 3. New York: Academic Press, p. 157 (1968).

Green, D. E., and Baum, H. *Energy and the Mitrochondrion*, New York: Academic Press (1970).

Hatzelbauer, G. L., and Adler, J. "Role of Galactose Binding Protein in Chemotaxis of *Escherichia Coli* Toward Galactose, *Nature* **230**:101–104 (1971).

Kaley, G., and Weiner, R. "Effect of Prostaglandin E_1 on Leucocyte Migration," *Nature (New Biol.)* **234**:114–115 (1971).

Kaley, G., and Weiner, R. "Prostaglandin E_1: A Potential Mediator of the Inflamatory Response," *Ann. N.Y. Acad. Sci.* **180**:338–350 (1971).

Katz, B. *The Release of Neural Transmitter Substances*. Liverpool: University Press (1969).

Katz, B. "Ueber die Freisetzung des Uebertragerstoffs aus den Nerven in Form von Quanten," *Angew. Chem.* **83**:818–821 (1971).

Kay, A. B., and Austen, K. F. "Chemotaxis of Human Basophil Leucocytes," *Clin. Exp. Immunol.* **11**:557–563 (1972).

Kishuo, S., Kunisuke, T., Shoji, K., Yoshisuke, K., Hideaki, K., and Shogo, F. *Gumna J. Med. Sci.* (Japan) **7**:91 (1958).

Konijn, T. M., van de Meene, J. G. C., Bonner, J. T., and Barkley, D. S. "The Acrasin Activity of Adenosine 3′,5′-Cyclic Phosphate," *Proc. Nat. Acad. Sci.* (U.S.A.) **58**:1152–1154 (1967).

Leahy, D. R., McLean, E. R., and Bonner, J. T. "Evidence for Cyclic 3′,5′-Adenosine-Monophosphate As Chemotactic Agent for Polymorphonuclear Leukocytes," *Blood* **36**:52–54 (1970).

Lichtenstein, L. M. "Characteristics of Leukocytic Histamine Release by Antigen and by Antiimmunoglobulin and Anticellular Antibodies." In *Cellular and Humoral Mechanisms in Anaphylaxis and Allergy*, H. Z. Movat, ed. Basel: Karger Verlag, pp. 176–186 (1969).

Lichtenstein, L. M., Levy, D. A., and Ishizaka, K. "In Vitro Reversed Anaphylaxis: Characteristics of Anti-IgE-Mediated Histamine Release," *Immunol.* **19**:831–842 (1970).

Miller, M. E., Oski, F. A., and Harris, M. B. "Lazy-Leucocyte Syndrome. A New Disorder of Neutrophil Function," *Lancet i*:665–669 (1971).

Mustard, J. F., Evans, G., Packham, M. A., and Nishizawa, E. F. "The Platelet in Intravascular Immunological Reactions." In *Cellular and Humoral Mechanisms in Anaphylaxis and Allergy*, H. Z. Movat, ed. Basel: Karger Verlag, pp. 157–163 (1969).

Ramsey, W. S. "Analysis of Individual Leukocyte Behavior During Chemotaxis," *Exp. Cell Res.* **70**:129–139 (1972a).

Ramsey, W. S. "Locomotion of Human Polymorphonuclear Leucocytes," *Exp. Cell Res.* **72**:489–501 (1972b).

Rifkind, J. M., and Eichhorn, G. L. "Specificity for the Interaction of Nucleotides with Basic Polypeptides," *Biochemistry* **9**:1753–1761 (1970).

Rossomando, E. F., and Sussman, M. "Adenyl Cyclase in *Dictyostelium discoidenum*. A Possible Control Element of the Chemotactic System," *Biochem. Biophys. Res. Commun.* **47**:604–610 (1972).

Sorkin, E., Stecher, V. J., and Borel, J. F. "Chemotaxis of Leukocytes and Inflammation," *Ser. Haematol.* **III** (1) 131–162 (1970).

Sorkin, E., Stecher, V. J., and Wissler, J. H. "The Anaphylatoxin-Related Leukotactic Binary Peptide System." In *A Reexamination of Nonspecific Factors Influencing Host Resistance.* Basel: Karger Verlag (in press—1972).

Symon, D. N. K., McKay, I. C., and Wilkinson, P. C. "Plasma-Dependent Chemotaxis of Macrophages Toward *Mycobacterium tuberculosis* and Other Organisms," *Immunol.* **22**:267–276 (1972).

Ward, P. A., Offen, C. D., and Montgomery, J. R. "Chemoattractants of Leukocytes with Special Reference to Lymphocytes," *Fed. Proc.* **30**:1721–1724 (1971).

Ward, P. A., and Zvaifler, N. J. "Complement Cleaved Leukotactic Factors in Inflammatory Synovial Fluids of Humans," *J. Clin. Invest.* **50**:606–616 (1971).

Wilkinson, P. C. "Chemotaxis of Phagocytic Cells Toward Proteins: The Effect of Protein Denaturation." In *The Reticuloendothelial System and Immune Phenomena*, N. R. Di Luzio, ed. New York: Plenum Press, pp. 59–70 (1971).

Wilkinson, P. C. "Characterization of the Chemotactic Activity of Casein for Neutrophil Leukocytes and Macrophages," *Experientia* **28**:1051–1052 (1972).

Wilkinson, P. C. Personal communication (1973).

Wilkinson, P. C., and McKay, I. C. "The Chemotactic Activity of Native and Denatured Serum Albumin," *Int. Arch. Allergy* **41**:237–247 (1971).

Wissler, J. H. "A New Biologically Active Peptide System Related to Classical Anaphylatoxin." *Experientia* **27**:1447–1448 (1971).

Wissler, J. H. "Chemistry and Biology of the Anaphylatoxin Related Serum Peptide System. I. Purification, Crystallization and Properties of Classical Anaphylatoxin from Rat Serum," *Eur. J. Immunol.* **2**:73–83 (1972a).

Wissler, J. H. "Chemistry and Biology of The Anaphylatoxin Related Serum Peptide System. II. Purification, Crystallization, and Properties of Cocytotaxin, a Basic Peptide from Rat Serum," *Eur. J. Immunol.* **2**:84–89 (1972b).

Wissler, J. H., and Sorkin, E. "Structural Requirements of Proteins for Their Chemotactic Recognition by Neutrophil Phagocytes." *Commun. XXIVth Int. Congress of Pure and Applied Chemistry*, Hamburg, 1973 (in press—1973).

Wissler, J. H., Stecher, V. J., and Sorkin, E. "Biochemistry and Biology of a Leukotactic Binary Serum Peptide System Related to Anaphylatoxin," *Int. Arch. Allergy* **42**:722–747 (1972a).

Wissler, J. H., Stecher, V. J. and Sorkin, E. "Chemistry and Biology of the Anaphylatoxin Related Serum Peptide System. III. Evaluation of Leukotactic Activity As a Property of a New Peptide System with Classical Anaphylatoxin and Cocytotaxin As Components," *Eur. J. Immunol.* **2**:90–96 (1972b).

Wissler, J. H., Stecher, V. J., and Sorkin, E. "Regulation of Chemotaxis by the Anaphylatoxin Related Peptide System." *Proc. XXth Colloquium Protides of the Biological Fluids*, H. H. Peeters, ed. Oxford: Pergamon Press, Abstr. Commun., p. 111 (in press—1972c).

Wissler, J. H., Stecher, V. J., Sorkin, E., and Jungi, Th. W. "Mechanism of Leucotaxis." Abstr. Commun. 4th., Tagung Gesellschaft für Immunologie, Bern, 1972, p. 61 (1972d).

Wissler, J. H., Stecher, V. J., and Sorkin, E. "Secondary Structural Properties of Anaphylatoxin Preparations and Chemotactic Activity for Neutrophils." Abstr. Commun. Vth, Int. Complement Workshop, La Jolla, 1973, *J. Immunol.* (in press—1973a).

Wissler, J. H., Stecher, V. J., and Sorkin, E. "Chemistry and Biology of the Anaphylatoxin Related Serum Peptide System. VI. On the Mechanism of Chemotaxis and of Cellular Recognition oof Neutrophil Leukocytes" (manuscript in preparation—1973b).

CHANGES IN THE LEVELS OF CYCLIC AMP IN HUMAN LEUKOCYTES DURING PHAGOCYTOSIS

B. H. Park, N. P. Beck, and R. A. Good

I. Introduction

Cyclic adenosine 3′,5′-monophosphate (cyclic AMP) has been postulated to be a "second messenger" of various hormonal effects in the appropriate target cells (1). We have previously shown that the intracellular concentration of cyclic AMP promptly increased in human leukocytes within the first 5 min of phagocytosis of latex particles (2). Further, it was postulated that the initial increase in cyclic AMP might be a triggering mechanism for the subsequent metabolic changes during phagocytosis and eventual fusion of the phagosomes and appropriate lysosomal granules.

The leukocytes of patients with chronic (fatal) granulomatous disease (CGD) have been shown to be defective in their bactericidal capacity of certain bacteria and metabolic changes during phagocytosis (3,4). We, therefore, investigated the changes in cyclic AMP levels in the leukocytes of patients with CGD.

II. Materials and Methods

Human leukocytes—between 75 and 90% polymorphonuclears—were isolated by dextran sedimentation of heparinized venous blood from healthy volunteers and patients with CGD. The diagnosis of CGD was established by standard methods as previously described.

Rabbit peritoneal exudates were induced by injection of 80 ml of 12% casein in phosphate-buffered saline (pH 7.0) into the peritoneum of a rabbit weighing 4 to 6 lb. The peritoneal exudate rich in polymorphonuclears (60 to 85%) was obtained by washing the peritoneum 18 hr after the injection of casein, whereas the exudate rich in mononuclears and macrophages (75 to 90%) was obtained by washing the peritoneum 48 hr after the injection of casein.

In vitro phagocytosis of latex particles by human leukocytes was carried out as previously described (2). The in vitro phagocytosis of latex particles by the leukocytes of rabbit peritoneal exudate were carried out similarly. The assay of cyclic AMP in human leukocytes was carried out using the method of Kaneko and Field (6), and the assay of cyclic AMP in leukocytes of rabbit peritoneal exudates was carried out using the method described previously (7). Adenyl cyclase activity in the disrupted human leukocytes was measured using the methods described by Bourne et al. (8).

III. Results

Table 1 shows the levels of cyclic AMP in the leukocytes of patients with CGD. The levels of cyclic AMP both in the resting and phagocytosing leukocytes were found to be significantly lower as compared to the control leukocytes.

Table 1. The Levels of Cyclic AMP in the Leukocytes of Patients with Chronic Granulomatous Disease (pmoles per 10^8 cells)

	Level of cAMP	
Subject	In resting leukocytes	In phagocytozing leukocytes
Control	261	984
CGD patient	142	205

Since the intracellular concentrations of cyclic AMP reflect the relative activities of at least two enzymes, the adenyl cyclase and phosphodiesterase and the amount of ATP, the lower concentration of cyclic AMP in the leukocytes of CGD patients might be due to (1) the increased degradation of cyclic AMP by phosphodiesterase, (2) a decreased amount of ATP available for the formation of cyclic AMP, or (3) a decreased activity of adenyl cyclase activity. We found the activity of phosphodiesterase and the amount of ATP in the leukocytes of CGD patients to be comparable to the control leukocytes. Therefore, we measured the activity of adenyl cyclase in the leukocytes of patients with CGD. Table 2 shows our preliminary results, which indicate that the leukocytes of patients with CGD might have a deficiency in the activity of membrane-bound adenyl cyclase.

Table 3 shows the levels of cyclic AMP in the leukocytes of rabbit peritoneal exudates during phagocytosis. The leukocytes rich in mononuclears and those rich in polymorphonuclears shows comparable levels of cyclic AMP in the resting state and during phagocytosis of latex particles.

IV. Discussion

According to the "second-messenger hypothesis" proposed by Sutherland et al. (1), a neurohormone or circulating hormone (first messenger) impinges upon target cells specifically at the adenyl cyclase system. This complex is called the first target system. As a result of the interaction between the hormone and its receptor, adenyl cyclase activity is affected in such a way as to bring about a change in the intracellular level of cyclic AMP. The information originally contained in the hormone is translated intracellularly through changes in cyclic AMP

Table 2. Adenyl Cyclase Activity in Human Leukocytes (pmoles of cyclic AMP formed per mg of protein per min)

	Adenyl cyclase activity	
Subject	Without NaF	With NaF
Controls[a]	8.9	36.2
Mother[b]	6.4	20.3
Patient J[b]	2.5	8.9
Patient D[b]	3.2	10.4

[a]Average of 12 experiments.
[b]Average of 3 experiments, patients with chronic granulomatous disease.

Table 3. Changes in Cyclic AMP Levels in the Peritoneal Macrophages of a Rabbit During Phagocytosis (pmoles of cyclic AMP per 10^6 cells)

	cAMP levels in cells		cAMP level in supernatant	
Time (min)	Resting	During phagocytosis	Resting	During phagocytosis
0	4.15	2.7	5.07	6.31
5	14.89	19.65	5.33	6.72
15	7.45	4.8	6.83	6.55
30	4.92	5.26	7.69	7.70

levels, and it is the altered level of intracellular cyclic AMP that changes the metabolic behavior of a cell so that it can be said to have "responded to the hormone."

Intracellular concentrations of cyclic AMP reflect the relative activities of at least two enzymes, which function in opposition to each other. Production of cyclic AMP is catalyzed by adenyl cyclase. This enzyme has been detected in almost every mammalian tissue in which it has been sought and is also present in the tissues of many lower organisms. Magnesium and ATP are required for its activity, so the substrate is probably Mg-ATP. In addition to cyclic AMP, 1 mole of pyrophosphate is formed per mole of ATP utilized.

Adenyl cyclase was first described in detail by Sutherland and Rall and their co-workers in 1962 (9). At that time, it was realized that the cyclase was not a simple soluble enzyme, but might be a component of a nonmitochondrial membrane. This suspicion has been borne out, of course, and with the single exception of a soluble enzyme from *Brevibacterium liquefaciens,* all adenyl cyclases that have been studied have been found to be associated with membranes. In the case of certain types of muscle, the cyclase system appears to be bound to the membranes of the sacroplasmic reticulum.

Regulation of the intracellular concentration of cyclic AMP is also controlled by one or more specific cyclic AMP phosphodiesterases. This enzyme, first described by Sutherland and Rall, hydrolyses the 3'-phosphate ester bond of cyclic AMP to yield 5'-AMP. Like adenyl cyclase, phosphodiesterase appears to be ubiquitous. It requires a divalent cation for catalytic activity, usually Mg^{2+}, and ethylene-diaminetetraacetic acid (EDTA) inhibits the enzyme activity.

In contrast to adenyl cyclase, phosphodiesterase appears to exist in the cell in either (or both) a soluble or particulate form. The enzyme has been purified by a number of investigators and has greater affinity for cyclic AMP than for other cyclic nucleotides. The Km's lie in the range of about 10^{-4} M cyclic AMP. In addition, enzymes with Km's of about 10^{-6} M were found.

Phosphodiesterase appears to be present in cells at far greater activity than is adenyl cyclase (10).

The role of cyclic AMP in the phagocytosis process has been studied by Borne et al. (8) and Weissmann et al. (11). Addition of exogenous cyclic AMP resulted in a decreased candicidal activity of leukocytes (8) and a decreased release of lysosomal hydrolases (11). Our preliminary studies show increased levels of cyclic AMP during the early stage of phagocytosis, and decreased levels of cyclic AMP were associated with a defective phagocytosis function. It is, therefore, tempting to postulate that the changes in the cyclic AMP levels during a certain stage of phagocytosis may be an essential step for the expression of phagocytosis function.

Manganiello et al. (12) reported that the levels of cyclic AMP in "purified human polymorphonuclears" showed much smaller increases in cyclic AMP content in response to latex particles as compared to those of the unfractionated leukocytes containing mixed populations. In our studies, the levels of cyclic AMP in the leukocytes of peritoneal exudate were comparable in both polymorphonuclear-rich leukocytes and leukocytes rich in macrophages and mononuclears. Therefore, it is most likely that the polymorphonuclears and macrophages are responding with increased levels of cyclic AMP at comparable magnitudes.

Further studies of the metabolism of cyclic AMP in the leukocytes of both human and experimental animals are in progress.

We thank Mrs. Sarah W. Reed for her excellent technical assistance.

Discussion of Dr. Park et al.'s Paper

DR. WEISSMANN: Am I correct in saying that in mixed populations of human leukocytes, you did not find the adenylate cyclase to be fluoride sensitive?

DR. PARK: There usually are small increases in enzyme activity with fluoride, and in some experiments, large increases are obtained, as Dr. Bourne has shown in a published paper.

DR. WEISSMANN: If, in separated leukocytes, the bulk of cyclic AMP that accumulates is not in polymorphonuclear cells, how does that affect the interpretation of the data?

DR. PARK: Polymorphonuclear leukocytes are easily affected by manipulation, and I wonder if the poor cyclic AMP response observed by other investigators in these cells might be due to nonspecific damage.

DR. WEISSMANN: There are cationic proteins in polymorphonuclear leukocytes that interfere with cyclic AMP protein-binding assays. Moreover, when you extract cyclic AMP from these cells with acid and pass the extract over a Dowex column, the peak where cyclic A emerges contains another dialyzable small molecule that interferes in the binding assay. These difficulties do not hamper chemical or immunoassay measurements of cyclic AMP, and apparently these other two methods have also indicated that these cells do not accumulate cyclic AMP after phagocytosis. How do you reconcile these observations with your theory?

DR. PARK: I was not aware of these findings.

DR. PARKER: At the risk of belaboring a point, I must ask you if the data you presented on patients with granulomatous disease involved cell preparations that contain lymphocytes?

DR. PARK: Yes.

DR. PARKER: Are there differences in the percentage of lymphocytes in preparations from patients with chronic granulomatous disease and normal controls? As already implied, the work of Manganiello and Vaughan would indicate that when mixed leukocytes are presented with a phagocytic stimulus, an increase in cyclic AMP is obtained, but Vaughan's data would further indicate that the increase is probably largely or entirely in the lymphocyte. One might, therefore, assume that your results are due to variation in numbers of lymphocytes or some quantitative change in lymphocyte function rather than changes in neutrophilic cell function.

DR. PARK: The cyclic AMP changes we observed in mixed leukocytes did not correlate with the percentages either of lymphocytes or monocytes. Rabbit peritoneal exudates rich in neutrophiles or macrophages have an easily demonstrable increase in cyclic AMP after exposure to the latex particles. Therefore, I think that increases in cyclic AMP levels may be involved in bactericidal function.

References

1. Sutherland, E. W., Oye, I., and Butcher, R. W. "The Action of Epinephrine and the Role of the Adenyl Cyclase System in Hormone Action," *Recent Progr. Horm. Res.* **21**:623 (1965).
2. Park, B. H., Good, R. A., Beck, N. P., and Davis, B. B. "Concentration of Cyclic Adenosine 3′,5′-Monophosphate in Human Leukocytes during Phagocytosis," *Nature* (London) **229**: 27 (1971).
3. Holmes, B., Quie, P. G., Windhorst, D. B., and Good, R. A. "Fatal Granulomatous Disease of Childhood: An Inborn Abnormality of Phagocytic Function," *Lancet* **1**:1225 (1966).
4. Holmes, B., Page, A. R., and Good, R. A. "Studies of the Metabolic Activity of Leukocytes from Patients with a Genetic Abnormality of Phagocytic Function," *J. Clin. Invest.* **46**:1422 (1967).
5. Good, R. A., Quie, P. G., Windhorst, D. B., Page, A. R., Rodey, G. E., White, J., Wolfson, J. J., and Holmes, B. H. "Fatal (Chronic) Granulomatous Disease of Childhood: A Hereditary Defect of Leukocyte Function." *Seminars Hematol.* **5**:215 (1968).
6. Kaneko, T., and Field, J. B. "A Method for Determination of 3′,5′-Cyclic Adenosine Monophosphate Based on Adenosine Triphosphate Formation," *J. Lab. Clin. Med.* **74**:682 (1969).
7. Beck, N. P., Reed, S. W., Murdaugh, H. V., and Davis, B. B. "Effects of Catecholamines and Their Interaction with Other Hormones on Cyclic 3′,5′-Adenosine Monophosphate of the Kidney." *J. Clin. Invest.* **51**:939 (1972).
8. Bourne, H. R., Lehrer, R. I., Cline, M. J., and Melmon, K. L. "Cyclic 3′,5′-Adenosine Monophosphate in the Human Leukocyte: Synthesis, Degradation, and Effects on Neutrophil Candidacidal Activity," *J. Clin. Invest.* **50**:920 (1971).
9. Sutherland, E. W., Rall, T. W., and Menon, T. "Adenyl Cyclase. I. Distribution, Preparation and Properties." *J. Biol. Chem.* **237**:1220 (1962).
10. Cheung, W. Y. "Cyclic Nucleotide Phosphodiesterase." In P. Greengard and E. Costa, eds. *Role of Cyclic AMP in Cell Function.* New York: Raven Press, pp. 51–65.
11. Weissmann, G., Dukor, P., and Zurier, R. B. "Effect of Cyclic AMP on Release of Lysosomal Enzymes from Phagocytes," *Nature New Biol.* **231**:131 (1971).
12. Manganiello, V., Evans, W. H., Stossel, T. P., Mason, R. J., and Vaughan, M. "The Effect of Polystyrene Beads on Cyclic 3′,5′-Adenosine Monophosphate Concentration in Leukocytes," *J. Clin. Invest.* **50**:2741 (1971).

THE ROLE OF CYCLIC AMP IN THE CONTROL OF CELL DIVISION

J. R. Sheppard

Evidence is rapidly accumulating in support of the thesis that cyclic nucleotides are important biological regulators of cellular proliferation. This chapter will be divided into four sections discussing various aspects of the association between cyclic nucleotides and cell division:

(1) review of the evidence that cyclic nucleotides are associated with the control of cell proliferation;

(2) enzymatic control of cyclic AMP levels in normal and transformed fibroblasts;

(3) possible mechanisms of cyclic AMP control of cell division;

(4) the biological signal that regulates basal cellular cyclic AMP levels.

I. Review of the Evidence that Cyclic Nucleotides Are Involved in the Regulation of Cell Proliferation

Cyclic AMP was discovered by Sutherland and Rall in 1957 to mediate the action of glucagon (1), and until recently, the biological significance of this nucleotide was generally relegated to its role in carbohydrate metabolism. One of the first reports of its possible inhibiting effect on cell proliferation was reported by Bürk in 1968 (2); simultaneous independent studies by Ryan and Heidrich (3,4) directly measured the inhibitory effect of cyclic AMP on the growth of cultured Hela and L cells. In 1969, an in vivo study by Gericke and Chandra reported the nucleotide's inhibition of tumor growth.

Johnson et al. (6) and Hsie and Puck (7) in 1970 showed that the dibutyryl derivation of cyclic AMP had an effect on the morphology of cultured animal cells. Soon after, a report from our laboratory showed that the dibutyryl cyclic AMP (DBcAMP) had a growth inhibitory effect on virus and spontaneous transformed cells growing in culture (8). Little effect of the same concentration of DBcAMP was seen on the growth of the normal, contact-inhibited cell line. We suggested at that time, because of the effect of the DBcAMP treatment on the agglutinability of the transformed cells by plant lectins, that the DBcAMP growth inhibitory effect might be membrane mediated.

Agents such as prostaglandin E_1 and cyclic nucleotide phosphodiesterase inhibitors, which increase the endogenous level of cyclic AMP in transformed cells, also inhibit the growth of those cells (9). Again, these agents have little effect on the growth of the normal cell. These data support the initial observation, which used exogenous addition of DBcAMP and theophylline.

It is interesting to note that only DBcAMP and not cyclic AMP will produce the observed effects. Possible explanations for this occurrence have been suggested (10,11), but none proved. In contradiction, Ryan and Heidrich report the exact opposite pattern in their early studies (3).

Following our initial observations, we measured the steady-state levels of cyclic AMP in the various cell lines derived from the 3T3 mouse fibroblast line (12). The cell lines that exhibit contact inhibition have higher basal levels of cyclic AMP than the cell lines that are not contact inhibited (Table 1). The data also suggest an inverse relationship between basal cyclic AMP levels and the saturation densities of different lines. This relationship, which is illustrated in Fig. 1, should not be accepted as definitive, since other variables such as substrate composition (i.e., glass versus plastic Petri dishes) are also important factors in determining cell saturation density. A relationship between growth rate and cyclic AMP levels has been shown by Otten et al. (13). A subsequent confirming study has shown similar differences in the cyclic AMP level between normal cells and virus-transformed cells derived from hamster, chicken, and monkey tissues (14).

Table 1. Steady-State Cyclic AMP Levels in Several Cell Lines

Cell line	Doubling time hr	pmole cyclic AMP mg protein	pmole cyclic AMP 10^6cells
3T3	24	22.4 ± 3.4	7.2 ± 0.8
3T3/Balb	24	23.9 ± 3.6	8.5 ± 1.0
3T6	18	9.7 ± 2.9	2.8 ± 0.5
3T12/Balb	20	11.8 ± 2.7	2.9 ± 0.5
py3T3	20	13.3 ± 3.4	3.9 ± 0.5
SV3T3	19	10.2 ± 3.1	2.4 ± 0.5
flpy	28	25.5 ± 6.0	8.9 ± 1.9
flSV	24	24.1 ± 5.2	8.5 ± 0.8

Note: The values shown are the means of 11 separate experiments; cells were assayed at the confluent stage of their growth, and each measurement was done in triplicate. Standard deviations are indicated for each value. The cells used were of Swiss mouse origin unless Balb origin is indicated. Reprinted from J. R. Sheppard, *Nature (New Biol.)* **236**:14 (1972).

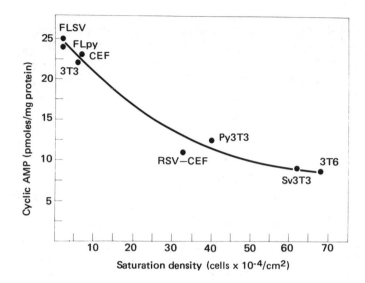

Fig. 1. Cells were grown and counted as described in References 8 and 14. The cyclic AMP levels and protein measured are as described in Reference 12. These results are the average of at least 12 separate determinations.

Once normal cells are transformed (e.g., by virus infection), a population of cells with a "normal phenotype" can be selected from the transformed cells. These cells, which still possess the integrated virus genome, are called revertants (15). We measured the cyclic AMP levels in the revertant lines, which have previously been shown to exhibit low saturation density, slow rate of growth, low tumorgenicity, and low agglutinability, all characteristics of the parent, contact-inhibited cell line (16). The cyclic AMP studies showed that the revertant lines have high basal levels of cyclic AMP, again similar to the parent normal line (Table 1).

Once normal cells reach confluency, they cease dividing, which is characteristic of contact inhibition, and will resume division only after direct stimulation. Several different agents have been described as stimulators, e.g., serum (17) and trypsin (18,19). Stimulation of confluent cellular monolayers to divide is accompanied by a transient depression of the cyclic AMP level. Soon after exposure to these growth stimulatory agents (within 5 min), the cyclic AMP level is decreased and remains low for about 120 to 180 min, after which the cyclic AMP returns to its normal elevated level (12). Furthermore, DBcAMP has been observed to inhibit or interfere with the growth stimulatory action of trypsin (20) and serum (21). In these experiments, DBcAMP was added simultaneously with the stimulating agent to confluent monolayers of cells and the resulting effect on cell division noted. In both cases, DBcAMP prevented the action of trypsin and serum, which normally stimulate cell division.

The recent studies by Hadden et al. (22) and Goldberg et al. (23) appear to support our thoughts concerning cyclic AMP and further extend them in a new direction showing that cyclic GMP levels rise in response to the growth stimulation of peripheral lymphocytes by phytohemagglutinin. These studies suggest that cyclic GMP may be an active signal to induce proliferation, while cyclic AMP may function to inhibit or limit cell division.

In summary, the following observations have been made concerning cyclic AMP and cell proliferation.

(1) Exogenous addition of DBcAMP or agents that increase the endogenous level of cyclic AMP prevent transformed cells from growing to high densities.

(2) Normal, contact-inhibited cells have higher basal levels of cyclic AMP than the corresponding virus and spontaneous transformed lines. Revertant cell lines, which have low saturation densities like the contact-inhibited parent cells, also have high basal cyclic AMP levels.

(3) Stimulation of confluent monolayers to divide is accompanied by a transient depression of the cyclic AMP level.

(4) Addition of DBcAMP with serum or trypsin (which alone stimulates the growth of confluent monolayers) prevents the stimulating effect of these agents.

These observations led to the hypothesis that elevated cellular levels of cyclic AMP are associated with the inhibition of cell growth and that depressed cyclic AMP levels are associated with stimulated or accelerated cell division.

II. Enzymic Control of Cellular Cyclic AMP Levels in Normal and Transformed Cells

After formulating the above hypothesis, we went on to investigate possible explanations for the difference in the cyclic AMP levels of the normal and transformed cells.

Two enzymes are responsible for the cellular levels of cyclic AMP— adenylate cyclase and cyclic nucleotide phosphodiesterase. Other factors may also affect the cyclic AMP level, e.g., transport between and out of cells. However, these two enzymes are today considered most important in the maintenance of the cell's cyclic AMP level.

Our studies (24) and those of Pastan's group (25, 26) on adenylate cyclase and phosphodiesterase have led to no solid evidence indicating the enzyme responsible for the difference in basal levels of cyclic AMP between normal and transformed cells.

Adenylate cyclase, which catalyzes the production of cyclic AMP from ATP, is found in a higher basal specific activity in transformed cell lines (Table 2). These data, derived from a broken cell assay, are paradoxical in view of the fact that transformed cells have lower levels of cyclic AMP. The observed decreased sensitivity of the transformed cell's adenylate cyclase to hormones reported by Perry et al. (25) has suggested an alternation in the cell's hormone receptor site or surface structure of the transformed cell. Other peripheral evidence for this possibility exists; changes in the transformed cell's plasma membrane chemical composition, antigenicity, transport, architecture, enzyme activity, and membrane potential have all been reported (27). Nevertheless, a broken cell plasma membrane preparation is a poor source for the measurement of an intact cell's basal or stimulated capacity to synthesize cyclic AMP.

Table 2. Adenylate Cyclase Specific Activity of Homogenates of Normal and Transformed Cells

	pmoles of cAMP formed per min per mg protein		
Cell type	Basal	F^- (8 mM)	PGE_1 (10 μg/ml)
3T3	6 ± 2	20 ± 4	10 ± 3
py3T3	12 ± 4	38 ± 7	29 ± 6
3T6	32 ± 7	94 ± 11	49 ± 7

Note: Adenylate cyclase was measured by the method of G. Krishna, B. Weiss, and B. B. Brodie, *J. Pharmacol. Exp. Ther.* **163**: 379 (1968), on cells grown as described in Reference 8.

Cyclic nucleotide phosphodiesterase has also been studied, but with less equivocal results. We have, again, observed an increased specific activity in the transformed cell's enzyme compared to the normal cell (Table 3). This difference in enzyme activity might account for the decreased cyclic AMP level of the transformed cells. A kinetic study of this enzyme and a comparison of these kinetic parameters from several

Table 3. Kinetic Studies of cAMP Phosphodiesterase in Normal and Transformed Cells

	V_{max} (pmoles cAMP degraded)	
Cell type	mg protein/min	K_m (μM)
3T3	3.0 ± 0.5	1.0
py3T3	6.8 ± 0.6	1.2
3T6	5.5 ± 1.4	1.0

Note: Phosphodiesterese was measured by the method of R. F. O'Dea, M. K. Haddox, and N. D. Goldberg, *J. Biol. Chem.* **246**: 6183 (1971), on cells grown as described in Reference 8.

cell lines suggest no large difference between normal and transformed lines in apparent substrate binding constants, variations in pH optima, or inhibition kinetics using theophylline, papaverine, and RO-20 (an experimental inhibitor from Dr. H. Sheppard, Hoffman LaRoche). Thus, the increase observed in transformed cell's phosphodiesterase activity may possibly be accounted for by differences in the actual amount of the enzyme or some activator substance.

At the present time, the data suggest that differences do exist in the enzymes responsible for synthesis and degradiation of cyclic AMP, but it is difficult to say whether these differences can account for the depressed cyclic AMP level of the transformed cell.

III. Cyclic AMP's Mechanism of Regulating Cell Division

The first problem we approached in an investigation into the biochemical mechanism of cyclic AMP's action on cell growth was *where in the cell's division cycle does cyclic AMP exert its effect?*

Initial studies with Dr. G. R. Shepherd suggested that transformed cells treated with DBcAMP did not accumulate in any one portion of the cell cycle (Table 4). The effect of DBcAMP was a generalized slowdown throughout the cell cycle. Other experiments showed that DBcAMP-tre-

Table 4. Cell Cycle Analysis Using Flow Microfluorimetry of DBcAMP-Treated CHO Cells[a]

Phase of cell cycle	% of cells in phase of cell cycle	
	−DBcAMP	+DBcAMP
G_1	72.1	78.1
S	15.0	9.3
$G_2 + M$	12.4	12.6

[a]Dr. G. R. Shepherd's unpublished results.
Note: Cell cycle analysis was performed by the method of M. A. Van Dilla, T. T. Trujillo, P. F. Mullaney, and J. R. Coulter, *Science* **163**: 1213 (1969). One mM dibutyryl cyclic AMP was added to experimental nutrient medium 48 hr prior to analysis. The cells in this experiment were Chinese hamster ovary cells originating in T. T. Puck's laboratory.

ated 3T6 cells continued to incorporate ^3H thymidine as measured by radioautography. Figure 2 shows that upon replating (in the absence of DBcAMP) transformed 3T6 cells previously arrested at a monolayer by DBcAMP, there is immediate incorporation of ^3H thymidine, indicating that a number of cells were stopped in the S phase of the cell cycle. A peak in the ^3H thymidine incorporation occurs 18 hr after plating, suggesting that some synchrony was achieved. However, comparison with normal confluent 3T3 cells indicated the synchrony of the DBcAMP-treated transformed cells is not nearly the degree as that which occurs at true contact inhibition. Thus, what we had originally thought to be contact inhibition is merely a pronounced slowdown of the cell's passage through the entire cell cycle. Willingham et al. have also recently re-

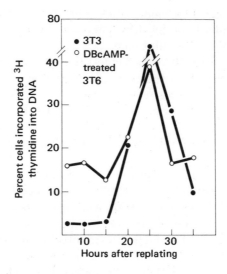

Fig. 2. Cells were grown to confluency and replated on circular cover slips as described in Reference 31. Slides with cells fixed after the appropriate interval were coated with K5 liquid emulsion (Ilford Ltd., Ilford, Essex, England) exposed for 7 days at 23° and developed in D-19 (Eastman Kodak Co., Rochester, N. Y.).

ported that cyclic AMP exerts inhibitory growth effects in at least two sites in the cell cycle (28).

While studying the cell cycle for the primary site of DBcAMP's action, we also studied the variation of cyclic AMP level throughout the cell cycle. These studies, done in collaboration with Burger et al. (20) and Prescott (29), showed that a depression of the cellular cyclic AMP level occurs during the mitotic phase. The studies with Prescott using CHO cells synchronized to at least 95% by mitotic shake-off also indicated a rebound of the cyclic AMP in very early G_1, then a stabilization late in G_1 and S. The depression of the cyclic AMP level during mitosis is paradoxical in light of Makman and Klein's (30) report of increased adenylate cyclase activity during mitosis.

The biological significance of the variation of cyclic AMP during the cell cycle is not presently clear. However, Burger et al. (20) have proposed a model for cell division control that incorporates this feature as well as the previously observed surface structural changes that occur during mitosis (31). This model suggests that the depressed cAMP level and the increased cellular agglutinability that occur during the mitotic phase in normal cells are necessary prerequisites for transient or permanent neoplastic transformation. The necessity of at least one round of cellular division (one mitosis) for the occurrence of transformation has been reported (32). The relationship between these two observed cellular changes during mitosis (depressed cyclic AMP and increased agglutinability) and their ultimate relationship to the cellular division that follows are all unknown.

IV. The Biological Stimuli Regulating Cellular Cyclic AMP Levels

Studies concerning the plasma membrane and the enzymes localized there led to our initial investigations concerning cyclic AMP and cell growth (8). The possibility existed that membrane interactions might stimulate the surface membrane-bound adenylate cyclase, which would lead directly to increased cyclic AMP levels and possibly inhibit or stimulate cell proliferation. Transformed cells, which have an altered membrane structure, might be insensitive to the cell contact signal. This was the working hypothesis of both ours (8) and the Pastan group (7). We tested it by measuring the level of cyclic AMP in normal and transformed cells before and after confluency (12). If cell-cell contact or membrane stimulation of adenylate cyclase was responsible for the inhibition of growth, which describes contact inhibition, cyclic AMP levels should rise in the contact-inhibited cells at confluency; the cyclic AMP level should not change in transformed cells at confluency. The results presented in Fig. 3 show that no observed change in the cyclic AMP level occurred in either the normal or transformed lines after reaching confluency. The cyclic AMP levels of the normal cells were higher than the transformed

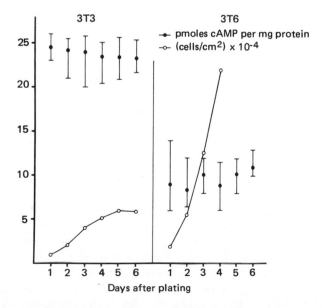

Fig. 3. Cyclic AMP levels at confluency. (Reprinted from Reference 12.)

cells before confluency and remained proportionately higher after confluency. Pastan's group has presented evidence supporting the hypothesis (13), which, although an attractive one, is not supported by our
data.

As an alternative to this initial hypothesis for the biological stimulus
to which cyclic AMP responds in its regulation of cell division, we suggest
a new model initially proposed at a recent symposium (33). The idea is
not original and had been mentioned earlier by Robisón et al. (34) in
regard to the bacterial system. These authors suggest that glucose, a
basic carbon source, which is known to depress *E. coli* cyclic AMP levels
(35), may function as a primitive hormone similar to action to that of insulin, some of the prostaglandins, and the alpha adrenergic
catecholamines in mammalian systems. In the absence of glucose, cyclic
AMP levels in the bacteria increase 10-fold. These elevated cyclic AMP
levels lead to the reversal of catabolite repression, the mechanism of
which Pastan and Perlman have investigated and elegantly described
(36).

Certain key nutrients such as glucose may also play a similar role in
the animal cell system. While these nutrients are available, cyclic AMP
levels are low, and cell proliferation is maximal. If a nutrient becomes
limiting in the medium or transport into the cell is impaired, cyclic AMP
levels would rise, which serves as a signal for the cell to slow its rate of
growth. In short, we suggest that the cellular cyclic AMP level responds
to the availability of nutrients. Artificially increasing the cyclic AMP
level (by exogenous addition of DBcAMP or stimulation of the endogenous cyclic AMP level by prostaglandin E_1 or phosphodiesterase inhibitors) tricks the cell into a response (inhibited growth) that normally
occurs during nutrient deprivation. This biological stimulus, which
results in elevated cellular cyclic AMP levels, is more basic than the previously conceived contact-inhibition, cell-interaction hypothesis. Elevated levels of cyclic AMP may be a prerequisite for contact inhibition,
but it appears that cyclic AMP does not directly participate in the contact-inhibition phenomenon. Furthermore, if the cyclic AMP level does
indeed rise in some instances after confluency occurs (13), the elevation
could be reconciled with this revised hypothesis by citing an effect of
confluency on the transport of nutrients (37).

The accelerated growth of transformed cells may result from a
membrane change, which leads to facilitated availability of nutrients-
stimulated transport—which would keep cyclic AMP levels low and the
rate of proliferation maximal. Aberrations in the cell membrane-
transport mechanism have been proposed as the primary control site
responsible for the growth of neoplasms (38, 39), and transformed cells
have been shown to possess increased rates of transport for amino acids
(40, 41), phosphate (42), and glucose (43, 44) in comparison to contact-
inhibited cells.

Two different laboratories recently reported that exogenous addition of DBcAMP inhibits cellular transport mechanisms (45, 46). If an elevated cyclic AMP level is a general cellular stress signal in response to the absence of a necessary nutrient, one of the fastest ways to stop further progress through the cell cycle would be to inhibit transport of other key nutrients. This would prevent unbalanced growth. The observations presented here that treatment of animal cells with DBcAMP does not affect one specific phase of the cell cycle (e.g., does not lead to a buildup of the cells in the G_1 phase, but results in a generalized slowdown) support the hypothesis that cyclic AMP levels serve as a potential stress signal in response to many biological stimuli and not as a specific mediator of the contact-inhibition phenomenom.

Discussion of Dr. Sheppard's Paper

DR. AUSTEN: What about cyclic GMP?

DR. SHEPPARD: Our working hypothesis could be described in Nelson Goldberg's scheme as a unitary bidirectional type-B model. We are now seriously considering a dualistic bidirectional type-B model incorporating his cyclic GMP observations.

DR. GOLDBERG: We have measured steady-state levels of cyclic GMP in 3T3 and 3T6 cells and have found that the levels of cyclic GMP are four- to fivefold higher in 3T6 cells (the spontaneously transformed fibroblasts) than in untransformed 3T3 cells. We have also shown that serum that is mitogenic increases cGMP levels in 3T3 cells. So there is a consistency; several mitogenic agents are associated with cellular cyclic GMP accumulation, and the levels of this cyclic nucleotide are higher in transformed versus untransformed cells. It is hard to explain what these differences in steady-state levels mean, because we have put forth the idea that the temporally discrete increases in cyclic GMP levels induced by mitogens represent a trigger type of signal. Of course, we are not working with synchronized cells, so it could be theorized that there are enough cells going through initiation to account for the increased level of cyclic GMP.

References

1. Sutherland, E. W., and Rall, T. W. *J. Amer. Chem. Soc.* **79**:3608 (1957).
2. Burk, R. R. *Nature* **219**:1272 (1968).
3. Ryan, W. L., and Heidrich, M. L. *Science* **162**:1484 (1968).
4. Heidrich, M. L., and Ryan, W. L. *Cancer Res.* **30**:367 (1970).
5. Gericke, D., and Chandra, P. *Hoppe-Seyler's Z. Physiol. Chem.* **350**:1469 (1970).
6. Johnson, G. S., Friedman, R. M., and Pastan, I. *Proc. U. S. Nat. Acad. Sci.* **68**:425 (1971).
7. Hsie, A. W., and Puck, T. T. *Proc. U. S. Nat. Acad. Sci.* **68**:358 (1971).
8. Sheppard, J. R. *Proc. Nat. Acad. Sci. U.S.*, **68**:1316, 1971.
9. Sheppard, J. R. (unpublished data).

10. Posternak, T., Sutherland, E. W., and Henion, W. F. *Biochim. Biophys. Acta* **65**:558 (1962).
11. Heersche, J. N. M., Fedak, S., and Aurbach, G. O. *J. Biol. Chem.* **246**:6770 (1971).
12. Sheppard, J. R. *Nature (New Biol.)* **236**:14 (1972).
13. Otten, J., Johnson, G. S., and Pastan, I. *Biochem. Biophys. Res. Commun.* **44**:1192 (1971).
14. Sheppard, J. R., and Lehman, J. M. In *Medical Aspects of Prostaglandins and Cyclic AMP,* R. Kahn, ed. New York: Academic Press (1972).
15. Pollack, R., Green, H., and Todaro, G. *Proc. U. S. Nat. Acad. Sci.* **60**:126 (1968).
16. Pollack, R., and Teebor, G. *Cancer Res.* **29**:1770 (1969).
17. Todaro, G. J., Lazar, G. K., and Green, H. *J. Cell Comp. Physiol.* **66**:325 (1965).
18. Burger, M. M. *Nature* **227**:843 (1970).
19. Sefton, B. M., and Rubin, H. *Nature* **227**:843 (1970).
20. Burger, M. M., Bombik, B. M., Breckenridge, B. M., and Sheppard, J. R. *Nature (New Biol.)* **239**:161 (1972).
21. Froelich, J. E., and Rachmeler, M. *J. Cell Biol.* **55**:19 (1972).
22. Hadden, J. W., Hadden, E. M., Haddox, M. K., and Goldberg, N. O. *Proc. Nat. Acad. Sci.* (U.S.A.) **69**:3024 (1972).
23. Goldberg, N. D., Haddox, M. K., Hartle, D. K., and Hadden, J. W. *Fifth International Congress on Pharmacology,* Vol. 5 (in press—1972).
24. Sheppard, J. R., and McGloughlin, Wm. *Life Sci.* (in press).
25. Peery, C. V., Johnson, G. S., and Pastan, I. *J. Biol. Chem.* **246**:5785 (1971).
26. d'Armiento, M., Johnson, G. S., and Pastan, I. *Proc. Nat. Acad. Sci.* (U.S.A.) **69**:459 (1972).
27. Burger, M. M. In *Current Topics in Cellular Regulation,* B. L. Horecker, ed., Vol. 3. New York: Academic Press, p. 135 (1971).
28. Willingham, M. E., Johnson, G. S., and Pastan, I. *Biochem. Biophys. Res. Commun.* **48**:743 (1972).
29. Sheppard, J. R., and Prescott, D. M. *Exp. Cell Res.* **75**:293 (1972).
30. Makman, M. H., and Klein, M. I. *Proc. Nat. Acad. Sci.* (U.S.A.) **69**:456 (1972).
31. Fox, T. O., Sheppard, J. R., and Burger, M. M. *Proc. Nat. Acad. Sci.* (U.S.A.) **68**:244 (1971).
32. Todaro, G. J., and Green, H. *Proc. Nat. Acad. Sci.* (U.S.A.) **55**: 302 (1966).
33. Sheppard, J. R. In *Membranes, Viruses and Immunopathology,* S. Day and R. A. Good, eds. New York: Academic Press (1972).
34. Robison, G. S., Butcher, R. W., and Sutherland, E. W. *Cyclic AMP.* New York: Academic Press, p. 427 (1971).
35. Makman, R. S., and Sutherland, E. W. *J. Biol. Chem.* **240**:1309 (1965).
36. Pastan, I., and Perlman, R. L. *Nature (New Biol.)* **299**:1 (1971).
37. Sefton, B. M., and Rubin, H. *Proc. Nat. Acad. Sci.* (U.S.A.) **68**:3154 (1971).
38. Pardee, A. B. *Nat. Cancer Inst. Mono.* **14**:7 (1964).
39. Wallach, D. F. H. *Proc. Nat. Acad. Sci.* (U.S.A.) **61**:868 (1968).
40. Foster, D. O., and Pardee, A. B. *J. Biol. Chem.* **244**:2675 (1969).
41. Isselbacher, K. J. *Proc. Nat. Acad. Sci.* (U.S.A.) **69**:585 (1972).

42. Cunningham, D. O., and Pardee, A. B. *Proc. Nat. Acad. Sci.* (U.S.A.) **64**:1049 (1961).
43. Hatananka, M., and Hanafusa, H. *Virology* **41**:647 (1970).
44. Martin, G., Venuta, S., Weber, M., and Rubin, H. *Proc. Nat. Acad. Sci.* (U.S.A.) **68**:2739 (1971).
45. Grimes, W. J. *J. Cell Biol.* (in press).
46. Hauschka, P. V., Everhart, L. P., and Rubin, R. W. *Proc. Nat. Acad. Sci.* (U.S.A.) (in press).

CONTROL OF NORMAL CELL PROLIFERATION IN VIVO AND IN VITRO BY AGENTS THAT USE CYCLIC AMP AS THEIR MEDIATOR

J. P. MacManus, J. F. Whitfield, and R. H. Rixon

I. Introduction

Over the past several years, while the mechanism whereby the hormones of the thyro-parathyroid complex control calcium metabolism in their classical target organs (bone and kidney) has been under intense investigation (1), another role for these hormones has been emerging. It can now be clearly shown that these hormones, parathyroid hormone and calcitonin, intimately control cell proliferation in bone marrow (13, 19–23,31), thymus (8,9,13,18,19,31), and liver (24,31). This retrospective report will give some indication of the physiological importance of this control by the calcium homeostatic system (31). Then two in vitro models will be discussed. Finally, our thoughts on the mechanism whereby cell proliferation is controlled will be presented.

II. In Vivo Experiments

The circulating plasma calcium concentration can be lowered by surgical removal of the parathyroid glands, provided the animals are also given a low-calcium diet [Fig. 1(a)]. When this occurs, there is an atrophy of the thymus gland, so severe that after nine days only one-tenth of the tissue mass remains [Fig. 1(b)]. The mitotic activity of the thymus also decreases (8,18). Thus, the atrophy is caused by the continuous normal loss of cells from the thymus (16,18) without concomitant replacement. Similarly, in the absence of the parathyroid glands, the bone marrow becomes hypoplastic [Fig. 1(c)] due to the decrease in marrow mitotic activity [Fig. 1(d)]. In this case, there is a selective nonrenewal of cells of the erythroid and lymphoid subpopulations, but the myeloid elements are unaffected (23). There is also a decrease in the synthesis of hemoglobin as evidenced by a severe depression (85%) of iron incorpora-

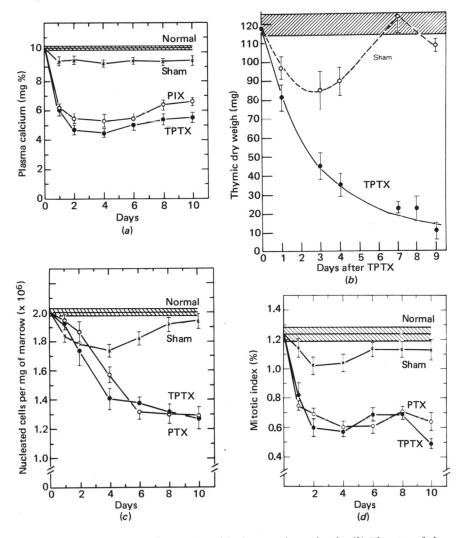

Fig. 1. Effect of parathyroidectomy on: (*a*) plasma calcium levels. (*b*) The size of the thymus gland. (*c*) The size of the total femoral bone cell population. (*d*) Bone marrow mitotic index. All animals were pair-fed a calcium-deficient diet. PTX = parathyroidectomy by electrocautery. TPTX = surgical thyroparathyroidectomy. Mean ± SEM are indicated.

tion into red blood cells (19,20). This malfunctioning of marrow activity in an aparathyroid rat leaves the animal unable to cope normally with a physiological stress such as hemorrhage (21).

This central influence of the calcium homeostatic system on cell proliferation is not limited to renewal tissues such as thymus or bone marrow. The absence of the parathyroid glands also severely reduces and delays the proliferative transformation of liver cells following partial hepatectomy (Fig. 2).

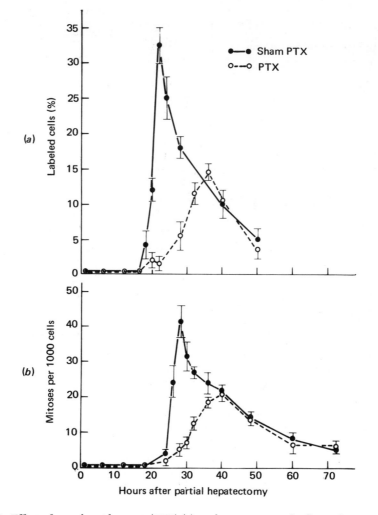

Fig. 2. Effect of parathyroidectomy (PTX) (*a*) on the proportion of cells synthesizing DNA or (*b*) entering mitosis at various times following partial hepatectomy. The parathyroid glands were removed 24 hr before surgery, and the animals were starved thereafter. Each value is the mean ± SEM of from five to nine rats. [Reproduced from Rixon and Whitfield, 1972 (24) with permission of the Society for Experimental Biology and Medicine, New York.]

All of these phenomena can be ascribed to a failure of potential progenitor cells to expand into a proliferative subpopulation in the presence of a low circulating plasma calcium concentration. This is confirmed by the ability of treatment that again increases the blood calcium level to restore proliferative responses (18,21,23), that is, by the stimulation of mitotic activity by calcium. An example of this is seen in rats serially injected with calcium chloride to increase extracellular calcium levels (Fig.

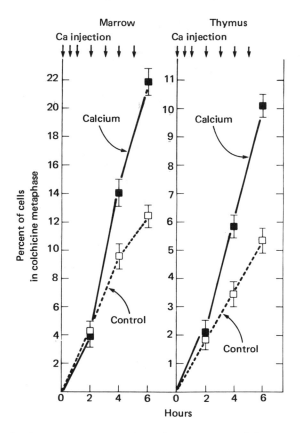

Fig. 3 Effect of calcium on mitotic activity in bone marrow and thymus. Arrows indicate serial injections of 1 ml of 62.5 mM calcium chloride. Each value is the mean ± SEM of at least six rats. [Redrawn from Perris, 1971 (19).]

3). Proliferating cells are trapped in metaphase by the mitotic poison, colchicine, and the accumulation of such metaphase cells indicates the mitotic activity of the tissue. It can be seen that animals treated with calcium have a higher mitotic rate in both marrow and thymus than normal animals (Fig. 3). Calcium is envisaged as governing the intensity of mitotic activity in a tissue, which, in turn, controls that particular tissue's response to a call for increased proliferative action.

III. In Vitro Models

The ability of an increase in extracellular calcium to stimulate cell proliferation can also be demonstrated in cells maintained in vitro. Thymic lymphocytes suspended in a medium free of serum continue to proceed into mitosis after isolation (Fig. 4). Raising the calcium concentration in such a medium, from 0.6 mM to 1.2 mM or more, stimulates a

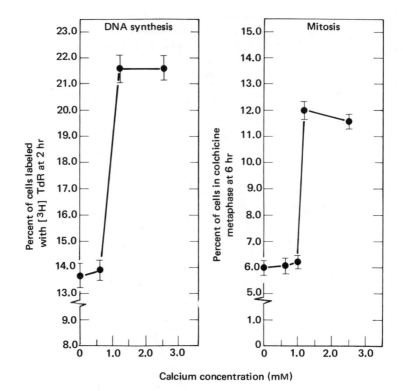

Fig. 4. Effect of calcium on DNA synthesis and mitotic activity in thymocytes suspended in MAC-1 medium free of serum. Cells were suspended in medium in the presence [³H]TdR (2 μCi/ml) and autoradiographs made after 2 hr (11). Colchicine was also present (0.06 mM), and the accumulated colchicine-metaphase cells were scored at 6 hr. Each value is the mean ± SEM of eight separate determinations.

new population of lymphoblasts to initiate DNA synthesis, and these stimulated cells continue into mitosis (Fig. 4). The ability of calcium to increase mitotic activity in vivo (Fig. 3) is, therefore, believed to be due to the ability of the cation to initiate the synthesis of DNA (Fig. 4).

This action of calcium can also be demonstrated in avian fibroblasts grown in culture. Normal, freshly isolated chicken fibroblasts maintained in a medium, containing homologous plasma rather than serum, and a calcium concentration approaching zero, do not synthesize DNA or multiply (Fig. 5). When such cells are exposed to a final calcium concentration of 1.5 mM, they begin to synthesize DNA and subsequently proceed into mitosis (Fig. 5). It is also of interest that such fibroblasts, when infected with *Rous sarcoma* virus, are no longer controlled by calcium (2). Thus, calcium switches on a "master reaction," which governs the cell's ability to proliferate, and the transformation to neoplasia may involve loss of this overall control.

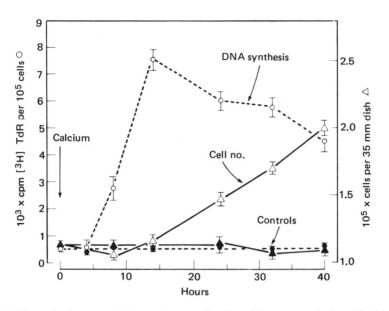

Fig. 5. Effect of calcium on DNA synthesis and cell proliferation in chicken fibroblasts maintained for three days in a medium containing homologous plasma and a very low con centration of calcium. At zero time, calcium was added to a concentration of 1.5 mM. Each value is the mean ± SEM from four culture dishes. [Redrawn from Whitfield et al., 1972 (31).]

IV. Mechanism

If calcium controls a master reaction, what are the elements of such a control system? Because cyclic AMP is involved in the cellular mediation of the hormones associated with calcium homeostasis (1), it was felt that the cyclic nucleotide might be involved in the control of cell proliferation. An increase in the calcium concentration of the suspension medium of thymocytes above 1.2 mM did lead to an increase in the cellular level of cyclic AMP (Fig. 6). This was seen not as an increase in absolute concentrations of the cyclic nucleotide, but as a relative increase in radioactive cyclic AMP in cells that had a radioactive ATP pool produced by preincubation with ^{3}H adenine (Fig. 6). In fact, there was a similar relation between the extracellular calcium concentration and all three parameters studied: cyclic AMP levels, DNA synthesis, and mitosis (Figs. 6 and 4). The cyclic AMP concentration was not increased by a stimulation of adenylate cyclase (12,28), but an inhibition of a phosphodiesterase was considered a possible mechanism (12,28). However, other mechanisms are now considered possible for calcium action (see below). Several hormones, e.g., parathyroid hormone, have also been found to stimulate DNA synthesis and cell proliferation in

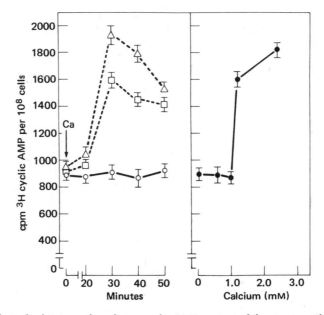

Fig. 6. Effect of calcium on the relative cyclic AMP content of thymocytes. The ATP pool of the cells was first labeled by incubating the cells in medium containing 5 μCi/ml of ^{3}H adenine (6 Ci/mmole) for 2 hr (11). The cells were then washed and incubated in medium containing the indicated concentration of calcium. The cells were ruptured and the ^{3}H cyclic AMP isolated by Dowex chromatography (11). Each value is the mean $\pm$ SEM of eight separate determinations.

thymocyte populations, provided calcium was present in the medium (13,26,28).

The ability of cyclic AMP to stimulate DNA synthesis and cell proliferation was investigated to substantiate this proposed role as mediator of calcium's mitogenic effects. Exposure to cyclic AMP at 10^{-7} M increased, in 1 hr, the proportion of cells in a thymocyte population synthesizing DNA, and these stimulated cells subsequently enter mitosis (Fig. 8). A dose-response curve indicated that the cyclic nucleotide can stimulate mitotic activity from 10^{-8} to 10^{-6} M; higher concentrations were found to be either ineffective or toxic (Fig. 7). Such mitogenic effects of cyclic AMP are specific since 5'-AMP gave no such stimulation (Fig. 7). Also, cyclic AMP exerts such a mitogenic action even in the complete absence of ionic calcium (27). Therefore, cyclic AMP is very probably the intracellular mediator of the mitogenic action of calcium.

These doses of cyclic AMP (10^{-7} to 10^{-6} M), which stimulate cell proliferation, do not increase the cellular content of the cyclic nucleotide (unpublished observations), nor does the cyclic AMP even enter the cell at this extracellular concentration (10). Specific binding sites for cyclic AMP have been demonstrated on the surface of these cells (14). An ac-

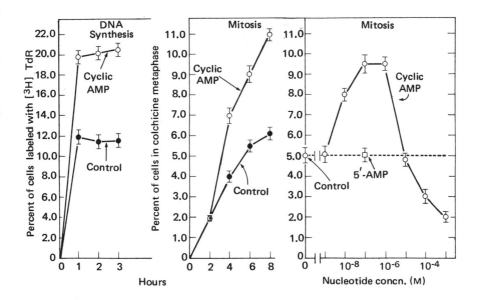

Fig. 7. Effect of cyclic AMP on DNA synthesis and mitosis in thymocytes. DNA synthesis and mitotic activity were measured as in Fig. 4. 10^{-7}M cyclic AMP was used in the time-course experiments on both DNA synthesis and mitosis. The total accumulation of cells in colchicine metaphase at 6 hr was used in experiments on the dose-response relation. Each value is the mean ± SEM of at least eight separate experiments.

tivation site containing adenylate cyclase, cyclic AMP receptors, and phosphodiesterase is visualized in the cell membrane (10,14,30,31), which can turn on the cellular proliferative machinery.

Further evidence implicating cyclic AMP in the control of cell proliferation was obtained with prostaglandins. Prostaglandin E_1 (PGE_1) and also PGA_1, stimulated adenylate cyclase from thymocytes (5), increased both the relative and absolute levels of cyclic AMP, initiated DNA synthesis, and increased mitotic activity (Fig. 8). The failure of $PGF_{1\alpha}$ (differing in only one H atom from PGE_1) to stimulate adenylate cyclase or raise cyclic AMP levels (5) made this prostaglandin unable to increase either DNA synthesis or cell proliferation (5). The involvement of cyclic AMP is also supported by the use of imidazole, a compound which stimulates cyclic AMP hydrolysis (12), and caffeine, which inhibits cyclic AMP hydrolysis (12). These compounds lead to inhibition and potentiation, respectively, of the effects of calcium (12) and other hormones (8,11,27) on mitotic activity.

Since cyclic AMP appears as an element in the control of proliferation of thymic lymphoblasts, evidence of its involvement in other proliferative models was sought. Therefore, we determined the changes in cyclic AMP concentrations in the liver remnant after partial (70%) hepatectomy, a procedure that initiates the proliferative transformation

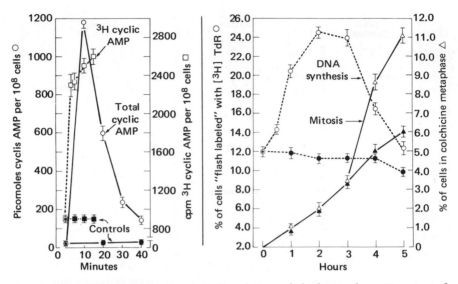

Fig. 8. Effect of PGE$_1$ (5 μg/ml) on both the relative and absolute cyclic AMP content of thymocytes, and also on DNA synthesis and mitosis. The relative cyclic AMP content was measured as in Fig. 6, and the absolute content by the method of Gilman described in Whitfield et al., 1972 (30). Synthesis and mitosis were measured as in Fig. 4. Each value is the mean ± SEM of at least 12 separate experiments.

of hepatocytes by a process controlled by the calcium homeostatic system (Fig. 2, References 24,31). It was found that this operation caused an unusual biphasic increase in cyclic AMP levels prior to the initiation of DNA synthesis (Fig. 9). Similarly, induction of DNA synthesis in *intact* liver by hormone infusion (25) is preceded by two waves of cyclic AMP accumulation (15). By using various adrenergic blocking agents, it has been possible to show that the first wave of cyclic AMP accumulation has nothing to do with the initiation of DNA synthesis, but delaying or reducing the second wave of accumulation correspondingly delays or reduces DNA synthesis (MacManus et al., in preparation).

Thus, the immediate pre-DNA synthetic events in liver following partial hepatectomy and in thymocytes treated with calcium or PGE$_1$ are essentially similar. Also, very interestingly, in established cell lines grown to confluency, which are stimulated to proliferate by addition of pronase or serum, an immediate pre-S rise in cyclic AMP levels has also been noted (4,17). However, it is the *rapid* drop in cyclic AMP levels soon after treatment that has hitherto been considered to be responsible for the subsequent initiation of DNA synthesis (4, 17).

V. Conclusion

It would appear from the foregoing that cyclic AMP is central to the control of cell proliferation (31) and probably mediates the mitogenic ac-

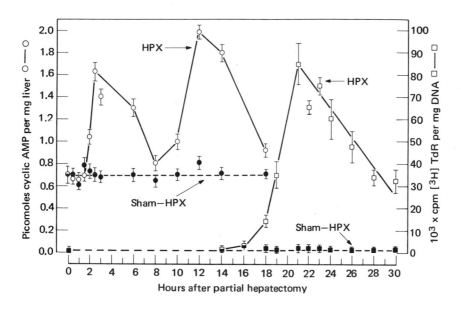

Fig. 9. Effect of 70% partial hepatectomy (HPX) on the cyclic AMP content and DNA synthesis in liver. The absolute content of cyclic AMP was measured as in Fig. 8. Each value is the mean ± SEM of between 10 and 30 separate determinations. [Reproduced from MacManus et al., 1972 (15) with permission of the Academic Press, New York.]

tion of calcium and several mitogenic hormones. However, although an immediate pre-S-phase burst of cyclic AMP accumulation is regarded as crucial to the initiation of DNA synthesis, it remained unclear how agents, such as calcium and various calcium-dependent hormones (12,-13,26,28), initiate events that lead to this increase in cyclic AMP and the eventual stimulation of DNA synthesis.

While it had been proposed that calcium and calcium-dependent hormones raise the cyclic AMP content of thymocytes by reducing the activity of a membrane-bound phosphodiesterase (12), we suspected that this initial reaction leading to proliferation was more complex. In faltering steps to clarify this reaction, we tried several years ago to stimulate cell proliferation with the other naturally occurring cyclic nucleotide, cyclic GMP (29). At concentrations where exogenous cyclic AMP was ineffective (10^{-11} to 10^{-10} M), exogenous cyclic GMP strongly stimulated thymic lymphoblast proliferation by a process which caused an elevation of the cellular cyclic AMP level (Fig. 10, Reference 29). This observation, plus recent reports on a calcium-dependent increase in cyclic GMP levels in lymphocytes treated with phytohemagglutinin (6), prompted us to present the scenario in Fig. 11.

Calcium, unlike its effect on adenylate cyclase, can stimulate guanylate cyclase (7). Thus, PHA in lymphocytes (32,33) or other hormones such as PTH (3) cause an influx of calcium into cells, which, in turn, leads

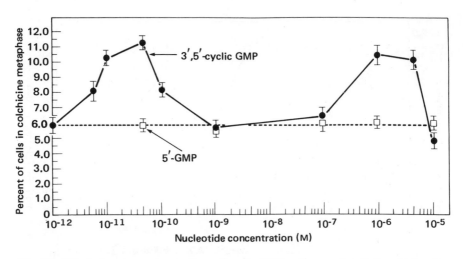

Fig. 10. Effect of various concentrations of cyclic GMP on thymocyte mitotic activity after 6 hr. The level of mitotic activity in untreated cell populations is represented by the broken line at 5.9 ± 0.1% (n = 40). Each value is the mean ± SEM of at least 10 separate determinations. [Reproduced from Whitfield et al., 1971 (29) with permission from the Society for Experimental Biology and Medicine, New York.]

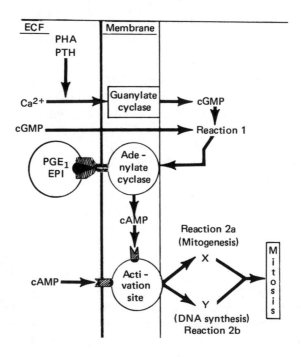

Fig. 11. Scenario presenting an integration of available data on the elements in the control of the initiation of DNA synthesis and cell proliferation. PHA = phytohemagglutinin; PTH = parathyroid hormone; PGE_1 = prostaglandin E_1, EPI = epinephrine, and ECF = extracellular fluid.

to an increase in cyclic GMP (Fig. 11). Thus, cyclic GMP can completely substitute for calcium and stimulate lymphoblast proliferation in serum-free medium without added calcium (Whitfield et al., in preparation).

Such an elevation of cyclic GMP is envisaged to lead to a heretofore unproved Reaction 1 (Fig. 11), which leads, in turn, to a stimulation of adenylate cyclase. The eventual stimulation of this enzyme causes an immediate pre-S-phase increase in cyclic AMP levels, either locally within the cell membrane or in the cytosol. This pre-S burst of cyclic AMP accumulation may occur rapidly as in thymic lymphoblasts (Fig. 8, References 5,30) or after many hours as in liver (15), established cell lines (4, 17), or peripheral lymphocytes. The newly synthesized cyclic AMP finally stimulates another group of reactions (Reactions 2a and 2b), which initiate events culminating in DNA synthesis and mitosis (Fig. 11).

Obviously, this cGMP-initiated cascade of reactions can be "short-circuited" by different agents at several points (Fig. 11). For example, this reaction sequence accounts for that adenylate cyclase stimulators such as epinephrine (11) or PGE_1 (5,30), as well as cyclic AMP itself, do not need calcium to promote DNA synthesis and cell proliferation. Despite the lack of evidence for Reaction 1, this scenario permits the integration of all of the available data, and it will facilitate the formulation of pertinent experiments.

Discussion of Dr. MacManus et al.'s Paper

DR. SHEARER: Have your cell proliferation studies in bone marrow and thymus been done only in rats or have mice and other species been used as well? There are genetic mutants in the mouse that control erythropoiesis. The genes involved control the differentiation of erythroid cells. It would be interesting to study these genetic models in your system.

DR. MACMANUS: Yes, it would.

DR. SHEARER: Do you known of anyone studying the effects of parathyroid hormone in these animal strains?

DR. MACMANUS: No.

DR. BLOOM: Is it your feeling that calcium acts on a different cell subpopulation in the thymus than PHA does?

DR. MACMANUS: Yes, I would think so. We are dealing with actively cycling cells, cells that make a G_1 to S commitment in the presence of calcium. We are not transforming them at all. I presume these cells are not affected by PHA.

DR. HIRSCHHORN: Do you known the relationship of your actively cycling cells to cortisol-resistant cells?

DR. MACMANUS: Very small doses (10^{-8} M) of cortisol will initiate DNA synthesis in the same cells that respond to calcium. The cortisol stimulation is also a calcium-dependent phenomenon. But we also get the expected cytolytic effect of cortisol at 10^{-6} M that everybody sees. Our

thymocytes do not appear to contain a cortisol-resistant population in as
much as all the cells are killed.

DR. HARDMAN: Would you comment further on the antiproliferative
effects of cyclic GMP that you have seen in the mid concentration range.

DR. MACMANUS: When we expose cells to PGE_1 and intermediate
concentrations of cyclic GMP at the same time, they will initiate DNA
synthesis, but they do not appear in mitosis. This is due to a block in G_2.
The inhibitory effect of cyclic GMP depends on extracellular calcium. If
calcium is not there, cyclic GMP does not arrest the cells in G_2, so we
have another point of control in the cell cycle. In our presentation, I was
specifically talking about an immediate pre-S-phase rise in cAMP being
correlated with the initiation of DNA synthesis. Increases or decreases in
cyclic AMP or cyclic GMP at other stages of the cell cycle lead to
different effects.

References

1. Aurbach, G. D., Marcus, R., Heersche, J. N. M., Winickoff, R. N., and Marx,
    S. J. "Cyclic Nucleotides in the Action of Native and Synthetic Parathyroid
    and Calcitonin Peptides." In *Calcium, Parathyroid Hormone, and the
    Calcitonins.* Amsterdam: Excerpta Medica, pp. 502–510 (1972).
2. Balk, S. D. "Calcium as a Regulator of the Proliferation of Normal, But Not
    Transformed, Chicken Fibroblasts in a Plasma-Containing Medium," *Proc.
    Nat. Acad. Sci.* (Wash.) **68**:271–275 (1971).
3. Borle, A. B. "Parathyroid Hormone and Cell Calcium." In *Calcium,
    Parathyroid Hormone, and the Calcitonins.* Amsterdam: Excerpta Medica,
    pp. 484–491 (1972).
4. Burger, M. M., Bombik, B. M., Breckenridge, B. M., and Sheppard, J. R.
    "Growth Control and Cyclic Alterations of Cyclic AMP in the Cell Cycle,"
    Nature (London) **239**:161–163 (1972).
5. Franks, D. J., MacManus, J. P., and Whitfield, J. F. "The Effect of
    Prostaglandins on Cyclic AMP Production and Cell Proliferation in Thymic
    Lymphocytes," *Biochem. Biophys. Res. Commun.* **44**:1177–1183 (1971).
6. Hadden, J. W., Hadden, E. M., Haddox, M. K., and Goldberg, N. D.
    "Guanosine 3′,5′-Cyclic Monophosphate: a Possible Intracellular Mediator
    of Mitogenic Influences in Lymphocytes," *Proc. Nat. Acad. Sci.* (Wash.)
    69:3024–3027 (1972).
7. Hardman, J. G., Beavo, J. A., Gray, J. P., Chrisman, T. D., Patterson, W. D.,
    and Sutherland, E. W. "The Formation and Metabolism of Cyclic GMP,"
    Ann. N.Y. Acad. Sci. **185**:27–35 (1971).
8. MacManus, J. P., and Whitfield, J. F. "Inhibition by Thyrocalcitonin of the
    Mitogenic Actions of Parathyroid Hormone and Cyclic Adenosine 3′,5′-
    Monophosphate on Rat Thymocytes," *Endocrinology* **86**:934–939 (1970).
9. MacManus, J. P., Perris, A. D., Whitfield, J. F., and Rixon, R. H. "Control of
    Cell Division in Thymic Lymphocytes by Parathyroid Hormone,
    Thyrocalcitonin and Cyclic Adenosine 3′,5′-Monophosphate." In *Proceed-
    ings of the Fifth Leukocyte Culture Conference.* New York: Academic
    Press, pp. 125–142 (1970).

10. MacManus, J. P., Whitfield, J. F., and Braceland, B. "The Metabolism of Exogenous Cyclic AMP at Low Concentrations by Thymic Lymphocytes," *Biochem. Biophys. Res. Commun.* **42**:503–509 (1971).

11. MacManus, J. P., Whitfield, J. F., and Youdale, T. "Stimulation by Epinephrine of Adenyl Cyclase Activity, Cyclic AMP Formation, DNA Synthesis and Cell Proliferation in Populations of Rat Thymic Lymphocytes," *J. Cell Physiol.* **77**:103–116 (1971).

12. MacManus, J. P., and Whitfield, J. F. "Cyclic AMP Mediated Stimulation by Calcium of Thymocyte Proliferation," Exp. Cell Res. **69**:281–288 (1971).

13. MacManus, J. P., Youdale, T., Whitfield, J. F., and Franks, D. J. "The Mediations by Calcium and Cyclic AMP of the Stimulatory Action of Parathyroid Hormone on Thymic Lymphocyte Proliferation." In *Calcium, Parathyroid Hormone, and the Calcitonins.* Amsterdam: Excerpta Medica, pp. 338–350 (1972).

14. MacManus, J. P., and Whitfield, J. F. "Cyclic AMP Binding Sites on the Cell Surface of Thymic Lymphocytes," *Life Sci.* **11**:837–845 (1972).

15. MacManus, J. P., Franks, D. J., Youdale, T., and Braceland, B. M. "Increases in Rat Liver Cyclic AMP Concentrations Prior to the Initiation of DNA Synthesis Following Partial Hepatectomy or Hormone Infusion," *Biochem. Biophys. Res. Commun.* **49**:1201–1207 (1972).

16. Metcalf, D. *The Thymus.* New York: Springer-Verlag (1966).

17. Otten, J., Johnson, G. S., and Pastan, I. "Regulation of Cell Growth by Cyclic Adenosine 3′,5′-Monophosphate," *J. Biol. Chem.* **247**:7082–7087 (1972).

18. Perris, A. D., Weiss, L. A., and Whitfield, J. F. "Parathyroidectomy and the Induction of Thymic Atrophy in Normal, Adrenalectomized and Orchidectomized Rats," *J. Cell Physiol.* **76**:141–150 (1970).

19. Perris, A. D. "The Calcium Homeostatic System as a Physiological Regulator of Cell Proliferation in Mammalian Tissues." In *Cellular Mechanisms for Calcium Transfer and Homeostasis.* New York: Academic Press, pp. 101–131 (1971).

20. Perris, A. D., and Whitfield, J. F. "Calcium Homeostasis and Erythropoietic Control in the Rat," *Can. J. Physiol. Pharmacol.* **49**:22–35 (1971).

21. Perris, A. D., MacManus, J. P., Whitfield, J. F., and Weiss, L. A. "Parathyroid Glands and Mitotic Stimulation in Rat Bone Marrow After Hemorrhage," *Amer. J. Physiol.* **220**:773–778 (1971).

22. Rixon, R. H. "Mitotic Activity in Bone Marrow of Rats and Its Relation to the Level of Plasma Calcium," *Curr. Mod. Biol.* **2**:68–74 (1968).

23. Rixon, R. H., and Whitfield, J. F. "Hypoplasia of the Bone Marrow in Rats Following Removal of the Parathyroid Glands," *J. Cell Physiol.* **79**:343–352 (1972).

24. ———. "Parathyroid Hormone: a Possible Initiator of Liver Regeneration," *Proc. Soc. Exp. Biol. Med.* **141**:93–97 (1972).

25. Short, J., Brown, R. F. Husakova, A., Gilbertson, J. R., Zemel, R., and Lieberman, I. "Induction of Deoxyribonucleic Acid Synthesis in the Liver of the Intact Animal," *J. Biol. Chem.* **247**:1757–1766 (1972).

26. Whitfield, J. F., Perris, A. D., and Youdale, T. "The Calcium Mediated Promotion of Mitotic Activity in Rat Thymocyte Populations by Growth Hor-

mone, Neurohormones, Parathyroid Hormone and Prolactin," *J. Cell Physiol.* **73**:203–212 (1969).

27. Whitfield, J. F., MacManus, J. P., and Gillan, D. J. "The Possible Mediation by Cyclic AMP of the Stimulation of Thymocyte Proliferation by Vasopressin, and the Inhibition of This Mitogenic Action by Thyrocalcitonin," *J. Cell Physiol.* **76**:65–76 (1970).

28. Whitfield, J. F., MacManus, J. P., Youdale, T., and Franks, D. J. "The Roles of Calcium and Cyclic AMP in the Stimulatory Action of Parathyroid Hormone on Thymic Lymphocyte Proliferation," *J. Cell Physiol.* **78**:355–368 (1971).

29. Whitfield, J. F., MacManus, J. P., Franks, D. J., Gillan, D. J., and Youdale, T. "The Possible Mediation by Cyclic AMP of the Stimulation of Thymocyte Proliferation by Cyclic GMP," *Proc. Soc. Exp. Biol. Med.* **137**:453–457 (1971).

30. Whitfield, J. F., MacManus, J. P., Braceland, B. M., and Gillan, D. J. "The Influence of Calcium on the Cyclic AMP Mediated Stimulation of DNA Synthesis and Cell Proliferation by Prostaglandin E_1," *J. Cell Physiol.* **79**:353–362 (1972).

31. Whitfield, J. F., Rixon, R. H., MacManus, J. P., and Balk, S. D. "Calcium, Cyclic AMP, and the Control of Cell Proliferation In Vitro" (in press—1972).

32. Whitney, R. B., and Sutherland, R. M. "Requirement for Calcium Ions in Lymphocyte Transformation Stimulated by Phytohemagglutin," *J. Cell Physiol* (in press—1972).

33. Whitney, R. B., and Sutherland, R. M. "Kinetics of Calcium Transport in Lymphocytes Before and After Stimulation by Phytohemagglutinin." In *Proceedings of the Seventh Leukocyte Culture Conference.* New York: Academic Press (in press—1972).

CYCLIC AMP AND MICROTUBULES

ELIZABETH GILLESPIE

I. Introduction

The variety of topics under consideration in this book attests to the importance of cAMP in immunology. Similarly, many of the functions ascribed to microtubules are important in a variety of immunologic situations. These functions all involve movement, either of constituents within cells or of cells themselves. They include secretion (exocytosis), chemotaxis, axonal flow, granule movement in melanocytes, movement (and maintenance of form) of cilia and flagella, and, of course, movement of chromosomes during cell division. The microtubule system is a logical point of control, and cyclic AMP a probable control molecule in many of these processes. The idea that cyclic AMP might affect the microtubule system, therefore, has become popular.

This chapter will briefly consider those functions known to involve both microtubules and cAMP, will review published work dealing more directly with the relationship between the two, and will also include new data on the effect of cAMP on microtubule assembly. It is probable that cAMP has multiple effects on microtubules. Emphasis will be placed on work from the author's laboratory, which favors the view that there is a direct effect of cAMP on microtubules and that this effect is to either break these structures into subunits or, more likely, to convert them to a less stable form.

Study of microtubule function has been greatly facilitated by the availability of compounds that interact with these structures. The most important of these is the drug, colchicine, which binds to the subunit of microtubules (tubulin) (Borisy and Taylor, 1967; Shelanski and Taylor, 1968) thereby leading to a gradual disappearance of the structure. While colchicine has been found in recent years to have other actions, these can be distinguished from its effect on microtubules. Unlike its other effects,

This is publication 83 from the O'Neill Research Laboratories of The Good Samaritan Hospital.

317

its action on microtubules (1) occurs at low concentrations (10^{-6} M), (2) increases with time of drug treatment, and (3) is irreversible within the context of the normal in vitro experiment, i.e., irreversible within a few hours. Other drugs that act much like colchicine include demecolcine (Colcemid), podophyllatoxin, griseofulvin, and the vinca alkaloids, vinblastine, and vincristine. These have not been studied as extensively as colchicine. It is known, however, that the effects of griseofulvin, unlike those of colchicine, are easily reversible (Malawista et al., 1968). It is also clear that vinblastine binds to a different site on tubulin than does colchicine (Bryan, 1972).

Deuterium oxide or heavy water affects microtubules in a manner opposite to that of colchicine in that it stabilizes them and thereby leads to their formation from subunits (Sato et al., 1966). D_2O has been studied in relation to several microtubule functions. While its effects on cells are undoubtedly multiple, most of its actions are inhibitory (Thompson, 1963). When it potentiates a function therefore, it seems probable that it is acting by promoting the formation of microtubules.

II. Functions of Microtubules

A. Secretion

The first evidence that microtubules might be important in secretory processes was provided in 1967 and 1968 when it was shown that colchicine inhibited the degranulation of lysosomes into the phagocytic vacuole (Malawista and Bodel, 1967), the release of insulin from isolated pancreatic islets (Lacy et al., 1968), and the release of histamine from rat peritoneal mast cells (Gillespie et al., 1968).

Shortly after these initial publications, it was shown that colchicine also inhibited antigen-induced release of histamine from the basophils of allergic individuals (Levy and Carlton, 1969). In this study, it was found that the action of colchicine was greatly enhanced by cold, a treatment known to favor the breakdown of microtubules. More recently, numerous other secretory processes have been shown to be inhibited by colchicine. These include thyroid hormone release (Williams and Wolff, 1970), free fatty acid release (Schimmel, 1972), amylase release from the parotid gland (Butcher and Goldman, 1972), collagen release from bone and fibroblasts (Diegelmann and Peterkofsky, 1972), and histamine release from platelets (White, 1969). Catecholamine release from the bovine adrenal medulla induced by acetylcholine is also inhibited by colchicine (Poisner and Bernstein, 1971). In this one case, however, the effect of colchicine is immediate in onset and instantly reversed on removal of the drug. It is unlikely, therefore, that colchicine is acting to disrupt microtubules. Nonetheless, D_2O potentiates release in this system, suggesting that microtubules are indeed involved in this release process.

The available evidence overwhelmingly supports the view that microtubules play a general role in secretion processes.

The role of cyclic AMP in controlling secretory events is not a uniform one. In many cases, cyclic AMP and/or compounds that raise cyclic AMP levels act to cause release. These cases include thyroid hormone release (Willems et al., 1970), amylase release from the parotid gland (Bdolah and Schramm, 1965) and the pancreas (Ridderstap and Bonting, 1969), and release of various hormones from the anterior pituitary (Schofield, 1967; Fleischer et al., 1969). In other cases, cyclic AMP is a potent inhibitor of the release process. It inhibits, for example, histamine release from leukocytes (Lichtenstein and DeBernardo, 1971) and rat peritoneal mast cells (Loeffler et al., 1971) and the release of cytotoxic substances from antigen-stimulated lymphocytes (Henney et al., 1972). In still other cases, cyclic AMP has no clear role to play. These cases include catecholamine release from the adrenal medulla, vasopressin release from the pituitary and insulin release from B cells of the pancreas.

A recent study describing the effects of D_2O on antigen-induced histamine release from leukocytes attempts to link cAMP and microtubules (Gillespie and Lichtenstein, 1972). This release process can be divided into two stages (Lichtenstein, 1971). The first stage involves the treatment of the cells with antigen in the absence of calcium. The cells are then washed free of excess antigen, and calcium is added to initiate the second stage during which secretion actually occurs. Compounds that raise cyclic AMP levels inhibit histamine release largely or only when they are present during the first stage (Lichtenstein and DeBernardo, 1971). If D_2O is present during the first stage and is removed before the second stage, it has no effect. At the same time, the presence of D_2O during the first stage almost completely prevented the inhibition due to cyclic AMP. As discussed above, D_2O stabilized microtubules. This experiment suggests, therefore, that cyclic AMP might have the opposite effect.

B. The Melanocyte Model

Both microtubules and cAMP are important for the functioning of melanocytes (melanophores). This cell type contains many pigmented granules, which move back and forth along armlike projections of the cell. Their location dictates whether the skin is light or dark. The role of microtubules in the functioning of melanocytes of the frog skin was elegantly dissected some years ago (Malawista, 1965). These structures are required for the inward movement of granules, but not for outward movement. Recently, it has been shown that outward movement can be inhibited by the drug cytochalasin B, and it has been suggested that microfilaments are involved here (McGuire and Moellmann, 1972). Cyclic AMP appears to mediate the effect of melanocyte-stimulating hormone and cause darkening (outward granule movement) of the skin (Bitensky and Burstein, 1965). While it has been suggested that this action is due to an effect of cyclic AMP on microfilaments, it is more consistent with other data to suppose that the true action of cyclic AMP in this system is the inhibition of microtubule function.

C. Chemotaxis

The movement of cells can be either random or directed. Available evidence suggests that an intact microtubule system is not required for random motion, but is essential for directed motion, i.e., chemotaxis (Caner, 1965; Vasiliev et al., 1970; Bhisey and Freed, 1971). Cyclic AMP is also involved in chemotaxis. It has been identified as the attractant, acrasin, in the slime mold *D. discoideum*, where a cAMP gradient directs the movement of the individual cells to aggregate with one another (Konijn, 1967). It is also possible that cAMP is important in the chemotaxis of human polymorphonuclear leukocytes. It is weakly chemotactic itself (Leahy, 1970), while prostaglandin E_1, which raises cAMP in these cells, also has chemotactic activity (Kaley and Weiner, 1971).

D. Other Functions of Microtubules

Other functions that involve microtubules include axonal flow (Dahlstrom, 1968), movement and the maintenance of form of cilia, and, of course, cell division. Cyclic AMP does not seem to have been studied in relation to the first two of these functions. It would be particularly interesting to know if it affected ciliary movement. While cyclic AMP is undoubtedly important in cell division, it is premature to try and relate cyclic AMP to the formation or functioning of the mitotic spindle.

III. Protein Kinase and Microtubules

Several investigators have directed their attention to the nature of the interaction between cyclic AMP and microtubules at the molecular level. Unfortunately, they reach different conclusions. In the first of these studies, it was found that tubulin purified from brain served as a substrate for cAMP-dependent brain protein kinase (Goodman et al., 1970). At the same time, there was considerable self-phosphorylation of tubulin, which was also cAMP dependent. Similar observations have been reported by Murray and Froscio (1971). In contrast, others have failed to confirm these claims (Rappaport et al., 1972). More recently, Soifer and his associates (1972) have found that purified tubulin will, itself, act as a protein kinase when histone is added as a substrate. In initial studies, the cAMP dependence of this phosphorylation varied markedly between preparations. By combining conventional purification procedures with vinblastine precipitation, however, material was prepared that is highly dependent on cyclic AMP for its activity (Soifer, 1972). It is of particular interest that GTP supports tubulin phosphorylation of histone almost as well as does ATP.

This idea that tubulin has protein kinase activity fits well with the observation that colchicine stimulates protein kinase activity (Lichtenstein et al., 1972). Figure 1 illustrates the effect of colchicine on partially purified beef heart protein kinase kindly supplied by D. A. McAfee and P. Greengard. The experiments illustrated were carried out at pH 7.0 in a medium containing 100 mM NaCl and 10 mM PIPES buffer. No effect of colchicine was noted at pH 6.0, the optimum pH for cAMP stimulation. As can be seen, colchicine increased the basal protein kinase activity and acted synergistically with cAMP. This observation is best interpreted by assuming that protein kinases and tubulin are inherently similar molecules.

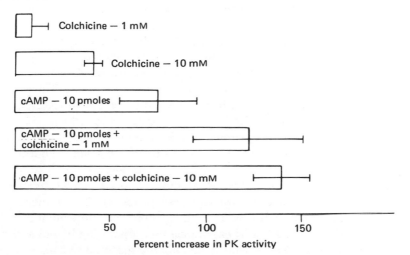

Fig. 1. Effect of colchicine on beef heart protein kinase activity. Average ± SD. The incubation mixture consisted of partially purified beef heart protein kinase, 3 μg; mixed histone, 40 μg; γ^{32}-P-ATP, 0.5 mμmole (1 to 3 × 10^6 CPM; NaCl, 12.4 μmoles; MgCl$_2$, 2.2 μmoles; piperazine-N, N-bis-2-ethane sulfonic acid (PIPES), 1.1 μmoles, pH 7.0; and cyclic AMP and colchicine as indicated in a final volume of 0.2 ml. Incubation was for 5 min at 30°C.

IV. Cyclic AMP and Microtubule Assembly

Recently, Weisenberg (1972) has shown that concentrated tubulin will reassemble into microtubules in the presence of Mg^{2+}, GTP, and ethylenebis (oxyethylenenitrilo) tetraacetic acid (EGTA). Since EGTA preferentially chelates calcium rather than magnesium, he concluded that Ca^{2+} causes the breakdown (disassembly) of microtubules. In a second related study, Borisy and Olmsted (1972) found that microtubules could be assembled in the supernatant of pig brain, again by the addition of GTP and EGTA. Under these conditions, doughnut-shaped structures, considered to be orienting or nucleating centers, were incorporated into the assembled microtubules. These observations have led to the development of a method for the partial purification of tubulin and orienting centers and permitted the study of microtubule assembly in a cell-free system. Cyclic AMP has been found to inhibit the assembly process.

A. Methods

Brain tissue from female Sprague-Dawley rats was homogenized in 100 mM PIPES, pH 6.5, containing 0.5 mM $MgCl_2$ and the homogenate centrifuged at 25,000 g for 30 min at 4°C. The supernatant was mixed with D_2O buffer, GTP, and EGTA so that the final concentrations were 50% D_2O, 1 mM GTP, and 1 mM EGTA. This mixture was incubated for 15 min at 37°, and then centrifuged at 15,000 g for 10 min at room temperature. The precipitate contained 40 to 55% of the total tubulin in the original supernatant as judged by colchicine binding. This material, presumably assembled microtubules, was resuspended in 100% D_2O buffer and recentrifuged at 15,000 g for 10 min three times. The precipitate was next resuspended in a buffer consisting of 100 mM KCl, 50 mM PIPES, pH 6.5, and 0.5 mM $MgCl_2$, and placed on ice for 10 min. This procedure led to complete breakdown of microtubules. Reassembly back to microtubules was accomplished by incubating the tubulin preparation at 37° for 10 min with GTP, EGTA, and other compounds as indicated. The material was again centrifuged at 15,000 g for 10 min at room temperature and the various precipitates resuspended in 0.24 M sucrose, 10 mM phosphate buffer, pH 7.0. Tubulin content was determined by incubating with 2.5×10^{-6} M ^{3}H colchicine and separating bound and free colchicine on Sephadex columns. Unincubated material was also assayed for colchicine binding activity.

B. Results

The effect of 1 mM GTP and various concentrations of EGTA on the assembly of microtubules is illustrated in Fig. 2. The proportion of tubulin that could be precipitated increased over a wide range of EGTA concentrations. The concentration of GTP also affects this process. Lowering the GTP level to 3×10^{-4} in the presence of 1 mM EGTA decreased assembly by 20 to 30%, while in the presence of 1×10^{-4} M GTP, assembly was only 10 to 20% of that seen in the presence of 1×10^{-3} M GTP. GDP substituted only very poorly for GTP. In four experiments, assembly in the presence of 1 mM EGTA and 1 mM GDP was greater than background, but was always less than 10% of that seen in the presence of 1 mM GTP.

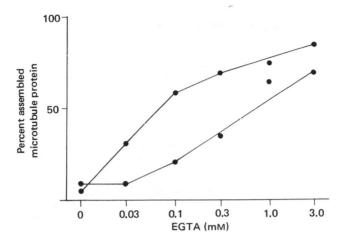

Fig. 2. Effect of EGTA on the assembly of microtubules. Microtubule protein was prepared and incubated as described in the text. Two different experiments are illustrated.

The assembly process can be inhibited by cyclic AMP (Fig. 3). Inhibition varied considerably from preparation to preparation. In eight different experiments, inhibition by 10^{-5} M nucleotide ranged from 0 to 70% with an average of 26%. It is not clear at present what factor or factors contributed to the variability of this effect. 5'-AMP has no effect on the assembly process at 10^{-5} M, while 3'-AMP and adenosine might have a modest effect.

C. Discussion

This finding that cAMP inhibits microtubule assembly is compatible with previous experiments dealing with the effects of Ca^{2+} and cAMP on the binding of colchicine to tubulin in tissue slices (Gillespie, 1971). Cyclic AMP and dibutyryl cAMP, alone and in combination, had a biphasic effect on this binding in slices of rat liver and spleen and in strips of pancreas in the presence of 2 mM Ca^{2+}, first decreasing then increasing colchicine binding. The biphasic nature of this effect was much less pronounced when Ca^{2+} was absent or present at a level of 20 mM. At the same time, Ca^{2+} (0 to 20 mM) decreased colchicine binding. These findings were interpreted as reflecting the relative availability of microtubule subunits. It was suggested that Ca^{2+} favored the formation of microtubules from subunits, while cAMP at low concentrations promoted the entry of Ca^{2+} into the tissue and at higher concentrations acted, by an unknown mechanism, to favor the breakdown of microtubules to subunits. This second action of cAMP is consistent with the present observation that cAMP inhibits assembly of microtubules.

In contrast to these results, cAMP appears to assemble microtubules in other systems. For example, dibutyryl cAMP will cause the transfor-

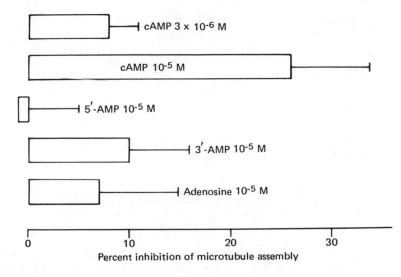

Fig. 3. Effect of cyclic AMP and related compounds on the assembly of microtubules. Average of four to eight experiments ± SD.

mation of several tumor cell lines from an epitheliallike form into a fibroblastlike form, while colchicine, Colcemid, and vinblastine cause the reverse effect (Puck et al., 1972; Johnson et al., 1971). The fibroblastlike cells contain many microtubules along the axes of their processes, and it has been suggested that cAMP acts to favor the assembly of microtubules.

Similar results have been reported in studies of neurite extension from embryonic ganglia. Here cAMP and dibutyryl cAMP increase both the length and number of processes, while colchicine and related compounds prevent elongation and cause the disruption of existing neurite extensions (Furmanski et al., 1971; Roisen et al., 1972; Hier et al., 1972). Synthesis of tubulin is not required for neurite extension, and it has been suggested that cAMP stimulates the assembly of microtubules from a preexisting subunit pool.

V. Discussion and Conclusions

It is clear that cAMP can affect microtubules under a variety of conditions and in a variety of ways. It is impossible to formulate one single action that will explain all the observed effects. It is probable, therefore, that cAMP can affect the assembly and function of the microtubule system in more than one way. Three possibilities will be considered: (1) a direct effect, (2) an effect through Ca^{2+}, and (3) an effect of cAMP on microtubule assembly.

A. *Direct Effect*

The finding that cAMP can inhibit the assembly of microtubules in a partially purified preparation and also the finding that purified microtubule protein has cAMP-dependent protein kinase activity both provide good evidence that cAMP does indeed interact directly with microtubules. The interaction prevents the assembly of microtubules, i.e., favors subunits over microtubules. At the same time, cAMP seems to influence the *activity* of microtubules in a variety of cases. While it is possible that the breakdown of microtubules into subunits is a necessary event in microtubule function, it is difficult to visualize this. It is perhaps more likely and equally compatible with the data to suppose that cAMP acts to convert microtubules into a form which is less stable than alternate forms and thereby breaks down to subunits. GDP is much less effective than GTP in supporting microtubule assembly, and GTP is almost as good as ATP in the phosphorylation of histone by microtubule protein. This suggests the possibility that cAMP stimulates the phosphorylation by microtubules of some associated protein using the GTP known to be incorporated in microtubules.

B. *Effect Through Calcium*

It is clear that calcium is important to microtubule function—all processes involving microtubules that have been studied to date have an absolute requirement for calcium. The precise role of this ion is less clear. In any event, any effect of cAMP on calcium metabolism will, in turn, have an effect on the microtubule system.

C. *Effect on Assembly*

As discussed above, microtubules will assemble in the presence of EGTA. This observation has most frequently been interpreted as meaning that calcium is responsible for the disassembly of microtubules. At the same time, calcium decreases rather than increases the binding of colchicine in tissue slices, indicating that fewer, rather than more, subunits are present. It should also be pointed out that no direct effect of calcium on microtubule morphology has been reported. It is possible, therefore, that EGTA assembles microtubules in its own right as a chemical and not because of its ability to chelate calcium. If this is the case, presumably it is mimicking the action of an as yet unidentified compound(s) present in cells. This provides a third locus for cAMP action; it could increase the formation of this unknown compound and thereby cause the assembly of microtubules.

Discussion of Dr. Gillespie's Paper

DR. MacMANUS: Does cyclic AMP bind to microtubules?

DR. GILLESPIE: No one has been able to demonstrate this. In fact,

Bryan has found that there is no effect of cyclic AMP on vinblastine precipitation of microtubule protein.

DR. WEISSMANN: After Dr. Gillespie presented data showing that cyclic AMP and colchicine increase kinase activity, Dr. Tsung in our laboratory added colchicine to beef heart protein kinase preparations and showed disaggregation of the R from the C unit in sucrose gradients, which fits very nicely. However, we have also found that the most highly purified protein kinase preparations we have from the human polymorphonuclear leukocytes are clearly distinct from any colchicine-binding protein.

DR. GILLESPIE: Agreed. I don't think that protein kinase necessarily binds colchicine the way tubulin does.

References

Bdolah, A., and Schramm, M. "The Function of 3',5'-Cyclic AMP in Enzyme Secretion," *Biochem. Biophys. Res. Commun.* **18**:452 (1965).

Bhisey, A. N., and Freed, J. J. "Ameboid Movement Induced in Cultured Macrophages by Colchicine or Vinblastine," *Exp. Cell. Res.* **64**:419 (1971).

Bitensky, M. W., and Burstein, S. R. "Effects of Cyclic Adenosine Monophosphate and Melanocyte Stimulating Hormone on Frog Skin in Vitro," *Nature* **208**:1282 (1965).

Borisy, G. G., and Olmsted, J. B. "Nucleated Assembly of Microtubules in Porcine Brain Extracts," *Science* **177**:1196 (1972).

Borisy, G. G., and Taylor, E. W. "The Mechanism of Action of Colchicine. Binding of Colchicine ^{3}H to Cellular Protein," *J. Cell Biol.* **34**:525 (1967).

Bryan, J. "Definition of Three Classes of Binding Sites in Isolated Microtubule Crystals," *Biochemistry* **11**:2611 (1972).

Butcher, F. R., and Goldman, R. H. "Effect of Cytochalasin B and Colchicine on the Stimulation of α-Amylase Release from Rat Parotid Tissue Slices," *Biochem. Biophys. Res. Commun.* **48**:23 (1972).

Caner, J. E. Z. "Colchicine Inhibition of Chemotaxis," *Arthritis Rheum.* **8**:757 (1965).

Dahlstrom, A. "Effect of Colchicine on Transport of Amine Storage Granules in Sympathetic Nerves of Rat," *Eur. J. Pharmacol.* **5**:111 (1968).

Diegelmann, R. F., and Peterkofsky, B. "Inhibition of Collagen Secretion from Bone and Cultures Fibroblasts by Microtubular Disruptive Drugs," *Proc. Nat. Acad. Sci.* (U.S.A.) **69**:892 (1972).

Fleischer, N., Donald, R. A., and Butcher, R. W. "Involvement of Adenosine 3'5'-Monophosphate in Release of ACTH, *Amer. J. Physiol.* **217**:1287 (1969).

Furmanski, P., Silverman, D. J., and Lubin, M. "Expression of Differentiated Functions in Mouse Neuroblastoma Mediated by Dibutyryl Cyclic AMP," *Nature* **233**:413 (1971).

Gillespie, E. "Colchicine Binding in Tissue Slices. Decrease by Calcium and Biphasic Effect of Adenosine 3',5'-Monophosphate," *J. Cell Biol.* **50**:544 (1971).

Gillespie, E., and Lichtenstein, L. M. "Histamine Release from Human Leukocytes: Studies with Deuterium Oxide, Colchicine and Cytochalasin B," *J. Clin. Invest.* **51**:2941 (1972).

Gillespie, E., Levine, R. J., and Malawista, S. E. "Histamine Release from Rat Peritoneal Mast Cells: Inhibition by Colchicine and Potentiation by Deuterium Oxide," *J. Pharmacol. Exp. Ther.* **164**:158 (1968).

Goodman, D. B. P., Rasmussen, H., DiBella, F., and Guthrow, Jr., C. G. "Cyclic Adenosine 3′,5′-Monophosphate Stimulated Phosphorylation of Isolated Neurotubule Subunits," *Proc. Nat. Acad. Sci.* **67**:652 (1970).

Henney, C. S., Bourne, H. R., and Lichtenstein, L. M. "The Role of Cyclic 3′5′-Adenosine Monophosphate in the Specific Cytolytic Activity of Lymphocytes," *J. Immunol.* **108**:1526 (1972).

Hier, D. B., Arnason, B. G. W., and Young, M. "Studies on the Mechanism of Action of Nerve Growth Factor," *Proc. Nat. Acad. Sci.* **69**:2268 (1972).

Johnson, G. S., Friedman, R. M., and Pastan, I. "Restoration of Several Morphological Characteristics of Normal Fibroblasts in Sarcoma Cells Treated with 3′5′-Adenosine Monophosphate and Its Derivatives," *Proc. Nat. Acad. Sci.* **68**:425 (1971).

Kaley, G., and Weiner, R. "Effect of Prostaglandin E$_1$ on Leukocyte Migration," *Nature (New Biol.)* **234**:114 (1971).

Konijn, T. M., van de Meene, J. G., Bonner, J. I., and Barkley, D. S. "The Acrasin Activity of Cyclic 3′5′-Adenosine Monophosphate," *Proc. Nat. Acad. Sci.* **58**:1152 (1967).

Lacy, P. E., Howell, S. L., Young, D. A., and Fink, C. J. "New Hypothesis of Insulin Secretion," *Nature* **219**:1177 (1968).

Leahy, D. R., McLean, E. R., and Bonner, J. T. "Evidence for Cyclic 3′,5′-Adenosine Monophosphate as Chemotactic Agent for Polymorphonuclear Leukocytes," *Blood* **36**:52 (1970).

Levy, D. A., and Carlton, J. A. "Influence of Temperature on the Inhibition by Colchicine of Allergic Histamine Release," *Proc. Soc. Exp. Biol. Med.* **130**:1333 (1969).

Lichtenstein, L. M. "The Immediate Allergic Response: in Vitro Separation of Antigen Activation, Decay and Histamine Release," *J. Immunol.* **107**:1122 (1971).

Lichtenstein, L. M., and DeBernardo, R. "The Immediate Allergic Response: in Vitro Action of Cyclic AMP-Active and Other Drugs on the Two Stages of Histamine Release," *J. Immunol.* **107**:1131 (1971).

Lichtenstein, L. M., Gillespie, E., and Bourne, H. "Studies on the Biochemical Mechanisms of IgE-Mediated Histamine Release." In *The Biological Role of the Immunoglobin E System.* Bethesda, Md.: National Institute of Child Health and Human Development, NIH (1972).

Loeffler, L. J., Lovenberg, W., and Sjoerdsma, A. "Effects of Dibutyryl Cyclic AMP, Phosphodiesterase Inhibitors and Prostaglandin E on Compound 48/80 Induced Histamine Release from Rat Peritoneal Mast Cells in Vitro," *Biochem. Pharmcol.* **20**:2287 (1971).

Malawista, S. E. "On the Action of Colchicine—the Melanocyte Model," *J. Exp. Med.* **122**:361 (1965).

Malawista, S. E., and Bodel, P. "Dissociation by Colchicine of Phagocytosis from Increased Oxygen Consumption in Human Leukocytes," *J. Clin. Invest.* **46**:786 (1967).

Malawista, S. E., Sato, H., and Bensch, K. G. "Vinblastine and Griseofulvin Reversibly Disrupt the Living Mitotic Spindle," *Science* **160**:770 (1968).

McGuire, J., and Moellmann, G. "Cytochalasin B: Effects on Microfilaments and Movement of Melanin Granules Within Melanocytes," *Science* **175**:642 (1972).

Murray, A. W., and Froscio, M. "Cyclic AMP and Microtubule Function—Specific Interaction of Phosphorylated Protein Subunits with a Soluble Brain Component," *Biochem. Biophys. Res. Commun.* **44**:1089 (1971).

Poisner, A. M., and Bernstein, J. "A Possible Role of Microtubules in Catecholamine Release from the Adrenal Medulla: Effect of Colchicine, Vinca Alkaloids and Deuterium Oxide," *J. Pharmacol. Exp. Ther.* **177**:102 (1971).

Puck, T. T., Waldren, C. A., and Hsie, A. W. "Membrane Dynamics in the Action of Dibutyryl Adenosine 3′,5′-Cyclic Monophosphate and Testosterone on Mammalian Cells," *Proc. Nat. Acad. Sci.* **69**:1943 (1972).

Rappaport, L., Leterrier, J. F., and Nunez, J. "Nonphosphorylation in Vitro of the 6S Tubulin from Brain and Thyroid," *FEBS Letters* **26**:349 (1972).

Ridderstap, A. S., and Bonting, S. L. "Cyclic AMP and Enzyme Secretion by the Isolated Rabbit Pancreas," *Pflugers Archiv. Eur. J. Physiol.* **313**:62 (1969).

Roisen, F. J., Murphy, R. A., Pichichero, M. E., and Braden, W. G. "Cyclic AMP Stimulation of Axonal Elongation," *Science* **175**:73 (1972).

Sato, H., Inoué, S., Bryan, J., Barclay, N. E., and Platt, C. "The Effect of D_2O on the Mitotic Spindle," *Biol. Bull.* **131**:405 (1966).

Schimmel, R. J. "Inhibition of Free Fatty Acid Mobilization by Colchicine," *Fed. Proc.* **31**:351 (1972).

Schofield, J. G. "Role of Cyclic AMP in the Release of Growth Hormone in Vitro., *Nature* **215**:1382 (1967).

Shelanski, M. L., and Taylor, E. W. "Properties of the Protein Subunit of Central Pair and Outer-Doublet Microtubules of Sea Urchin Flagella," *J. Cell Biol.* **38**:304 (1968).

Soifer, D. "Effect of Precipitation by Vinblastine on the Protein Kinase Activity of Tubulin from Porcine Brain," *J. Cell Biol.* **55**:245a (1972).

Soifer, D., Laszlo, A. H., and Scotto, J. M. "Enzymatic Activity in Tubulin Preparations. I. Intrinsic Protein Kinase Activity in Lyophilized Preparations of Tubulin from Porcine Brain," *Biochim. Biophys. Acta* **271**:182 (1972).

Thompson, J. F. *Biological Effects of Deuterium*. New York: Macmillan (1963).

Vasiliev, J. M., Gelfand, I. M., Domnina, L. V., Ivanova, O. Y., Komm, S. G., and Olshevskaja, L. V. "Effect of Colcemid on the Locomotory Behavior of Fibroblasts," *J. Embryol. Exp. Morph.* **24**:625 (1970).

Weisenberg, R. C. "Microtubule Formation in Vitro in Solutions Containing Low Calcium Concentrations," *Science* **177**:1104 (1972).

White, J. G. "Effects of Colchicine and Vinca Alkaloids on Human Platelets. Influence on Primary Internal Contraction and Secondary Aggregation," *Amer. J. Pathol.* **54**:467 (1969).

Willems, C., Rocmans, P. A., and Dumont, J. E. "Stimulation in Vitro by TSH, Cyclic AMP, Dibutyryl Cyclic AMP and Prostaglandin E_1 of Secretion by Dog Thyroid Slices," *Biochim. Biophys. Acta* **222**:474 (1970).

Williams, J. A., and Wolff, J. "Possible Role of Microtubules in Thyroid Secretion," *Proc. Nat. Acad. Sci.* (U.S.A.) **67**:1901 (1970).

TISSUE LEVELS OF CYCLIC AMP AND TUMOR INHIBITION

WAYNE L. RYAN and JAMES E. MCCLURG

I. Summary

The hypothesis that stimulated this investigation is that cyclic AMP may be a regulator of cell division and therefore may control the growth of tumors. Some inhibition of the 6C3HED, L5178Y, and L1210 tumors results from daily administration of 3 μmoles of cyclic AMP or 5.5 μmoles of theophylline. Treatment of the mice with cyclic AMP 48 hr prior to tumor inoculation produced the greatest tumor inhibition. 5'-AMP, if given during or before tumor transplantation, accelerates tumor growth. Intraperitoneal injection of 3 μmoles of cyclic AMP elevates the cyclic AMP level in the tumor for only 2 hr. The reversibility of the inhibition found in cell culture suggests that this is probably inadequate for producing tumor inhibition. The stimulation of tumor growth by 5'-AMP, a metabolite of cyclic AMP, makes multiple injections of cyclic AMP ineffective in tumor inhibition.

The data presented suggest that tumor inhibition by cyclic AMP is a host-mediated response. To test the original hypothesis, other means for elevating the level of cyclic AMP are needed.

A consideration of the disciplined nature of cell division in normal organs and tissues has suggested the presence of naturally occurring regulators of cell division to a great number of investigators. The presence of such inhibitors seems particularly apparent in processes of wounding and cell damage where cell division is initiated by the injury and then returns to a normal rate after the tissue is repaired. Our own search for such compounds lead to the observation that adenosine 3',5'-monophosphate (cyclic AMP) is an inhibitor of cell division (Ryan and Heidrick, 1968).

There are several characteristics of the inhibition of cell division by cyclic AMP that seem unique. For example, as the concentration is increased, cell division slows and then stops. If further increments of cyclic AMP are added to the cells, no toxicity, such as detachment from the glass, results. Replacement of the culture medium and removal of cyclic AMP permit the cells to divide again in a normal manner. For a compound to function as a regulator of cell division, there must be a variety of means for controlling the tissue level of the regulator. The many agents capable of activating adenylate cyclase and the inducibility of phosphodiesterase indicate that means are available for elevating and lowering the concentration of the inhibitor (D'Armiento et al., 1972); Maganiello and Vaughan, 1972).

The availability of such an inhibitor has suggested experiments in diverse areas such as psoriasis and cancer where control of cell division is desirable (Webb et al., 1972; Chandra et al., 1972; Gericke and Chandra, 1969; Voorhees and Duell, 1971). However, the rapid rate of metabolism of cyclic AMP and the inducibility of phosphodiesterase would suggest that limited success will be achieved in in vivo studies because of the problems of maintaining adequate tissue levels of cyclic AMP for any appreciable time. Because of this probable difficulty, we have examined the tissue levels in the liver, spleen, and tumor following the administration of cyclic AMP, N^6-2'-O-dibutyryl-adenosine 3',5'-monophosphate (dibutyryl cyclic AMP) and theophylline. In addition, in vivo experiments were conducted to determine the effects of different methods of administration of cyclic AMP on tumor growth.

II. Materials and Methods

A. Tumors in Mice

Female C3H/He mice weighing 15 to 20 g were obtained from Flow Laboratories, Dublin, Virginia. The 6C3HED tumor was obtained from Jackson Laboratory, Bar Harbor, Maine, and carried in C3H/He mice. To carry the 6C3HED tumor, 7- to 10-day-old tumor was removed, prepared, and injected according to NCI procedures, (CCNSC Protocol, 1962). Both trypan blue and eosin stains were used to determine cell viability. Only cell preparations with 95% or greater viability were used. The cells were diluted to 10×10^6 cells/ml with sterile phosphate-buffered saline $(0.1 \text{ M NaH}_2\text{PO}_4)$ (PBS). Subcutaneous inoculation of 2×10^6 cells along the axillary line resulted in a tumor that was lethal in 18 to 21 days.

Female BDF_1 SCH mice weighing 15 to 20 g were obtained from ARS/Sprague-Dawley, Madison, Wisc. The murine leukemias $L5178Y_f/407$ and $L1210$ were obtained from Arthur D. Little, Inc., Acron Park, Cambridge, Mass., and were carried in the BDF_1 SCH mice by intraperitoneal injection. The ascitic fluid containing L1210 or L5178Y cells was removed, prepared, and injected according to NCI procedures (CCNSC Protocol, 1962). Both trypan blue and eosin stains were used to determine viability. The cells were diluted to 5×10^5 cells/ml with sterile PBS. Intraperitoneal inoculation of 1×10^5 cells resulted in death in 14 to 16 days for L5178Y and 10 to 12 days for L1210.

B. Cyclic AMP Determinations

The mice were killed by cervical dislocation. The liver, spleen, and tumor were removed, placed in liquid nitrogen, and weighed while frozen. The frozen tissue was homogenized in cold 1.0 N perchloric acid in 25% ethanol (100 mg tissue per milliliter). For estimation of percent recovery, tracer cyclic AMP, 0.009 μc of (H^3) cyclic AMP, was added per milliliter of 1.0 N perchloric acid. The precipitated protein was removed by centrifugation and the supernatant chromatographed according to the method of Brooker (1968). Cyclic AMP was estimated by the method of Gilman (1970), and the protein was estimated by the method of Lowry (1951).

III. Results

Table 1 indicates the effect of incubation of the cells in cyclic AMP and theophylline on subsequent growth of each of the tumor cells in their respective hosts. There is a slight increase in the average day of death in each instance. However, only the BDF$_1$ SCH mice receiving the L5178Y lived significantly longer ($p < 0.05$).

Table 1. In Vitro Effects of Cyclic AMP and Theophylline on Subsequent Tumor Growth in Vivo

Tumor	Average day of death		Percent change
	Control	Exp.	
L1210	10.5	12.0	+14
6C3HED	15.9	17.0	+ 7
L5178Y	15.6	19.3	+24

Note: Tumor cells at a cell density of 50,000 per milliliter were incubated in McCoy's containing 10% fetal calf serum. Experimental media contained 3mM cyclic AMP and 1 mM theophylline. Cells were incubated 24 hr, counted, and viability estimated (90 to 95% in each group). Each mouse received 50,000 cells intraperitoneally (10 mice per group).

Because of the expected rapid metabolism of cyclic AMP, multiple injections of cyclic AMP and its metabolite 5'-AMP were given to C3H/He mice in order to maintain tissue levels. The injections were started three days after tumor inoculation. No significant increase in life was obtained with cyclic AMP when 12 μmoles and 24 μmoles of cyclic AMP were given (Table 2). Somewhat surprising is the significant decrease in life in mice treated with 5'-AMP ($p < 0.05$). Because multi-

Table 2. Effects of Acute Administration of Cyclic AMP to C3H/He Mice Bearing 6C3HED Tumor

Treatment (μmoles)	Average day of death		Percent change
	Control	Exp.	
5'-AMP (12)	17	15	−12
(24)	17	13	−34
c-AMP (12)	17	16.4	−4
(24)	17	16.2	−5

Note: 3 μmoles of cyclic AMP or 5'-AMP in 0.1 M tris, pH 7.2, were injected every 90 min until the total dosage was given. Controls were sham-injected with buffer (20 mice per group). Injections given intraperitoneally three days after the tumor (1 × 10^6 cells) was transplanted subcutaneously in the groin.

ple injections of 5'-AMP appeared to decrease the life of the mice, daily injections of smaller doses of theophylline and cyclic AMP were investigated (Table 3). This mode of administration produced an increase in life with all three tumors using either cyclic AMP or theophylline. In the preceding experiments, it is difficult to determine if the increase (or decrease) in life of the mice is due to the nucleotide acting on the tumor or affecting the host response in some manner. For this reason, the mice were treated with cyclic AMP and 5'-AMP 48 hr before inoculation of

Table 3. Effects of Daily Administration of Cyclic AMP and Theophylline on L5178Y, L1210, and 6C3HED Tumors

Tumor	Average day of death		Percent change
	Control	Exp.	
cAMP			
L1210	9.9	11.9	+20
6C3HED	19.6	24.0	+22
L5178Y	13.5	16.7	+24
Theophylline			
L1210	9.9	12.9	+30
6C3HED	19.6	27.0	+38
L5178Y	13.5	17.4	+29

Note: Animals were injected twice a day with 3 μmoles of cyclic AMP and 5.5 μmoles of theophylline in 0.1 M tris, pH 7.4. Control mice were sham-injected with buffer (10 mice per group).

mice with the tumor cells (Table 4). The 5'-AMP injection produced a decrease in life just as when given to mice with a tumor. This effect of the 5'-AMP must be a consequence of the 5'-AMP acting upon the host, since it would have been completely metabolized in 48 hr. Theophylline also has a similar effect causing a significant decrease in the life of the mice. On the other hand, administration of cyclic AMP produced a significant increase in the life of the mice ($p < 0.05$). Again, it is concluded

Table 4. Effects of Administration of Cyclic AMP and Theophylline to C3H/He Mice Before 6C3HED Tumor

Treatment (μmoles)		Average day of death		Percent change
		Control	Exp.	
5'-AMP	(12)	20	19.8	−1
	(24)	20	15.4	−33
cAMP	(12)	20	26.0[a]	+30
	(24)	20	24.4	+22
Theophylline	(11)	20	14.0	−30

Note: 3 μmoles of 5'-AMP and cyclic AMP per 0.2 ml of 0.1 M tris, pH 7.2, were given intraperitoneally every 90 min until the total dosage was administered. Theophylline was given in two 5.5-μmole injections 2 hr apart in 0.1 M tris, pH 7.2. Controls were sham-injected with buffer. Tumor (1×10^6 cells) was inoculated subcutaneously in the groin 48 hr later.
[a]Two mice survived to 40 days.

that this is an effect of cyclic AMP upon the host and not on the tumors, since all of the cyclic AMP would presumably have disappeared in 48 hr.

The marked difference in response of tumor growth to cyclic AMP, theophylline, and 5'-AMP suggested an investigation of the level of cyclic AMP in the tumor, spleen, and liver following the injection of cyclic AMP, dibutyryl cyclic AMP, and theophylline (Fig. 1).

The 6C3HED tumor (solid), rather than the L1210 and L5178Y, was employed for the measurements of cyclic AMP so that the drug could be given intraperitoneally without coming into direct contact with the tumor. However, the untreated levels of the ascitic form of 6C3HED, L1210, and L5178Y were found to be 12.9, 6.0, and 18.0 pmoles/mg protein. By comparison, the untreated levels of the spleen, liver, and solid form of 6C3HED were 4.4, 2.7, and 15.2 pmoles/mg protein, respectively.

Following injection of 3 μmoles of cyclic AMP, the highest cyclic AMP level was found in the spleen followed by the tumor and liver. However, in the case of dibutyryl cyclic AMP, the highest level was found in the liver followed by the spleen and the tumor. Particularly interesting was the elevation of cyclic AMP in the liver 4 hr after injection of dibutyryl cyclic AMP. The most surprising finding was that 11 μmoles of theophylline did not increase the cyclic AMP level in any of the tissues during the time periods investigated.

IV. Discussion

Cyclic AMP and theophylline produce a slight but significant inhibition of each of the three tumors investigated if they are given daily. A more marked inhibition is produced by cyclic AMP if it is given to the mice before inoculation of the tumor. Using the dosages described here, the evidence suggests that the inhibition of tumor growth may be a consequence of cyclic AMP affecting the host rather than the tumor.

The injection of 5'-AMP into mice either before or after receiving the tumor resulted in an increased rate of death. To our knowledge, no one has reported this effect previously, and it suggests a limited usefulness of cyclic AMP for tumor inhibition, since 5'-AMP will result from its

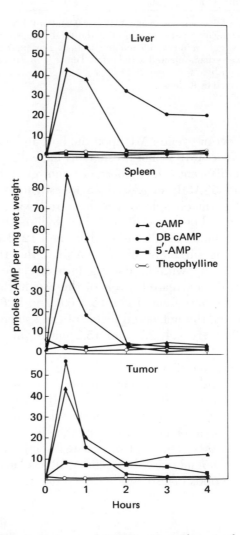

Fig. 1. Cyclic AMP levels in tissues of C3H/He mice. The animals were injected intraperitoneally with cyclic AMP (3 μmoles), dibutyryl cyclic AMP (3 μmoles), 5'-AMP (3 μmoles), and theophylline (11 μmoles), killed, tissues extracted, and assayed for cyclic AMP at 0.5, 1, 2, 3, and 4 hr. Zero time was not injected.

use. 5'-AMP causes vasodilation, which may be responsible for the increased tumor growth that follows its use. In experiments with cyclic AMP and poly I:C on growth of Friend Leukemia in mice, Gericke et al., (1970) observed decreased effectiveness of poly I:C when used with cyclic AMP. This response might result from rapid metabolism of cyclic AMP to 5'-AMP.

The effectiveness of pretreatment of the mice with cyclic AMP would seem to indicate an effect on the immune system of the host. Webb et al., (1972) examined the antitumor effects of polynucleotides and theophylline, which are believed to be mediated by cyclic AMP, on Rauscher leukemia. Inhibition was apparent even in irradiated mice, which would seem to rule out an effect in the immune system. Tissue levels attained following injection of cyclic AMP or dibutyryl cyclic AMP do not persist long enough to permit a test of the hypothesis that cyclic AMP will control cell division in the tumor, particularly in view of the rapid reversibility of the inhibition by cyclic AMP. Further, multiple injections are prohibited because 5'-AMP enhances tumor growth. Consequently, it would seem that some other method for continuous elevation of cyclic AMP is needed to determine the validity of the hypothesis. The failure of theophylline to elevate the level of cyclic AMP in any of the tissues assayed indicates that caution is essential in evaluating the response of a tumor to agents which may, under in vitro conditions, affect the cyclic AMP system.

Discussion of Dr. Ryan and Dr. McClurg's Paper

DR. CHANDRA: We did some studies with lymphosarcoma and were not able to get any effect after intraperitoneal application of cyclic AMP. I would like to ask whether you did experiments by applying (the compounds) intratumorally, because we were able to get much better results by applying cyclic AMP intratumorally.

DR. RYAN: We have not tried injecting it into the tumor.

DR. PARKER: From previous discussions, it seems likely that cholera toxin would provide a much more sustained elevation of cyclic AMP in tumor cells than the agents used in your studies. It would obviously be of interest to look at the effect of cholera enterotoxin. Do you know of any one who has attempted this as yet? I would wonder if the failure to demonstrate a theophylline effect is because the first determination was made at 30 min. In some tissues you can't sustain a theophylline effect for very long, particularly if the concentration of theophylline is not maintained at a high level.

DR. RYAN: You mean we should have elected shorter periods of time?

DR. PARKER: It is possible you would have seen a response if a shorter time interval had been used.

DR. LICHTENSTEIN: There are theoretical problems connected with attempting to control tumor growth with agents such as cholera enterotoxin or similar agents. You would increase cyclic AMP and perhaps stop the growth of the tumor, but this would also turn off immunologic surveillance. Presumably, lymphocytes have something to do with causing the tumor rejection, and as Dr. Henney has shown, increasing the cyclic AMP would stop their ability to kill the tumor cells.

DR. RYAN: At present, there is not really a good answer to this question. The idea is to get an experimental model in which you can demonstrate an elevation in vitro which does not affect other normal systems.

DR. HIRSCHHORN: I believe that theophylline is a mutagen in mice, which may explain some of your results.

DR. SHEPPARD: A couple of comments with regard to theophylline. We also have tried in vain to look for an increase in cyclic AMP levels in cultured cells treated with theophylline, but other inhibitors, such as papaverine, show very significant increase in the cyclic AMP level. In addition, we have looked at a mouse transplantable melanoma and a sarcoma; both were treated with cyclic AMP intratumorally, and have not been able to inhibit the growth of these tumors.

DR. J. HADDEN: I would like to add to the list of observations about theophylline. We have done some work with PHA-stimulated lymphocytes employing a broad concentration range of theophylline, and we have obtained curves that look very much like Whitfield and MacManus' curve for the effect of cyclic G in their thymus model; that is, there is a stimulation of DNA synthesis with a peak in the 10^{-12} to 10^{-11} M range and another peak at a higher concentration (10^{-6} M range). This may be a cyclic GMP mimicking effect, but whatever it is, it is another observation that makes the interpretation of theophylline data more difficult.

References

Brooker, G., Thomas, L. J., and Appleman, M. M. "The Assay of Adenosine 3',5'-Cyclic Monophosphate and Guanosine 3',5'-Cyclic Monophosphate in Biological Materials by Enzymatic Radioisotopic Displacement," *Biochemistry* 7 :4177–4181 (1968).

Cancer Chemotherapy National Service Center (CCNSC) "Protocols for Screening, Tumor Transplantation," *Cancer Chemotherapy Reports* 25:33–34 (1962).

Chandra, P., Gericke, D., and Becker, B. "Effect of Some Adenosine 3',5'-Monophosphate Analogues on Cell Proliferation, Tumor Growth and Macromolecular Synthesis in Various Organs of the Mouse," *Hoppe-Seyler's Z. Physiol. Chem.* 353:1506–1507 (1972).

D'Armiento, M., Johnson, G. S., and Pastan, I. "Regulation of Adenosine 3',5'-Cyclic Monophosphate Phosphodiesterase Activity in Fibroblasts by Intracellular Concentrations of Cyclic Adenosine Monophosphate," *Proc. Nat. Acad. Sci.* **69**:459–462 (1972).

Gericke, D., and Chandra, P. "Inhibition of Tumor Growth by Nucleoside Cyclic 3′,5′-Monophosphates," *Hoppe-Seyler's Z. Physiol. Chem.* **350**:1469–1471 (1969).

Gericke, D., Chandra, P., and Wacker, A. "Some Effects of Adenosine 3′,5′-Monophosphate and Double Stranded RNA, Poly I:Poly C, on Friend Leukemia in Mice," *Hoppe-Seyler's Z. Physiol. Chem.* **351**:411–412 (1970).

Gilman, A. G. "A Protein Binding Assay for Adenosine 3′,5′-Cyclic Monophosphate," *Proc. Nat. Acad. Sci.* **67**:305–312 (1970).

Lowry, O. H. Rosebrough, N. J., Farr, A. L., and Randall, R. J. "Protein Measurement with the Folin Phenol Reagent," *J. Biol. Chem.* **193**:265–275 (1951).

Maganiello, V., and Vaughan, M. "Prostaglandin E_1 Effects on Adenosine 3′,5′-Cyclic Monophosphate Concentration and Phosphodiesterase Activity in Fibroblasts," *Proc. Nat. Acad. Sci.* **69**:269–273 (1972).

Ryan, W. L., and Heidrick, M. L. "Inhibition of Cell Growth in Vitro by Adenosine 3′,5′-Monophosphate," *Science* **162**:1484–1485 (1968).

Voorhees, J. J., and Duell, E. A. "Psoriasis as a Possible Defect of the Adenyl Cyclase-Cyclic AMP Cascade," *Arch. Derm.* **104**:352–358 (1971).

Webb, D., Braun, W., and Plescia, O. J. "Antitumor Effects of Polynucleotides and Theophylline," *Cancer Res.* **32**:1506–1507 (1972).

CYCLIC AMP AND MICROTUBULAR DYNAMICS IN CULTURED MAMMALIAN CELLS

THEODORE T. PUCK and CAROL JONES

The discovery, announced simultaneously by the laboratory of Dr. Pastan (1) and by ourselves (2) of profound change in cell habitus produced by dibutyryl cyclic AMP, appears to hold important implications for the field of cellular interactions generally and the nature of malignancy specifically. The reaction under consideration here resembles in many respects the changes produced when a cell is transformed to malignancy by an oncogenic virus or chemical agent. However, in contrast to the irreversible nature of the latter set of changes, that produced by cyclic AMP is completely reversible. It thus appears to furnish an ideal tool for study of the molecular nature of the underlying effects.°

The properties that are altered as a result of the addition of dibutyryl cyclic AMP to our standard clone of Chinese hamster ovary cells (CHO-K1) are listed Table 1 (3).

We define as "reverse transformation" or "F" conditions the presence of 10^{-3} M dibutyryl cyclic AMP alone, or 1 to 2×10^{-4} M DBcAMP plus testosterone, prostaglandin E, or one of the other synergizing agents whose action we have described.

The reaction has a number of interesting features as follows.

First, the morphological changes are most clearly observed in single cells, which are sparsely distributed over the surface of a petri dish or in colonies arising from single cells, which are not allowed to become too dense (Fig. 1). At high densities, CHO-K1 cells tend to lose their epitheliallike morphology even in the absence of dibutyryl cyclic AMP

This investigation is a contribution from the Rosenhaus Laboratory of the Eleanor Roosevelt Institute for Cancer Research and the Department of Biophysics and Genetics (Number 530), University of Colorado Medical Center, and was aided by a USPHS Grant No. 5-PO1 HD02080 and by an American Cancer Society Grant No. VC-81C.

°The experiments described here have been carried out by Abraham Hsie, Carol Jones, Fa-Ten Kao, A. J. Kauvar, Ryushi Nozawa, David Patterson, Keith Porter, Theodore T. Puck, Carol Smith, Charles Waldren, and Leonor Wenger.

Table 1. Properties of the CHO Cell Altered by Addition
of Dibutyryl Cyclic AMP

	"F" or "reverse transformation" agents	
Epitheliallike cell	$\xrightarrow{\hspace{3cm}}$ $\xleftarrow{\hspace{3cm}}$	Fibroblastlike cell
	E agents	

Epitheliallike cell	Fibroblastlike cell
1. Compact cell morphology	Spindle-shaped morphology
2. Growth in multilayers	Contact inhibition producing only monolayered growth
3. Random cell orientation	Cell growth in close association, parallel to one another's long dimension
4. No morphological effect of drugs like Colcemid and vincristine which inhibit microtubular assembly	Conversion of fibroblastic morphology prevented by these drugs
5. Knoblike projections around the cell's periphery	Smooth cell periphery except for occasional knoblike structures at the pointed ends of the cell
6. Little or no collagen synthesis	Definite collagen synthesis
7. Strongly agglutinated and rounded by wheat germ agglutinin; also rounded up by specific cell antibody	Weakly agglutinated and rounded by wheat germ agglutinin; weakly rounded up by specific cell antibody

and achieve either a clearly fibroblastlike or an ambiguous morphology.

Second, while 10^{-3} M DBcAMP alone significantly reduced the growth rate, the morphologic effect of dibutyryl cyclic AMP and testosterone (or better still, testololactone) can be carried out without significant change in the generation time. Care is required in the interpretation of such experiments because of the presence of contact inhibition when dibutyryl cyclic AMP is present. Since the fibroblastlike cells obtained under these circumstances tend to inhibit one another's growth when a sufficiently high cell density is achieved, comparison of growth rates in the presence and absence of the reverse transformation conditions should be limited to the first three or four days of cultivation. Reverse transformation, accomplished by means of 2×10^{-4} M DBcAMP + 8 μg/ml of testololactone added to single cells growing as an attached monolayer, can be achieved with a generation time of approximately 12 hr, which is indistinguishable from that obtained by the epitheliallike cells which grow in the basal medium alone.

Third, another interesting feature is the specificity of synergising hormones. The synergistic action of testosterone is highly specific. Of a variety of natural steroids studied, 17α-hydroxyprogesterone, androstenedione, and 5α-dihydrotestosterone exhibited a similar degree

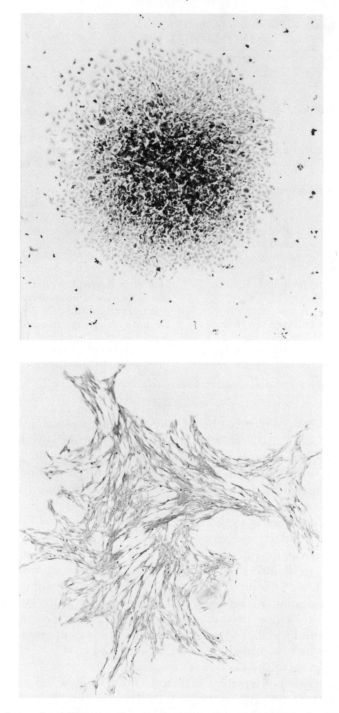

Fig. 1. Colonies derived from single cells that have grown up in basal medium (above) and in basal medium plus dibutyryl cAMP (below).

of activity. All of these compounds occur as members of a continuous metabolic chain (Fig. 2). Of special interest is the compound testololactone, which is fully as active as testosterone propionate, but appreciably less toxic. Testosterone propionate exerts recognizable inhibitory action on cell growth and toxic morphologic manifestations in concentrations of 6 μg/ml or higher. Testololactone, which seems to be equally active in producing the synergising action on dibutyryl cyclic AMP, can be used in concentrations as high as 20 μg/ml without evidence of any toxicity. Prostaglandins E_1, E_2, A_1, and A_2 are also active.

Fourth, an unexpected feature of the knobs, which are readily visualized by time-lapse photography studies, is their motion (4). The knoblike regions on the cell surface are continuously erupting and returning to the main body of the cell. Their expulsion and retraction appears to be fairly uniformly distributed over the cell surface. Extrusion of the knobs may extend to about 1/5 of the cell diameter. The minimum time required for a complete extension and retraction of a particular knob is about 15 sec.

Following the addition of 0.2 mM dibutyryl cAMP and 15 μM testosterone propionate, the first observable effect is the disappearance of the knobs from most of the cell periphery, a process that begins within a few minutes at 37°. A smooth structure is produced over all or most of the membrane within 15 min (Fig. 3). With the concentrations of reagents indicated, the cells do not begin to elongate to achieve the fibroblastlike condition until 5 hr later, and the majority of the cells are not fully elongated until about 8 hr have elapsed. Thus, the processes of knob disappearance and cell elongation that accompany reverse transformation appear to be dissociable. The smooth membrane produced by disappearance of the knobs is not completely quiescent, however, but has a slow ruffling or undulating movement that appears to be different in kind from the violent, localized movements exhibited by the cell membrane in normal growth medium.

An important exception to the disappearance of the knobs after addition of the reverse transformation agents is noteworthy. After mitosis, an intense period of knob formation regularly occurs and lasts about 1 hr in the cells that are already fibroblastlike in the presence of the reverse transformation agents. This is in contrast to the epitheliallike cells in which knob activity continues over a substantial part of interphase.

Several epitheliallike cells, like S3-HeLa cells, did not demonstrate the extensive elongation characteristic of CHO-K1 cells, and were, therefore, considered to be relatively unreactive to the fibroblast reagents. However, experiments demonstrated:

(1) HeLa cells in the native state exhibit knoblike structures similar to those of CHO-K1 cells; the proportion of cells exhibiting these knobs varies with the state of the culture, from about 0 to 25%.

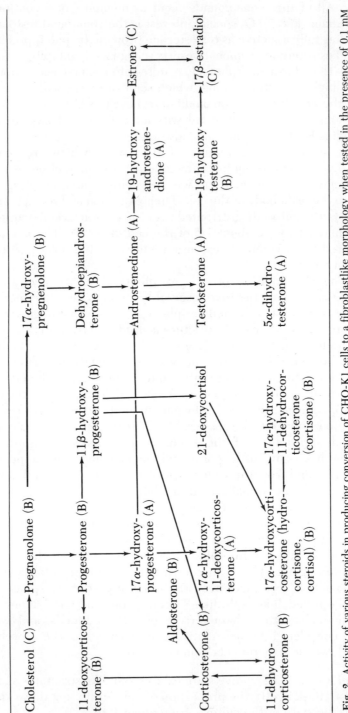

Fig. 2. Activity of various steroids in producing conversion of CHO-K1 cells to a fibroblastlike morphology when tested in the presence of 0.1 mM dibutyryl cyclic AMP. (A) represents high activity, (B) some activity, and (C) no activity. The diagram is arranged to show the biosynthetic relationship of the steroids.

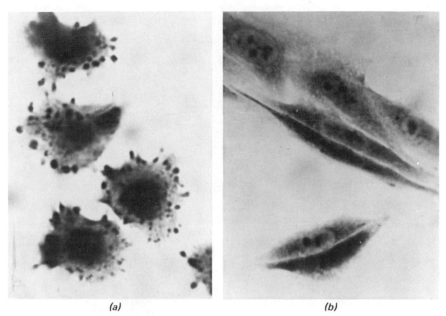

(a)   (b)

Fig. 3. (a) Light microscopic picture of cells grown in basal medium alone showing their compact structure and the knoblike protuberances. (b) The same cells in medium containing "reverse transformation" or "fibroblast" reagents.

(2) For such cultures, the proportion of knobbed cells can be increased to nearly 100% by the addition of 0.27 to 1.34 μM of Colcemid.

(3) The knobs produced through the action of Colcemid on HeLa cells also display the intense eruptive and regressive activity described for CHO-K1 cells.

(4) Addition of dibutyryl cAMP plus testosterone causes complete disappearance of the knobs whether they arise spontaneously or through Colcemid action. This disappearance of knobs is recognized within 30 min after addition of the reverse transformation agents (Fig. 4). Thus, HeLa cells also demonstrate dissociability of knob disappearance from the elongation process.

Studies were also done on permanent fibroblastlike V79 Chinese hamster cells. In basal medium, this cell already possesses the elongated form and smooth membranes typical of the fibroblastlike state. Addition of 0.27 μM Colcemid produces a rapid morphological change toward compactness to approximate the epitheliallike form and a set of typical knoblike structures in most of the cells within 30 min after addition of the drug. Finally, introduction of the agents that produce reverse transformation antagonizes this action of Colcemid, maintaining an elongated, smooth-membraned morphology.

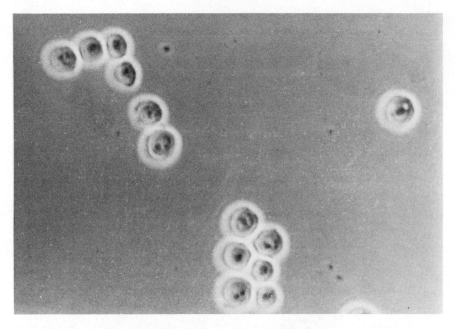

Fig. 4. *(a)* HeLa cells grown in standard growth medium for 16 hr.

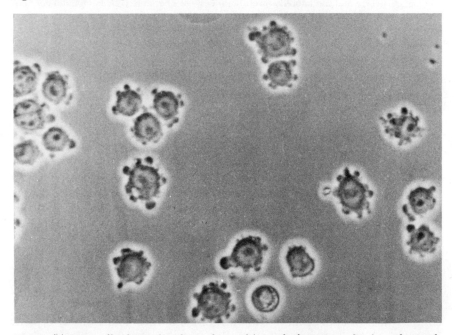

Fig. 4. *(b)* HeLa cells identical to those of Fig. 3(a), to which 0.27 μM of Colcemid was added and which were then incubated at 37° for 30 min. Note the resemblance between these knobbed cells and those of Fig. 2(a). Since virtually all the cells of the culture are affected within 30 min by Colcemid, the effect is not due to entrapment of cells in mitosis, but is a direct action of the drug on cells throughout interphase.

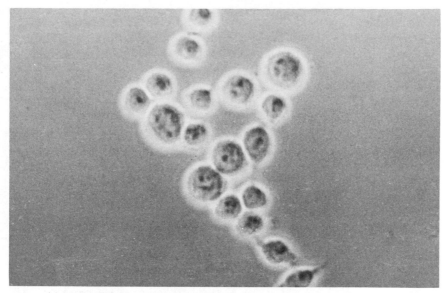

Fig. 4. (c) HeLa cells identical to those of Fig. 3(a) but to which was then added a mixture of 0.27 μM Colcemid plus 1 mM dibutyryl cAMP plus 15 μM testosterone. The reverse transformation conditions have completely prevented the production of knobs as in Fig. 3(b). Cytochalasin B acts similarly to Colcemid.

The inference that knob production involves disorganization of the cellular microtubular (MT)-microfibrillar (MF) system can be entertained only if this effect is produced specifically by these but by few or no other agents. Therefore, it was necessary to test a variety of toxic agents to determine whether knob production is triggered nonspecifically. CHO-K1 cells were treated with pactamycin, actidione, p-fluorophenylalanine, 5-fluorouracil, and actinomycin D in concentrations extending well into the toxic range. None of these agents produced the knobs that characterize the action of Colcemid or vinblastine and cytochalasin B.

The *fifth* interesting feature is the inhibitory action of estradiol. Estradiol prevents the reverse transformation reaction promoted by testosterone and dibutyryl cyclic AMP and so behaves like Colcemid and Cytochalasin. Estradiol also causes human and mouse fibroblastlike cells to become converted to epitheliallike, randomly growing forms, which are also more reactive to Concanavalin A. This action is also reversible.

Sixth, the addition of agents inhibiting either RNA synthesis (actinomycin D) or protein synthesis (puromycin, pactamycin, cyclohex-imide) had no effect either on the rate of the reverse transformation reaction induced by dibutyryl cyclic AMP plus testosterone or on the reversal of this reaction when the latter agents were removed. Neither reaction then appears to require new RNA or protein synthesis.

The *seventh* feature is the electron micrographic studies: role of the microtubules. Because agents like Colcemid and vinblastine, which disorganize microtubules, and cytochalasin B, which disrupts microfibrils, act to prevent or reverse assumption of the fibroblastlike condition, it was postulated that dibutyryl cyclic AMP acts to promote organization of the microtubular microfibrillar system inside the cell and that this action underlies the characteristic morphologic interconversions here considered. In order to test this hypothesis, electron microscopic studies were carried out on CHO-K1 cells in the basal medium and after treatment with testosterone and dibutyryl cyclic AMP, in the hope that the expected change in microtubular structure would be revealed. These studies required quantitative determination of the number and orientation of microtubules per unit volume of cytoplasm. The results of these determinations revealed that the reverse transformation agents produce a small increase in the total number of microtubules and a large increase in their degree of parallel alignment inside the cell. The microtubules become oriented in the direction of the long axis of the fibroblastlike cell in contrast to the more nearly random orientation of these structures in the compact multipolar epitheliallike cell.

Of particular interest was the ultrastructural studies of the knobs, which we found to characterize the CHO-K1 cell in its basal condition, but which disappeared under the influence of the F reagents. Knoblike and bleblike structures have been described in the past in a variety of different cells and have been designated by a variety of different names. Some of these protuberances develop as a result of treatment of the cells with a variety of apparently nonspecific toxic agents. The knobs that are described here are characterized by the following properties:

(1) If they preexist in cells, they can be caused to vanish as a result of addition of the F reagents.

(2) When cells are converted from the E to the F form, the first step observable is the disappearance of the knobs that occurs within a very few minutes, and the cell elongation occurs as a subsequent step at least in the CHO-K1 cells.

(3) Cells that do not display these knobs can be induced to produce them by the addition of E reagents like Colcemid or cytochalasin B.

(4) These knobs, when they are present, exhibit highly active oscillatory movements on the cell surface. The movement of the knobs appears to be a separate mode of membrane motion, which requires careful study.

Eighth, studies have been performed measuring cell adhesivity in the E and F conditions. The affinities of a large variety of cells for plastic surfaces and for one another is much greater in the F than in the E condition. This appears to be true regardless of whether the F condition is the one that obtains in basal medium or whether it is induced by means of dibutyryl cyclic AMP and testosterone. It is tempting to speculate that this property may influence the metastatic property of malignant growth in the body.

The tentative hypothesis that we use to explain these findings is as follows: In the epitheliallike cell, microtubular protein is present, some of it in the polymeric structure, but much of it still in small-molecular-weight units. Those complete microtubules, which do exist, are randomly oriented along cell diameters so that the resulting morphology is compact, approximately circular or multipolar. Under the influence of the fibroblast agents, more microtubules form, and these become organized in parallel fashion so that an elongated cell is produced in the course of this action. New cell membrane structures are exposed so that new cell orientation and associational recognition patterns become established. Since these changes can occur in either direction without new protein synthesis, the presumption is that only activation or inactivation of preexisting proteins is required to change the cell morphology. The detailed biochemistry underlying these changes, the genetic control of these reactions, and the role of these reactions in normal differentiation processes and in cancer are the subjects of continuing study.

Examination of genetic aspects of this phenomenon is now in progress. Cells with altered genomes are being produced by treatment with conditions that produce single-gene mutations, changes in chromosomal number and ploidy, and changes like translocations and deletions in chromosomal structure. One particularly interesting set of cells are hybrids which contain the standard CHO-Kl chromosomal complement plus particular normal human chromosomes. These clonal strains, which have been produced in this laboratory by methods described earlier, are being examined for changes in the properties discussed. For example, mutants of the CHO-K1 cell have been obtained that adopt the F condition. Other variants have been produced that fail to respond to the F agents, and still other clonal strains have been produced that display an exaggerated response. These experiments are being pursued in the hopes of obtaining genetic biochemical insight into the mechanisms underlying these phenomena.

Discussion of Dr. Jones and Dr. Puck's Paper

Dr. GILLESPIE: Those knobs are fascinating. I am sure you must have pictures of their fine structure in ordinary electron micrographs. Do they have a defined structure or any interesting structural features?

Dr. JONES: The only identified structure so far seen in the knobs is an unusually high density of ribosomes.

Dr. WILLINGHAM: I am interested that you mentioned time-lapse pictures, because I was wondering if these knobs represent motility on the part of the cells. In other studies, we had shown that motility of cells is markedly decreased by agents that effect cyclic AMP.

DR. JONES: Despite the great morphological changes associated with knob activity when dibutyryl cyclic AMP or Colcemid are added to our cultures, we could not document any clear effect on cell locomotion.

DR. WILLINGHAM: Did they show ruffling edges?

DR. JONES: The fibroblast form shows ruffling edges.

References

1. Johnson, G. S., Friedman, R. M., and Pastan, I. "Restoration of Several Morphological Characteristics of Normal Fibroblasts in Sarcoma Cells Treated with Adenosine 3',5'-Cyclic Monophosphate and Its Derivatives," *Proc. Nat. Acad. Sci.* (U.S.A.) **68**:425–429 (1971).
2. Hsie, A. W., and Puck, T. T. "Morphological Transformation of Chinese Hamster Cells by Dibutyryl Adenosine Cyclic 3',5'-Monophosphate and Testosterone," *Proc. Nat. Acad. Sci.* (U.S.A.) **68**:358–361 (1971).
3. Hsie, A. W., Jones, C., and Puck, T. T. "Further Changes in Differentiation State Accompanying the Conversion of Chinese Hamster Cells to Fibroblastic Form by Dibutyryl Adenosine Cyclic 3',5'-Monophosphate and Hormones," *Proc. Nat. Acad. Sci.* (U.S.A.) **68**:1648–1652 (1971).
4. Puck, T. T., Waldren, C. A., and Hsie, A. W. "Membrane Dynamics in the Action of Dibutyryl Adenosine 3',5'-Cyclic Monophosphate and Testosterone on Mammalian Cells," *Proc. Nat. Acad. Sci.* (U.S.A.) **69**:1943–1947 (1972).

CYCLIC AMP CHANGES AS A CONSEQUENCE OF TUMOR–NORMAL CELL INTERACTIONS

WERNER BRAUN and CHIAKI SHIOZAWA

Our present concentration on the role of cAMP-mediated events in immune responses is the direct consequence of our earlier discovery and analysis of the effects of nucleic acid breakdown products on population changes in bacteria and on antibody formation (1,2). Critical turning points in establishing a basis for the stimulatory activity of DNA breakdown products were: first, the discovery that double-stranded synthetic polynucleotides, including poly A:U, poly G:C, and poly I:C, were even better and more consistent stimulators than the natural nucleic acids previously employed, and second, the discovery of cAMP and its role in regulating cell function. In these studies of ours, the polynucleotide of choice was poly A:U, because it is neither pyrogenic nor toxic compared to poly I:C.

Part of the mystery of how poly A:U affects immune responses disappeared when it was discovered that the response of immunocompetent cells to extracellular activating signals is, like most hormone-dependent activations of cells, mediated and modified by endogenous cAMP levels and that poly A:U and other oligo- and polynucleotides enhance the activity of adenyl cyclase. A modifier of cAMP, in addition to specific antigen, seems to be required for full activation of antibody-forming cells.

Immunoenhancing agents that function as modifiers of the cAMP system also show antitumor effects. Among these agents are the double-stranded synthetic polyribonucleotides.

It is important to note that *in the activation of immunocompetent cells*, changes in endogenous cAMP levels can lead either to enhanced or reduced responses: A modest stimulation of the cAMP system causes enhancement, whereas an excessive stimulation causes reduced or inhibited responses (3). In contrast to a biphasic response in the course of activation of immature or resting cells, *already activated* immunocompe-

---

This work was supported in part by grants from the NIH (AI-09343) and the N. Y. Cancer Research Institute.

tent cells tend to be inhibited in their functions when exposed to antigen in the presence of cAMP stimulators. Studies by Lichtenstein and associates, by Parker's and by Austen's groups (reviewed in this book) have provided a good deal of information on this phenomenon.

A suppression of functions of already activated cells by agents elevating endogenous cAMP levels also seems to occur in the case of tumor cell populations. We became aware of this in our work with chemically or virus-induced mouse tumors implanted into syngeneic hosts (4). We confirmed that many tumor cell lines can be retarded in their rate of growth by the ip injection of poly I:C, and we showed that a similar, but somewhat less pronounced, retardation can be produced by treatment with poly A:U. Next, we found that the addition of theophylline at time of injection with poly A:U increased the antitumor effect of the polynucleotides and that even theophylline alone retarded the growth of a number of tumor cell lines.

Other modifiers of endogenous cAMP levels, such as isoproterenol and cAMP itself, produced comparable effects which we, like other investigators, believed to be due to an alteration of appropriate host immune responses. But then we discovered that essentially identical effects were produced when the mice were irradiated just prior to tumor implantation, using an X-ray dose that should have sufficed to impair immune mechanisms. Subsequent tests showed that an in vitro exposure of tumor cells to poly A:U or theophylline, prior to the cells' implantation into untreated hosts, sufficed to retard tumor growth (4).

We thus recognized that agents influencing intracellular cAMP levels can have an inhibitory effect on tumor cells that is independent of any stimulation of the conventional host immune responses. In addition, and in conformance with the results of others with BCG and fractions of tubercle bacilli (5), we found that a direct injection of poly A:U, theophylline, isoproterenol, or cAMP itself into a growing tumor produced better antitumor effects than systemic treatment. In fact, with a number of virus-induced tumor cell lines that failed to respond to systemic treatment, fair therapeutic effects were obtained by the direct injection of the agents into the growing tumor six to eight days after the i.D. implantation of 10^6 tumor cells.

Also, we have measured cAMP levels directly in RLV-induced tumor cells which, in the form of a cell suspension, were exposed to poly A:U or theophylline in vitro. No changes were observed. However, when we added normal, syngeneic spleen cells to the tumor cell suspension, a rapid and significant temporary increase in endogenous cAMP levels was detected in both tumor and spleen cells, and this increase was further elevated by the addition of poly A:U.

Incubation of normal spleen cells alone for up to 120 min resulted in either no change or a decrease in intracellular cAMP levels, and incubation of tumor cells alone resulted in an initial slight, but probably insignificant, increase in cAMP concentration, no change or a decrease. However, when tumor cells and syngeneic normal spleen cells were incubated together at a ratio of either 2:1 (2×10^7 tumor cells: 1×10^7 spleen cells) or 1:1 (2×10^7 of each population), a striking increase in endogenous cAMP levels, expressed as cAMP per 10^6 cells or cAMP per culture, was detectable at the earliest feasible assay time, i.e., 10 min after start of cultures. These results are summarized in Fig. 1 and Tables 1, 2. Appropriate controls revealed that the higher levels in mixed-cell suspensions were not due to the higher concentration of cells during incubation; mixed cultures containing the same total number of cells per milliliter yielded congruent results.

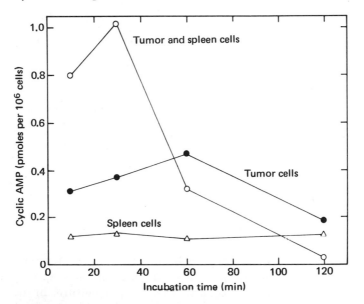

Fig. 1. Cyclic AMP levels (pmoles per 10^6 cells) in suspensions of normal CDF_1 spleen cells, Rauscher leukemia virus-induced tumor ascites cells (MCDV-10), and in mixed suspensions (1:2) thereof, after various periods of incubation in MEM. (All values are averages from duplicate cultures.)

To determine whether both tumor cells (Tu) and normal spleen cells (Sp) contribute to the increase in cAMP, Tu or Sp, or both, were sonicated at 0°C for 5 min. The results indicated that both cell populations contribute to the increase in cAMP during 30 min of incubation, since sonication of Tu reduced the cAMP level to 51% of base line, and sonication of Sp reduced it to 57%. Sonication of both reduced the level to 12%.

On the basis of information available from studies on allogeneic lymphocyte interactions, it was suspected that only a certain proportion of cells in either population (Tu vs. Sp) might participate in the interaction leading to altered cAMP levels. We tested, therefore, the effect of

Table 1. Cyclic AMP Levels (pmoles x 10/5 ml) in CDF_1 Spleen Cell Suspensions (I), Suspensions of Virus-Induced LSTRA Tumor Cells (II), and 1:1 Mixtures Thereof (III) After Various Periods of Incubation in MEM[a]

Time (min)	(I) CDF_1 spleen cells (2×10^7)[b]	(II) LSTRA (2×10^7)	(III) CDF_1 + LSTRA (2×10^7) + (2×10^7)	(III) − (I + II)
10	44.0	16.6	80.0	19.4
30	24.0	13.7	51.0	13.7
60	20.1	18.6	48.5	9.8

[a]All figures are averages from duplicate cultures.
[b]Number of cells per 5 ml.

Table 2. Cyclic AMP Levels (pmoles $\times$ 10/5 ml) in C57 Bl Spleen Cell Suspensions (I), Suspensions of Methylcholanthrene-Induced Tumor Cells (II), and 1:1 Mixtures Thereof (III) After Various Periods of Incubation in MEM[a]

Time (min)	(I) C57 Bl spleen cells (2×10^7)[b]	(II) MC-16 (2×10^7)	(III) C57 Bl + MC-16 (2×10^7) + (2×10^7)	(III) − (I + II)
10	11.4	26.7	59.7	21.6
30	9.8	16.6	52.2	25.8
60	7.0	20.0	42.2	15.2
90	4.9	17.1	37.2	15.2
120	5.7	16.8	43.0	20.5

[a]All figures are averages from duplicate cultures.
[b]Number of cells per 5 ml.

changes in the ratio of Tu and Sp cells on cAMP levels in mixed cultures, using different numbers of Sp cells in mixtures of Tu + Sp in which Tu concentration remained constant at 2×10^7 per culture. In the case of MCDV-10/CDF_1 spleen cell mixtures, 30 min after start of culture, increases in cAMP per 10^6 cells did not occur at 1/10, 1/20, or 1/40, but became apparent at a ratio of 1/5, and were pronounced at 1/1; in LSTRA/CDF_1 spleen cell mixtures, there was no significant change at 1/5, a modest change at 1/2, and a pronounced change at 1/1. Such results suggest that only a small proportion of cells in the Sp population is responsible for the triggering of cAMP changes in the mixed-cell population.

Addition of various amounts of poly A:U to the cultures failed to affect cAMP levels in pure Tu (MCDV-10 or LSTRA) or Sp (CDF_1) populations. However, in mixed Tu + Sp cultures (ratio 2:1), the addition of 0.01γ poly A:U per milliliter enhanced cAMP levels beyond that produced by the cell mixture per se in MCDV-10/CDF_1 mixtures. In LSTRA/CDF_1 mixtures, 1 to 10γ poly A:U per milliliter were required for additional amplification of cAMP levels.

In the course of these studies, it became apparent that cAMP levels declined more rapidly in normal spleen cells held in suspension than in tumor cells similarly maintained. This observation led to the testing of differences of phosphodiesterase (PD) activities in Sp and Tu cells, but, contrary to what might have been expected, PD activity, using cAMP as substrate, turned out to be higher in Tu cells than in Sp cells (Fig. 2).

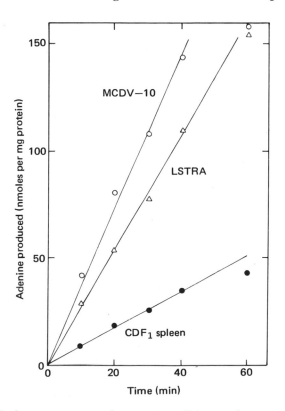

Fig. 2. Phosphodiesterase activity of two tumor cell lines and syngeneic normal spleen cells, using cAMP as substrate, as a function of time. The reaction was carried out at 37°C, using 0.4 mg protein.

Furthermore, pretreatment of Tu or Sp cells, or both, with theophylline resulted in a secondary increase in cAMP levels between 1 and 2 hr after start of mixed Tu + Sp cell suspensions, which was greatest following theophylline treatment of T cells, much less after treatment of Sp cells, and nil when no theophylline treatment was used (Fig. 3). These data suggest that, despite differences in PD activity in Tu and Sp cells, the more rapid loss of cAMP from Sp cells must be attributed to something other than PD activity, possibly diffusion through the membrane.

These data indicate that contact between virus- or chemically transformed tumor cells and syngeneic normal spleen cells leads to a

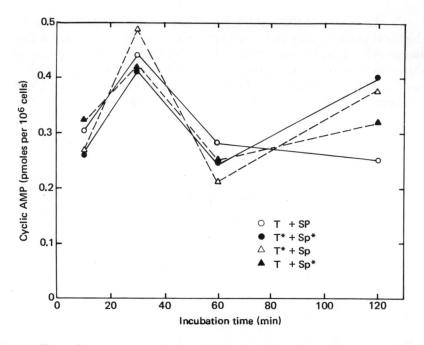

Fig. 3. Effects of pretreatment with theophylline (1 mM for 1 hr at 37°C) on changes in cAMP levels in mixed suspensions of tumor cells (MCDV-10) and syngeneic normal CDF_1 spleen cells (2×10^7 tumor cells + 1×10^7 spleen cells/5 ml). The pretreated component of the cell mixture is indicated by an asterisk (°). (All data are averages from duplicate determinations.)

rapid elevation of endogenous cAMP levels in members of both cell populations, an event that is reminiscent of, but more pronounced than, the changes in cAMP levels that occur following interactions among allogeneic spleen cells (6). Vanky et al. (7) have previously reported that human tumor cells stimulated autochthonous lymphocytes to increased DNA synthesis in a mixed lymphocyte target-interaction test. It is conceivable that the appearance of new antigenic sites on the surface of transformed cells may, in fact, render them equivalent to allogeneic cells. Since it is now believed that cAMP levels play a critical role in the initiation of mitotic events (8), it is tempting to speculate that the observed increases in cAMP levels, following interactions between Tu and Sp cells, may play a role in triggering the proliferation of transformed cells. Also, since changes in cAMP levels are one of the early events in lymphocyte transformation (9), the occurrence of cAMP changes in Sp cells after interaction with Tu cells may account for previously noted decreases in mitogen-activatable T cells in tumor-bearing animals and man. Although it is of interest, we have had no opportunity to determine whether the events occurring after interaction between Tu and Sp cells might also occur following an interaction of Tu cells with other types of normal syngeneic cells and what relationship these might have to tumorigenesis.

Discussion of Dr. Plescia's Presentation
of Dr. Braun and Dr. Shiozawa's Paper

DR. CHANDRA: In a recent paper, Schroeder and Plagemann (*Cancer Res.* Vol. 32) reported that in normal and hepatoma cells, there are at least two types of phosphodiesterases; one is a very low K_m enzyme, 1 to 2 mM, and the other 100 to 200 mM. Levels of the low K_m enzyme were about the same in all the cell lines studies. However, the high K_m enzyme varied markedly in its activity. Normal cells had 30 to 40 times more of high K_m enzyme than hepatoma cells. It is interesting that only the high K_m enzyme was competitively inhibited by theophylline. So I think that before any conclusion is made about phosphodiesterase activity, one should determine both the low and the high K_m enzyme. Secondly, do you believe that polynucleotides alter tumor growth as a consequence of cyclic AMP stimulation, and if so, do you have any data on poly I:C or poly A:U where you have measured cyclic AMP levels?

DR. PLESCIA: I would agree completely that one has to interpret the data on phosphodiesterase activity with considerable reservation, because these are multiple enzyme with different K_m's and possible differences in pH and metal ion requirements for optimal activity. As far as the mechanism of action of polynucleotides is concerned, it would be foolhardy on my part to say that the effect of polynucleotides on tumor growth is mediated through the cyclic AMP system. I don't think that Werner Braun himself would have insisted that this is necessarily the way the polynucleotides work. I think what he did say was that polynucleotides can affect the cyclic AMP system, and inasmuch as many of the biological systems reviewed here and reported in literature are dependent upon cyclic AMP, at least part of their affect probably does involve cyclic AMP.

DR. SHEARER: You showed a model in which Dr. Braun presented his interpretations of the cell-to-cell interaction necessary for antibody production. In that model, Dr. Braun depicted the sites of action of poly A:U as the B cell. I believe this contrasts directly with the results obtained by Dr. Johnson. Do you know if Dr. Braun had any direct evidence for B cells being the site of poly A:U stimulation?

DR. PLESCIA: No, this was based on the fact that one can get enhanced responses in terms of T-independent antigens using poly A:U. But there is, as far as I know, no direct evidence that the poly A:U, in the complete absence of T cells, will affect the function of the B cell.

DR. JOHNSON: This was a point of lively discussion between Dr. Braun and myself. The only pertinent data I know of are that of Campbell and Kind, and her system was quite different from ours.

DR. SHEARER: Dr. Braun, Edna Moses, and myself, studying controlled immune responses to synthetic polypeptides, have evidence that the B cell population is affected by poly A:U and that poly A:U stimulates immune responses in genetic low responders. This system may be different than that studied by Cone and Johnson. In our system, the overall immune responses of the high responders were not affected. It was the response of the selected low-responder strain that was modified in this particular example, and we detected an effect only on B cells. I would also raise the possibility that poly A:U may not directly work on any one cell population, but may facilitate cell-to-cell interaction, or work at the level of the macrophage.

DR. JOHNSON: Could you briefly tell us what the data were which led you to conclude that the effect was selectively on the B cells?

DR. SHEARER: It was exactly the same kind of reconstitution experiment you carried out using cell titrations. We injected animals with a varying number of T cells while maintaining B cells in excess. The recipients were given poly A:U 24 hr later at the same doses you have used. There was no change in these dilution curves. In contrast, B cell dilution curves were altered dramatically by the poly A:U treatment. We have also done a two-step experiment in which we have selectively treated the T cell recipient only with poly A:U in the presence of antigen. In that particular case, there was no effect on the education of T cells.

DR. PLESCIA: I think that in any definitive experiment, one would have to exclude T cells completely from the interacting system in order to prove definitively whether the poly A:U was acting on the B cells alone. When dealing with a low-responder strain, there is always the possibility that the low response is obtained because there are too few functioning T cells. All of these systems have to be looked at from the point of view of whether or not one has excluded completely one of the essential components. I would agree with Dr. Shearer that the evidence suggests that in the apparent absence of essential T cells, one can substitute for them with polynucleotides. In that regard, one can reasonably say that the poly A:U might well have its effect on B cells directly. This is what led Dr. Braun to postulate that perhaps B cells do have a site which can interact with the polynucleotides and cause the cells to amplify the signal given to it by the specific antigen.

References

1. Braun, W., and Firshein, W. *Bacteriol. Rev.* **31**:83 (1967).
2. Braun, W., Ishizuka, M., Yajima, Y., Webb, D., and Winchurch, R. In: *Biological Effects of Polynucleotides*, R. F. Beers and W. Braun, eds. New York: Springer-Verlag (1971).
3. Braun, W., and Ishizuka, M. *Proc. Nat. Acad. Sci.* **68**:1114 (1971).
4. Webb, D., Braun, W., and Plescia, O. J. *Cancer Res.* **32**:1814 (1972).
5. Zbar, B., Rapp, H., and Ribi, E. E. *J. Nat. Cancer Inst.* **48**:831 (1972).

6. Braun, W., and Shiozawa, C. In *Nonspecific Factors Influencing Host Resistance*, W. Braun and J. Ungar, eds. Basel: S. Karger (1973).
7. Vanky, F., Stjernsward, J., Klein, G., and Nilsonne, U. *J. Nat. Cancer Inst.* **47**:96 (1971).
8. Sheppard, J. R. *Nature (New Biol.)* **236**:14 (1972).
9. Smith, J. W., Steiner, A. L., and Parker, C. W. *J. Clin. Invest.* **50**:442 (1971).

EFFECTS OF NUCLEOSIDE CYCLIC MONOPHOSPHATES ON SOME ENZYMATIC PROCESSES, ANTIBODY SYNTHESIS, AND TUMOR GROWTH IN MICE

P. Chandra, D. Gericke, and B. Becker

The metabolic activity of a cell is dependent on its enzymatic architecture. Factors disturbing this architecture can influence the synthesis and breakdown of macromolecules, needed for the functional integrity of the cell. The result is cell death, increased rate of cell proliferation, or an alteration of its biological character such as cell surface. The synthesis and the catalytic activity of enzymes are regulated by genes, which are under the influence of environmental signals, namely, hormones, metabolites, various types of ions, and so on.

The basic mechanisms underlying this regulation are the influences on DNA to control its template functions in RNA synthesis. In large part, these mechanisms involve a direct action of the regulatory substances upon DNA. However, substances are known which may catalyze the chemical modification of a regulatory substance, so that the regulator may no more act on DNA, or this interaction is weakened. One such substance is cyclic AMP (cAMP), which was first shown by Langan and Smith (1967) and Langan (1968, 1969) to stimulate the phosphorylation of histones by histone kinase. It has been known for several years that histones inhibit RNA synthesis in isolated nuclei (Allfrey, 1961, 1962, 1963) and that they suppress the RNA polymerase reaction in pea-seedling chromatin fractions (Huang and Bonner, 1962). Langan (1968) found that the phosphorylated histones were much less inhibitory to DNA-directed RNA synthesis than free histones. This suggested the role of cAMP on the selective alteration of genetic expression by chemical modification of histones.

These studies indicated that cAMP might be involved in the regulation of multiplication of cells of animal origin. Bürk's (1968) studies on the effect of endogenous cAMP on two virus-transformed cells of a line of hamster kidney fibroblasts in tissue culture showed that cAMP was able to inhibit the growth of the original as well as the transformed cells.

In the same year, Ryan and Heidrick (1968) reported that exogenous cAMP inhibited the growth of HeLa cells and a line of chick fibroblasts, whereas equimolar concentrations of 5′-AMP and the dibutyryl derivative of cAMP were inactive, suggesting that these cells might lack the ability to deacylate the derivative. These observations encouraged us to look for the role of cAMP in tumor growth.

I. Regulation of Tumor Growth by Nucleoside Cyclic Monophosphates

In the first series of experiments, mice were treated with cAMP, cUMP, and cIMP. All the cyclic monophosphates were administered at a dose of 1 mg per 20 g body weight, and the treatment was carried out for 10 days. The growth of tumors was measured during the treatment and after the cessation of treatment.

Figure 1 plots the tumor growth at various stages of treatment with cyclic monophosphates. Measurements made on the 7th day of experiment (beginning from the day of tumor transplantation) show that cAMP and cIMP inhibit the growth of this tumor to about 50%, whereas cUMP has only a slight effect. On prolongation of the treatment, one finds a

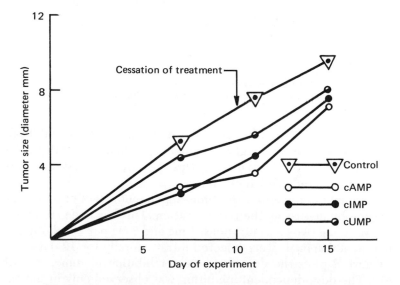

Fig. 1. Inhibition of tumor growth by nucleoside cyclic monophosphates. The studies were carred out on male and female albino mice weighing about 20 to 30 g. NKL lymphosarcoma was transplanted in the ascites form subcutaneously in the neck. The original tumor suspension was diluted 1:10 with Hank's balanced salt solution, and 0.1 ml of the material was injected subcutaneously. The compounds were dissolved in phosphate-buffered saline and injected intratumorally. Each experiment consisted of 10 treated and 10 untreated mice.

maximum inhibition of tumor growth by cAMP, followed by cIMP and cUMP respectively. Measurements made on the 11th day, that is, one day after the cessation of the treatment, show more than 50% inhibition of tumor growth by cAMP. It is interesting to note that soon after the cessation of treatment (see arrow), the tumors begin to grow very rapidly. We have observed that measurements made on the 21st day, that is, 11 days after the cessation of the treatment, do not exhibit any effect at all. This was irrespective of the type of cyclic monophosphate used in our studies.

The above results show that the duration of treatment is an important factor for the action of cyclic monophosphates on tumor growth. This has been confirmed in another series of experiments. The experiments were carried out under similar conditions as reported above. Animals were divided into three groups of 10 each. One group received intratumoral injections of cAMP (1 mg per 20g body weight) for the first 5 days, the second group was treated for 10 days, and the third group served as control. Table 1 shows the percentage of tumor growth as compared with controls, which were measured on the 11th day.

Table 1. Rate of Tumor Growth (NKL Lymphosarcoma) After Treatment with Cyclic AMP for Various Periods

Days of treatment	Tumor growth on 11th day (%)
0 (control)	100[a]
5	84.5
10	49.2

[a]100% designates a tumor diameter of 10.4 mm. Experimental details are the same as those described in Fig. 1.

Animals treated for the first five days only show an inhibition of about 16%, whereas on longer treatment, the tumor growth can be inhibited to more than 50%. This confirms again that the cessation of treatment leads to rapid growth of tumors.

One might ask whether the time-dependent effect of cAMP can be overcome by increasing the concentration. This has been studied by us for cAMP using two concentrations: 1 mg and 3 mg per 20 g body weight. Experimental animals were injected intratumorally for 12 days.

Figure 2 plots the dose-dependent inhibition of tumor growth by cAMP. The dose-dependent inhibition was observed only in the case of cAMP, whereas the other cyclic phosphates failed to show this response. The inhibition of tumor growth by cAMP at 3 mg per 20 g body weight is almost double that of the 1-mg dose. We have also observed that the growth of tumors after the cessation of treatment is slower in the group of animals treated at higher doses.

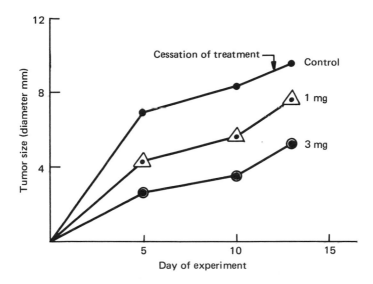

Fig. 2. Dose-dependent inhibition of tumor growth by cyclic AMP. Experimental conditions are the same as those described in Fig. 1.

The studies on the antitumor effects of cAMP were extended to other types of tumors. Figure 3 shows the effect of cyclic AMP on methylcholanthrene-induced tumors in mice. In these experiments, AKR mice were injected subcutaneously with 0.5 mg of methylcholanthrene. After 15 weeks the animals were divided into two groups of 30 each; one group received 20 μg per mouse of cAMP intratumorally daily, while the other served as control. The measurement of tumors after 10 days of treatment revealed that cAMP was stimulating the growth of these tumors. The growth-enhancing effect of cAMP on methylcholanthrene-induced tumors increased gradually with the duration of treatment.

The in vivo effect of cAMP on methylcholanthrene-induced tumors is different from its effect on cells transformed by methylcholanthrene. Thus, Johnson et al. (1971) have reported that L-929 cells (a mouse fibroblast line chemically transformed by 3-methylcholanthrene), during treatment with N-but. cAMP (1 mM), regain several of the morphological and growth characteristics of normal fibroblasts. On the other hand, Brown et al. (1969) have reported a very high level of adenyl cyclase activity in a chemically induced breast carcinoma. It is at present difficult to correlate the in vitro and in vivo effects of cAMP on chemical carcinogenesis.

The effect of cyclic AMP on the growth of transplanted Ridgeway-Osteo-Sarcoma (ROS) is shown in Fig. 4. In these experiments, the animals were transplanted with Ridgeway-Osteo-Sarcoma, and divided into two groups. One group received 1 mg cAMP per mouse (first 9 days subcutaneously and the remaining 9 days intratumorally), and the other

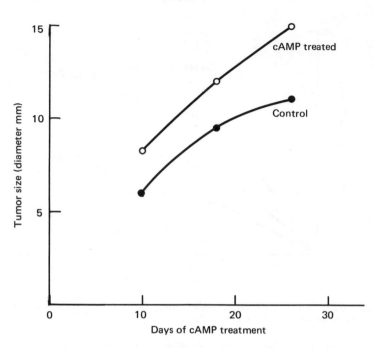

Fig. 3. Effect of cyclic AMP on chemical carcinogenesis by methylcholanthrene in mice. Experiments were carried out on male and female albino mice of AKR strain, weighing 20 to 30 g. Methylcholanthrene was dissolved in 2% Triton X-100 and injected subcutaneously (0.5 mg per mouse) for 15 weeks. Other details are described in text.

served as control. No beneficial effect was observed during the treatment with cAMP by the subcutaneous route. However, the intratumoral application was found to be quite effective. Thus, on 18th day there was more than 60% (average) inhibition of tumor growth in the cAMP-treated animals. The mean survival period in the treated group was higher than that of the animals in the control group.

The studies reported above indicate that the cytostatic activity of cAMP is tumor specific, i.e., the effect is seen only on transplantable tumors. Tumors induced by chemical carcinogens such as methylcholanthrene cannot be inhibited by cAMP. Secondly, the cAMP effect is seen only by the intratumoral application of the compound, and this effect is strictly dependent on the duration of treatment. It was therefore decided to study the cAMP action in combination with compounds that influence the metabolism of cAMP, such as theophylline, or that themselves exhibit cytostatic activity, e.g., poly r (I:C).

The effect of theophylline alone, and in combination with cAMP, is shown in Table 2. The strain AKD_2 was chosen specifically because according to our experience they could tolerate higher doses of cAMP. The scheme of application and duration of treatment are shown in the table. Each experimental group consisted of 20 mice. As follows from results,

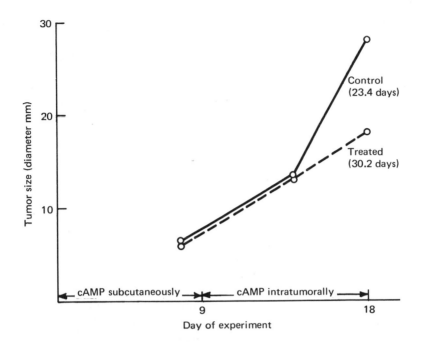

Fig. 4. Effect of cyclic AMP on the growth of transplanted Ridgeway-Osteo-Sarcoma (ROS) in mice. Figures in parentheses designate mean survival period. Other details are described in text.

cAMP and theophylline applied separately are able to significantly inhibit the growth of tumors. However, the application of both the compounds simultaneously almost doubles the inhibition of tumor growth. It is not clear whether the theophylline effect is due to inhibition of phosphodiesterase, which is responsible for the in vivo degradation of cAMP.

Schröder and Plagemann (1972) have recently reported two phosphodiesterases in Novikoff rat hepatoma and L cells, one with a low apparent K_m of 1 to 2 μM and another with an apparent K_m of 100 to 400 μM. Only the high-K_m phosphodiesterases were shown to be inhibited by theophylline in a competitive manner. Though the specific activities of low-K_m enzymes were found to be similar in L cells and hepatoma cells, the specific activities of high-K_m enzymes were found to be 30 to 40 times higher in L cells than in hepatoma cells (Schröder and Plagemann, 1972). This explains the observations of Brown et al. (1970), who have found that various Morris hepatomas have a higher basal activity of adenyl cyclase than the control liver tissue. Significantly, the increase in activity of adenyl cyclase was found to be correlated with the growth rate of the hepatomas, the fastest growing tumor having the highest activity.

Table 2. Effect of Theophylline and Cyclic AMP on the Growth of
Transplanted ROS Tumor in AKD$_2$ Mice

Group	Treatment	Duration of treatment	Tumor weight (g) on 14th day
I	cAMP intratumorally (6 mg per 20 g mouse)	10 days (daily) after tumor transplantation	2.42 ∓ 0.52
II	Theophylline subcutaneously (2 mg per 20 g mouse)	3 days before tumor transplantation and 10 days thereafter	2.64 ∓ 0.88
III	Theophylline and cAMP	As under II As under I	1.30 ∓ 0.42
IV	Phosphate-buffered saline intratumorally	As under I	4.06 ∓ 0.74
V	No treatment	—	3.84 ∓ 0.56

The effect of theophylline alone, or in combination with polynucleotides on the rate of intradermal growth of RLV-induced ascites tumor cells in syngeneic Balb/c mice, has been studied by Webb et al. (1972). The combination of polynucleotides and theophylline was found to be more effective than either agents alone. They have interpreted this effect as due to an involvement of theophylline in cAMP metabolism. If the observations of Schröder and Plagemann (1972) are a general feature of tumor cells, then it is questionable that the theophylline effect is really due to its action on phosphodiesterases.

Double-stranded synthetic polyribonucleotides, particularly poly r(I:C), is known to induce interferon and possess antitumor activity. Friedman and Pastan (1969) reported that the antiviral efficacy of interferon increases on preincubation with cAMP (10^{-2} M). The question arises whether this potentiating effect of cAMP is restricted to exogenous interferon, or whether it can also be achieved by using cAMP in combination with an interferon inducer. Previous studies in our laboratory have shown (Gericke and Chandra, unpublished results) that poly r(I:C) injected intraperitoneally (30 μg per mouse) into mice prior to infection with Friend leukemia virus (FLV) has a protective effect. We, therefore, studied the effect of cAMP in combination with poly r(I:C) on leukemia induced by FLV.

The animals were divided into eight groups of 8 to 10 each. Groups 1 to 5 received FLV suspension without any modification, whereas groups 6 to 8 received virus suspensions modified by various treatments in vitro. The latter experiments are designated as "in vitro" in the table. All the groups were infected with FLV suspensions on October 5, 1969. Animals in the groups 2,3,4, and 5 were pretreated with poly r(I:C) and cAMP as shown in Table 3; group 1 served as control. Three months later 90% of the animals in the control group had died. The animals treated with

Table 3. Effect of Poly r(I:C) and Cyclic AMP on the Mortality of Mice
Infected with Friend Leukemia Virus (FLV)

	Treatment				No. animals alive		Survival (%)
Group	2/10/ 1969	3/10	4/10	5/10	6/10/ 1969	6/1/ 1970	
In vivo							
1	—	—	—	FLV[a]	10	1	10
2	—	—	I:C[b]	FLV	10	5	50
3	—	I:C	—	FLV	10	5	50
4	—	I:C	cAMP[c]	FLV	8	2	25
5	I:C	cAMP	cAMP	FLV	8	1	12.5
In vitro							
6	FLV (preincubated 1 hr at 37°C)				10	4	40
7	FLV + cAMP (1 mg/ml) preincubated 1 hr 37°C				10	3	30
8	FLV + cAMP (3 mg/ml) preincubated 1 hr 37°C				8	0	0

[a]FLV = Friend leukemia virus suspension (1:20, ID_{90}) injected intraperitoneally.
[b]I:C = poly r(I:C) complex, prepared in 0.15 M NaCl; injected intraperitoneally (30 μg per mouse).
[c]cAMP = cAMP dissolved in phosphate-buffered saline and injected intraperitoneally (1 mg per 20 g body weight).

poly r(I:C) 24 hr (group 2) or 48 hr (group 3) before infection showed a 50% survival. A single dose of cAMP inhibits the protective action of poly r(I:C) (group 4); further treatment with cAMP increased this inhibition (group 5). It is not clear whether cAMP inhibits the action of poly r(I:C) complex or if it stimulates the synthesis of a viral component in the host. To study the effect of cAMP on virus, some in vitro experiments were designed. Animals infected with preincubated virus (group 6) showed a higher survival compared to those infected with freshly prepared suspensions (group 1). FLV incubated with cAMP (1 mg/ml) regained its infectivity; virus incubated with higher concentrations of cAMP (3 mg/ml) was quite lethal.

Winchurch et al. (17) have shown that the double-stranded synthetic polynucleotides stimulate adenyl cyclase activity in mouse lymphocytes, thus influencing the endogenous level of cAMP. Our results with cAMP and poly r(I:C) combination indicate that (1) immune system may not be involved during poly r(I:C) effect on viral-induced leukemia, though we have found that the immune response in FLV-treated animals is depressed (Chandra and Gericke, unpublished results); or (2) cAMP may inhibit the production of the active component induced by poly r(I:C), which is needed for the antitumor action. The latter possibility is strongly supported by the recent findings of Dianzani et al. (1972). Treatment of mouse L or rat embryo cells with dibutyryl cAMP was found to prevent, or strongly reduce, the production of interferon induced by poly r(I:C).

II. Effect of Nucleoside Cyclic 3′,5′-Monophosphates on Antibody Synthesis by Spleen Cells

To study the mechanism of inhibition of tumor growth by nucleoside cyclic monophosphates, we investigated their effect on the antibody synthesis in vivo and in vitro by spleen cells (Gericke et al., 1970; Chandra et al., 1971; Chandra, 1972).

Dissociated spleen cells obtained from normal mice can produce antibodies in vitro under a certain set of conditions (Mishell and Dutton, 1966). This has been demonstrated on spleen cell suspensions cultured with heterologous red cells. Mouse spleen cells incubated with sheep red blood cells (sRBC) develop hemolytic plaque-forming cells (PFC). The effect of various nucleoside cyclic monophosphates on the development of PFC is shown in Fig. 5.

Incubation mediums containing spleen cells, sRBC, and various cyclic monophosphates (0.01 mg/ml) were assayed on the fourth day. Among the various compounds tested, cGMP, cIMP, and cTMP do not show any significant inhibition; cUMP is ineffective, and cAMP and cCMP inhibit the development of plaque-forming cells to about 50%. To study the concentration dependency of these effects, experiments were done with concentrations varying between 0.01 to 1.00 mg/ml in the incubation medium. Under these conditions, cTMP did not have any effect; cUMP and cGMP caused 50% inhibition at 1 mg/ml. The other three com-

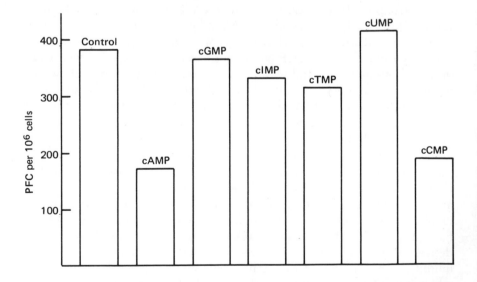

Fig. 5. In vitro effect of nucleoside cyclic monophosphates on antibody synthesis by spleen cells. Spleen cells were incubated with various compounds (10 μg/ml). On fourth day the incubated samples were analyzed by Jerne's plaque technique. The results are reported as plaque-forming cells (PFC) per 10^6 spleen cells.

pounds, namely, cCMP, cIMP, and cAMP inhibited the reaction (see Table 4). At 1 mg/ml, an almost total inhibition was observed for cAMP and cIMP, and about 90% inhibition for cCMP. The inhibitory effect of cIMP at this concentration is very striking, since no appreciable inhibition was observed at 0.01 mg/ml.

Table 4. Dose-Dependent Inhibition by Nucleoside Cyclic Monophosphates of Antibody Synthesis in Vitro

Compound added	Concentration (mg/ml)	PFC (10^6 cells)	% of control
Compound	—	236	100
cCMP	0.01	119	50.5
	0.10	50	21.2
	1.00	25	10.5
cIMP	0.01	206	87.5
	1.00	3	1.3
Control	—	380	100
cAMP	0.01	176	46.2
	1.00	7.5	1.98

Note: Experimental details are those described in Fig. 5.

The in vitro immunization studies show that various cyclic phosphates inhibit the development of PFC. Comparative studies have shown that the in vitro system is of only slight less magnitude than that which occurs in vivo. The fact that various reactions leading to protein synthesis are the same in in vivo and in vitro systems is now well documented. However, the in vitro action of cyclic monophosphates may not be the same as its in vivo effect (Chandra, Gericke, and Becker, unpublished results). One of the reasons for this is, perhaps, the mediation of in vivo effects through some intermediates, which under in vitro conditions are not formed. Therefore, we investigated the in vivo effects of cyclic monophosphates on antibody synthesis.

Mice immunized against sRBC were treated with various cyclic monophosphates (1 mg per 20 g body weight) intraperitoneally for five days. On the fifth day mice were killed, spleens aseptically removed and assayed by Jerne's plaque technique (Fig. 6). Surprisingly, none of the cyclic monophosphates inhibits the development of plaques; on the contrary, cAMP and cGMP exert a slight stimulatory effect.

In the in vitro experiments, cAMP and cCMP at concentrations as low as 0.01 mg/ml inhibited the antibody synthesis by more than 50%. Higher concentrations of these compounds caused a total inhibition. The inability of cyclic phosphates to inhibit the in vivo synthesis of antibody could be due to a rapid hydrolysis of cyclic phosphates by the phosphodiesterases. To test this, we studied the in vivo action of cyclic monophosphates at higher doses.

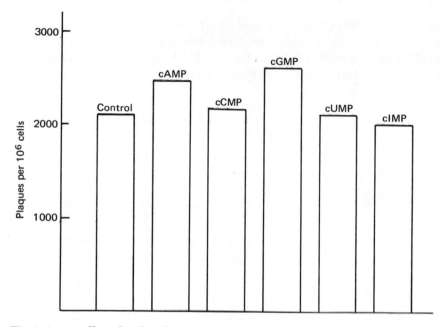

Fig. 6. In vivo effect of nucleoside cyclic monophosphates on the antibody synthesis. About 20 to 30 g heavy mice (AKR strain) were immunized against sRBC on the first day. Test compounds (1 mg per 20 g body weight) were injected intraperitoneally starting on the first day and continuing till the fifth day. On the fifth day spleens were aseptically removed and assayed by Jerne's plaque technique.

The effect of various cyclic monophosphates (3 mg per 20 g body weight) on the development of plaques and antibody titers is shown in Fig. 7. Under these conditions, all cyclic phosphates inhibit antibody synthesis. The maximum inhibition is observed by cCMP followed by cIMP and cAMP. The magnitude of inhibition is much less than that observed in the in vitro system. Nevertheless, cCMP, cAMP, and cIMP are still the most potent inhibitors.

The effect of cAMP on antibody formation has also been reported by Ishizuka et al. (1970) and Braun and Ishizuka (1971). The enhancement of antibody formation by cAMP, reported by these authors, was limited to a certain dose range; higher doses were reported to be inhibitory. This explains the slight stimulation of in vivo systems at lower doses observed by us. Under in vitro conditions, Braun and Ishizuka (1971) have reported a stimulation of antibody responses between 0.1 to 1 μg/ml, whereas no effect was seen at 10 μg/ml. However, in our experiments, 10 μg/ml produced a significant inhibition of antibody synthesis in vitro.

III. Studies with Cyclic AMP Analogues

To study the specific role of cAMP in tumor growth, immune response and cell proliferation analogues of cAMP, comparatively resis-

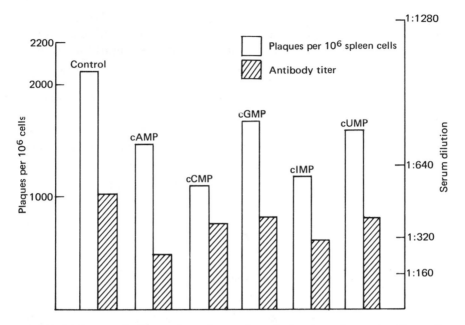

Fig. 7. Inhibition of plaque formation and antibody titers by nucleoside cyclic monophosphates. Experimental conditions were the same as under Fig. 6. Test compounds (3 mg per 20 g body weight) were injected intraperitoneally. On the fifth day spleens were removed aseptically and blood collected. Spleens were assayed by Jerne's plaque technique (left ordinate), and serums were analyzed by the hemagglutination test (right ordinate).

tent to phosphodiesterases (see Table 5), were used. These derivatives are: *N*-benzoyl cAMP, 8-methyl-mercapto DBcAMP, 8-mercapto DBcAMP, and 8-bromo cAMP.

Tumor suspensions (NKL lymphosarcoma) were incubated with the test compound (10^{-3} M) at 37°C for 2 hr and transplanted into experi-

Table 5. Rate of Hydrolysis of Cyclic AMP and Its Structural Analogues by Phosphodiesterase [a]

Derivative	Hydrolysis as % of cAMP (= 100)
N^6-benzoyl-A-3,5-MP	0.2—0.5
8-bromo-A-3,5-MP	8.0—10.0
8-mercapto-A-3,5-MP	0.5—0.8

[a]0.65 mM of cyclic phosphates were hydrolyzed by phosphodiesterase isolated from heart. Data taken from Michal et al. [*Z. Anal. Chem.* **252**:189 (1970)].

mental animals (15 animals per group); the control animals received preincubated suspensions at the same dilution. Under these conditions, 8-methyl-mercapto DBcAMP and 8-bromo GMP were ineffective, whereas the *N*-benzoyl and 8-mercapto DB derivatives showed a weak effect. Surprisingly, none of the animals transplanted with suspensions containing 8-bromo cAMP showed any signs of tumor until the 8th day; however, from the 13th day onward, the tumor growth was rapid. See Fig. 8.

We have previously observed that under the above conditions (preincubation in vitro followed by transfer in vivo), cAMP does not exhibit any antitumor activity. However, by using phosphodiesterase-resistant analogues of cAMP, one can inhibit the growth of lymphosarcoma.

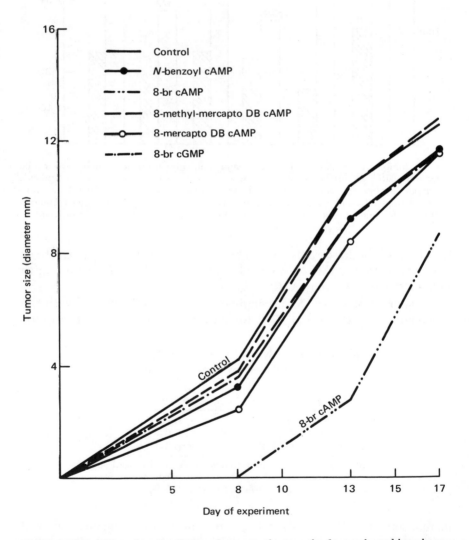

Fig. 8. In vitro effect of cyclic AMP analogues on the growth of transplanted lymphosarcoma (NKL). Experimental details are described in Fig. 1 and in text.

Though all the derivatives used are phosphodiesterase resistant, it is not understandable why only 8-bromo cAMP is effective. The acceleration of tumor growth after about 13 days in the 8-bromo cAMP group indicates that a part of the compound is already degraded; the compound is approximately 90% resistant toward phosphodiesterase (Table 5). In vivo studies are in progress to investigate these effects in details.

To study the mechanism of inhibition of growth of lymphosarcoma, we have studied the effect of these analogues on the macromolecular synthesis under in vitro conditions.

The effect of various cAMP analogues on the in vitro RNA synthesis by nuclei isolated from lymphosarcoma is shown in Table 6. The tumor

Table 6. Effect of Cyclic AMP Analogues on the in Vitro RNA Synthesis by Nuclei Isolated from Lymphosarcoma (NKL)

	^{3}H AMP incorporation into RNA (cpm/mg DNA) (% of control)	
Compound added	Mg^{2+}-dependent reaction	Mn^{2+} and $(NH_4)_2SO_4$-dependent reaction
Control	100	100
N-benzoyl cAMP	84	74
8-mercapto DBcAMP	79.5	61
8-methyl-mercapto DBcAMP	81.0	73
8-bromo cAMP	79	61

Note: The reaction mixture of the Mg^{2+}-activated polymerase was adjusted to a final volume of 0.65 ml and contained 0.15 ml of nuclear suspension, 50 μmoles tris/HCl (pH 8.2), 5 μmoles $MgCl_2$, 1 μmole of ATP (2 μCi), and 0.5 μmole (each) of CTP, UTP, and GTP. The reaction mixture for Mn^{2+}/$(NH_4)_2SO_4$-activated polymerase was identical with the Mg^{2+}-containing assay system, except that 2.2 μmoles of $MgCl_2$ and 0.2 μmole of amm. sulfate were substituted for the $MgCl_2$, and the pH of the system was 7.5.

tissue was homogenized with a hand homogenizer in 3 vol of a 0.08 M K-phosphate buffer (pH 7.4). The homogenate was passed through a nylon gauze, and the filtrate was centrifuged at 800 × g for 20 min. The resulting supernatant was decanted off, and the sediment resuspended in 10 vol of 2.3 M sucrose containing 3.3 mM $CaCl_2$. This suspension was centrifuged (Spinco) at 25,000 rpm for 90 min. The nuclear pellet was then suspended in 0.34 M sucrose (1 g tissue per milliliter). The nuclear suspension was preincubated at 37°C either with the solvent alone, or cAMP analogues (200 μg/ml) for 20 min. These suspensions were directly used in the assay system, described under Table 6. These nuclear preparations were active 2 to 3 hr after their purification, after which a sharp decrease in their RNA-synthesizing activity was observed.

As follows from results in Table 6, the nuclei preincubated with cAMP analogues, irrespective of the activating ions in the assay system, show a partial loss of their ability to synthesize RNA. However, the ability of nuclei preincubated with the analogues to synthesize RNA in the presence of Mn^{2+}ions supplemented with ammonium sulfate is preferentially depressed. The highest inhibition was seen in nuclei preincubated with 8-mercapto DBcAMP and 8-bromo cAMP. It is interesting to note that these two derivatives were most active against the growth of lymphosarcoma, but that the 8-bromo derivative was far superior to 8-mercapto DBcAMP in this system.

Table 7 shows the effect of cAMP analogues on protein synthesis in supernatants obtained after the removal of nuclei (800 × g). These experiments were done under identical conditions, i.e., the supernatant fraction was preincubated at 37°C for 20 min in the presence of cAMP analogues or the solvent alone. These fractions were directly added to the assay system, described under Table 7. As follows from results, no significant inhibition of amino acid incorporation into proteins was seen by any of the analogues used.

Table 7. Effect of Cyclic AMP Analogues on the Protein Synthesis in 800 g Supernatant of Lymphosarcoma (NKL)

Compound added	^{3}H-leucine incorporation into protein (cpm/mg protein) (% of control)
Control	100
N-benzoyl cAMP	84
8-mercapto DBcAMP	87
8-methyl-mercapto DBcAMP	94
8-bromo cAMP	84

Note: Protein synthesis was carried out by incubating 0.5 ml portions of 800 × g supernatant (control or preincubated with test compounds) with ^{3}H-leucine (2 μCi) in an ATP-regenerating buffer, as reported by DeChatelet and McDonald (1968). After incubation for 30 min at 37°C, the reaction was terminated by adding 3 ml of 10% trichloroacetic acid.

The studies reported in Tables 6 and 7 indicate that the inhibition of tumor growth by cyclic monophosphates is at the transcription level. We were unable to detect any in vitro effect (under the above conditions) of cAMP on RNA synthesis. This may be due to its fast degradation during the preincubation period by the phosphodiesterases.

In order to compare the biological activity of cAMP and its analogues, we have studied their activities in other systems. Figure 9 shows the in vivo effect of cAMP and some analogues on the antibody synthesis by mouse spleen.

The animals were divided into five groups of 10 each. Group 1 served as control (injected with saline), and other groups were treated with cyclic monophosphates (1 mg per mouse, ip, on days 1 to 4 after antigenic stimulus). The experimental conditions are described under Fig. 6. As follows from Fig. 9, the 8-bromo derivative shows a good stimulatory effect, followed by N-benzoyl cAMP and cAMP.

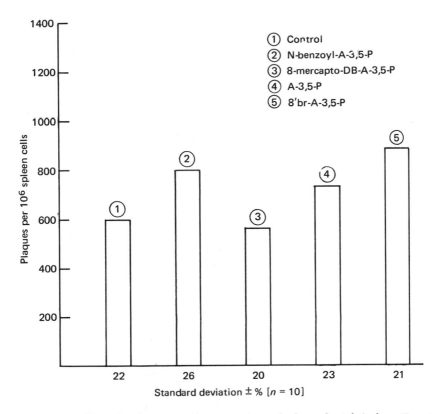

Fig. 9. In vivo effect of cyclic AMP analogues on the antibody synthesis by spleen. Experimental details are described in Fig. 6.

Figure 10 shows the growth rate of L cells in the presence of cAMP and its 8-bromo derivative. Sets of trypsinized triplicate cultures with an initial inoculum of 200×10^4 cells were incubated with cAMP or its 8-bromo derivative ($50 \mu g/ml$) for various periods, and cell counts were made every 24 hr. In all these studies, heat-inactivated serum was used. Under these conditions, both the compounds were able to depress the growth rate of L cells.

IV. Conclusions and Future Problems

The role of cyclic AMP in cancer is at present very confusing. The results reported in this chapter do not exhibit a uniformity in its antitumor activity. It shows an antitumor activity against some types of tumors, whereas the growth of several other types of tumors is not influenced, or even accelerated by cAMP. We believe that the antitumor effect of cAMP is not due to its stimulatory action on immune response. One reason for this belief is that cAMP concentrations needed to inhibit

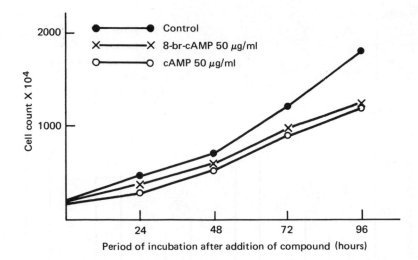

Fig. 10. Growth rate of L cells in the presence of cyclic AMP and 8-bromo cAMP. The cells were maintained in Eagle's medium supplemented with 5% fetal calf serum, penicillin 100 IU per milliliter, and 100 μg/ml of streptomycin. Cells were detached by trypsinization with 0.25% trypsin solution, suspended by agitation and counted by haemocytometer.

tumor growth are very high, and these concentrations actually inhibit antibody formation. The action of cAMP on antibody formation is strictly concentration dependent, i.e., at low concentrations it stimulates, whereas at higher concentrations it strongly depresses the immune system. It is not possible to interpret which action of cAMP is a specific one, the stimulation or the suppression of immune response. The recent report of Dianzani et al. (1972) and our own data (Table 3) indicate that interferon is not involved in the antitumor activity of cAMP.

In view of the very important metabolic role of cAMP, as well as the modifying influence on tissue cyclic nucleotide levels produced by hormonal stimuli, Chandra and Gericke (1972) have proposed a hypothesis to correlate malfunctions in cAMP metabolism as a direct or indirect cause to the abnormal behavior of cancer cells. This hypothesis is, however, applicable to endocrine tumors whose growth is hormonal dependent. Most of the endocrine tumors exhibit a very high intracellular level of cAMP. The mechanism by which exogenous cAMP inhibits growth of transplantable tumors remains still to be elucidated. The use of phosphodiesterase-resistent analogues of cAMP with similar biological properties opens new horizons to envisage such problems. Our studies with eight substituted derivatives, particularly 8-bromo cAMP, indicate that a direct effect on transcription may be one of the reasons for the inhibition of growth of lymphosarcoma. However, more in vivo studies are necessary to clarify the exact mechanism of tumor inhibition by exogeneous cyclic monophosphates.

The second possibility to embark on the specific role of cAMP in cancerogenesis is the use of compounds that regulate the endogenic concentration of cAMP. Our studies with theophylline, an inhibitor of cAMP phosphodiesterase, show that the stimulation of endogenic cAMP concentration leads to tumor inhibition. This means that cAMP is directly involved in the process of cancerogenesis. This belief is supported also by studies with phosphodiesterase-resistent analogues. Hilz et al. (1971) believe that the antitumor activity of cAMP is due to its extracellular degradation to adenosine. However, we failed to observe any cytostatic activity of exogenous adenosine in our systems. It may be useful to carry out more studies on the endogenic regulation of cAMP concentration and its implication on tumor growth. The studies on the role of prostaglandins, which stimulate adenyl cyclase, on cancer growth is presently being investigated in our laboratory.

Acknowledgment

The authors are grateful to Dr. Günther Weimann for the synthesis of cAMP analogues. It is a pleasure to acknowledge the skilled technical assistance of Mrs. A. Götz and Mrs. M. Scheerer.

References

Allfrey, V. G. In *Functional Biochemistry of Cell Structures*, O. Linberg, ed., Vol. VI-2. Oxford: Pergamon Press, p. 127 (1961).

Allfrey, V. G. In *15th Annual Symposium on The Molecular Basis of Neoplasia*. Austin: the Univ. of Texas Press, p. 581 (1962).

Allfrey, V. G., Littau, V. C., and Mirsky, A. E. "The Role of Histones in Regulating RNA Synthesis in Cell Nucleus," *Proc. Nat. Acad. Sci.* (U.S.A.) 49:414–421 (1963).

Braun, W., and Ishizuka, M. "Antibody Formation: Reduced Responses After Administration of Excessive Amounts of Nonspecific Stimulators," *Proc. Nat. Acad. Sci.* (U.S.A.) 68:1114–1116 (1971).

Brown, H. D., Chattopadhyay, S. K., and Mathews, W. S., et al. "Adenyl Cyclase Activity in Morris Hepatoma 7777, 7794A and 9618A, *Cancer Res. Biochim. Biophys. Acta* 192:372–375 (1969).

Brown, H. D., Chattopadhyay, S. K., and Morris, H. P., et al. "Adenyl Cyclase Activity in Morris Hepatoma 7777, 7794 A and 9618A, *Cancer Res.* 30:123–126 (1970).

Bürk, R. R. "Reduced Adenyl Cyclase Activity in a Polyoma Virus Transformed Cell Line," *Nature* (London) 219:1272–1275 (1968).

Chandra, P. "Zyklisches Adenosinmonophosphat in der Klinik und Biochemie der zellulären Prozesse, der Immunantwort und Kanzerogenese," *Med. Welt* 23:9–15 (1972).

Chandra, P., Gericke, D., and Wacker, A. "Effect of Nucleoside (3′,5′)-Monophosphates on Tumor Growth and Immunological Response in Mice." VIIth International Congress of Chemotherapy, Prague (1971).

Chandra, P., and Gericke, D. "Regulation des Tumorwachstums durch Adenosin-3′,5′-monophosphat," *Naturwissenschaften* 59:205–209 (1972).

DeChatelet, L. R., and McDonald, H. J. "Effect of in Vivo Administration of Oral
    Hypoglycemic Agents on Hepatic Protein Synthesis," *Proc. Soc. Exp. Biol.
    Med.* **127**:415–418 (1968).
Dianzani, F., Neri, P., and Zucca, M. "Effect of Dibutyryl Cyclic AMP on Inter-
    feron Production by Cells Treated with Viral or Nonviral Inducers," *Proc.
    Soc. Exp. Biol. Med.* **140**:1375–1378 (1972).
Friedman, R. M., and Pastan, I. "Interferon and Cyclic 3',5'-Adenosine
    Monophosphate: Potentiation of Antiviral Activity," *Biochem. Biophys.
    Res. Commun.* **36**:735–740 (1969).
Gericke, D., Chandra, P., Haenzel, I., and Wacker, A. "Studies on the Effect of
    Nucleoside Cyclic 3',5'-Monophosphates on Antibody Synthesis by Spleen
    Cells," *Hoppe-Seyler's Z. Physiol. Chem.* **351**:305–308 (1970).
Hilz, H., Nolde, S., and Kaukel, E. "The Cytostatic Action of Cyclic Nucleotides
    and Pyridine Nucleotides." VIIth International Congress of
    Chemotherapy, Prague. B-1.2/9 (1971).
Huang, R. C., and Bonner, J. "Histone, a Suppressor of Chromosomal RNA Syn-
    thesis," *Proc. Nat. Acad. Sci.* (U.S.A.) **48**:1216–1222 (1962).
Ishizuka, M., Gafni, M., and Braun, W. "Cyclic AMP Effects on Antibody Forma-
    tion and Their Similarities to Hormone-Mediated Events," *Proc. Soc. Exp.
    Biol. Med.* **134**:963–967 (1970).
Johnson, G. S., Friedman, R. M., and Pastan, I. "Restoration of Several
    Morphological Characteristics in Sarcoma Cells Treated with Cyclic AMP
    and Its Derivatives," *Proc. Nat. Acad. Sci.* (U.S.A.) **68**:425–429 (1971).
Langan, T. A., and Smith, L. K. "Histone Phosphorylation," *Fed. Proc.* **26**:603
    (1967).
Langan T. A. "Histone Phosphorylation: Stimulation by Adenosine 3',5'-
    Monophosphate," *Science* **162**:579–580 (1968).
Langan, T. A. *Phosphorylation of Proteins of the Cell Nucleus,* 3rd Kettering
    Symposium, A. S. Pietro, M. R. Lamborg, and F. T. Kenney, eds. New York:
    Academic Press, pp. 101-118 (1968a).
Langan, T. A. "Action of Adenosine 3',5'-Monophosphate-Dependent Histone
    Kinase in Vivo," *J. Biol. Chem.* **244**:5763–5765 (1969).
Mishell, R. I., and Dutton, R. W. "Immunization of Normal Mouse Spleen Cell
    Suspensions in Vitro," *Science* **153**:1004–1006 (1966).
Ryan, W. L., and Heidrick, M. L. "Inhibition of Cell Growth in Vitro by
    Adenosine 3',5'-Monophosphate," *Science* **162**:1484–1485 (1968).
Schröder, J., and Plagemann, P. G. W. "Cyclic 3',5'-Nucleotide
    Phosphodiesterases of Novikoff Rat Hepatoma, Mouse L, and HeLa Cells
    Growing in Suspension Culture," *Cancer Res.* **32**:1082–1087 (1972).
Webb, D., Braun, W., and Plescia, O. J. "Antitumor Effects of Polynucleotides
    and Theophylline," *Cancer Res.* **32**:1814–1819 (1972).
Winchurch, R., Ishizuka, M., Webb, D., and Braun, W. "Adenyl Cyclase Activity
    of Spleen Cells Exposed to Immunoenhancing Synthetic Oligo- and
    Polynucleotides," *J. Immunol.* **106**:1399–1403 (1971).

THE EFFECT OF CYCLIC AMP AND RELATED AGENTS ON SURVIVAL OF TUMOR-CHALLENGED ANIMALS

PERRY G. RIGBY, FRED McCURDY, and JERRY McCRERY

The effect of cyclic AMP on the immunologic system has been widely explored, especially in the context of its effect on the lymphocyte itself, the fundamental cell of immunologic function in man. That cyclic AMP is involved in the response of the lymphocyte to challenges such as mitogens, antigens, and hormones has received considerable experimental support. (2,4–7). Many questions remain, including very fundamental ones; i.e., how the lymphocyte works, what goes wrong in lymphocytes of those developing proliferative disorders, and why doesn't the lymphocyte respond when the malignant challenge appears from outside the immunologic system?

Previous studies in this laboratory have been concerned with the responses of normal human lymphocytes to added cyclic AMP in varying concentrations (4); a similar investigation has begun on lymphocytes from patients with lymphoproliferative disorders. Cyclic AMP in low concentrations (10^{-7} to 10^{-8} M) increases, and cyclic AMP in high concentrations (10^{-2} M) suppresses the tritiated thymidine uptake of normal lymphocytes in tissue culture (4–7). GMP (10^{-8} M) increases, but AMP and CMP do not significantly affect the tritiated thymidine uptake of normal lymphocytes (5). Cyclic AMP (10^{-7} M) and caffeine increase the tritiated thymidine uptake of PPD-stimulated human lymphocytes in tissue culture and PHA-stimulated human lymphocytes similarly (4). The 10^{-2} M dose of cyclic AMP suppresses lymphocytes challenged by mitogens and antigens as might be expected (4–7).

In consideration of these findings, many of which have been similarly reported by other authors, it was felt worthwhile to consider at least briefly the effect of added cyclic AMP concentrations on lymphocytes from patients with lymphoproliferative disorders. The clinical sources of these lymphocytes are listed in Table 1. We have divided these into three groups: those considered normal, those persons who were considered to have a chronic disease (listed to the middle column), and those

Table 1. Lymphocyte Sources Studied

Normal (NL)	Chronic disease (CDL)	Lymphproliferative (LPL)
2 attending physicians	Adenocarcinoma	Hodgkin's
4 medical students	Chronic active hepatitis	Lymphosarcoma
4 lab. techs.	Leprosy	Lymphoproliferative
	Emphysema	disorder
	Schmidt's syndrome	Chronic lymphocytic
	Dawson's disease	leukemia

considered to have lymphoproliferative disorders (listed in the final column). The clinical diagnosis were made at the University of Nebraska Medical Center employing the usual criteria confirmed by laboratory studies including biopsies of marrow and lymphnodes where appropriate.

The method of study was utilized as reported previously by McCrery and Rigby (5). This method is based on that reported by Hirshhorn et al. and later from this laboratory. Briefly, heparinized human blood was withdrawn and sedimented at 37° in screw cap glass tubes 1 to 3 hr. One to two million cells in 0.2 ml plasma were pipetted from the supernatant into tissue culture tubes containing 2 ml of MEM spinner media supplemented with 20% fetal calf serum, penicillin, streptomycin, and L-glutamine. Hepes media with similar additions was utilized alternatively in some studies as reported by Bach. The end point of lymphocyte cultures was evaluated by the uptake of tritiated thymidine and monitored by morphologic examination. Cell cultures were done in triplicate, cell counts in duplicate. Radioactive counting technique has been previously described utilizing 2 to 3 μCi of tritiated thymidine added to the tissue culture tube 49 to 72 hr after onset. Cells were harvested 24 hr later and, after washing, the DNA precipitated with cold 5% trichloroacetic acid. The material was collected on a filter (0.45 μ pores). These were dissolved in Bray's scintillant after drying and counted on a Unilux II scintillation counter. The data are expressed as a percentage of the corresponding control. Statistics were computed using the student t test.

The findings are presented as a profile of the lymphocyte studied. (See Table 2.) The normal profile indicates that cultured lymphocytes respond to small doses of cyclic AMP with increased uptake of tritiated thymidine, and large doses of cyclic AMP suppress this phenomenon. These same findings after added cAMP can be observed with phytohemagglutinin in the appropriate directions. Not all human lymphocytes consistantly do this, as these findings are influenced by viral disorders as well as many other factors. Additionally, there is variability in individual patients on a day-to-day basis and in comparing one person with another.

The lymphocytes from patients with chronic disorders showed a somewhat different profile. (See Table 3.) These lymphocytes appeared

Table 2. Profile Comparisons

Treatment	NHL	LPL	CDL
Control	—	—	—
cAMP 10^{-2} M	↓	—	↑↓
cAMP 10^{-7} M	↑	—	↑↓
PHA	↑↑↑	↑	↑↓
PHA + cAMP 10^{-2} M	↓	↓	↑↓
PHA + cAMP 10^{-7} M	↑↓	↑	↑↓

to be unable to additionally respond to small doses of cyclic AMP and on occasion did not suppress by large doses. In each patient studied, there was a change in at least one of these parameters, and thus some response was evident.

Table 3. Chronic Disease Lymphocyte (CDL) Profile[a]

Disease	cAMP 10^{-2} M	cAMP 10^{-7} M	PHA	PHA + cAMP 10^{-2} M	PHA + cAMP 10^{-7} M
Schmidt's syndrome	55	66	571	487	390
Chronic active hepatitis	14	82	2074	122	2791
Adenocarcinoma of eye lid	126	88	272	291	484
Hansen's disease	224	172	1557	328	2163
Pulmonary interstitial fibrosis	28	119	1458	251	1767
Hypogamma-globulinemia	18	102	5448	421	6861
Dawson's disease	366	127	1687	294	1172

[a]Expressed as % of control tritiated thymidine uptake, which in every case was 100%.

On the other hand, the profile of the lymphocytes from patients with lymphoproliferative disorders characteristically showed little change with either small or large doses of cyclic AMP, although the same directional change in smaller amount was seen with the addition of these agents after phytohemagglutinin. (See Table 4.) This latter finding might be interpreted as a change in the normal cell population existing in concert with the abnormal clone. This profile is, therefore, contrary to what would be considered the normal profile of lymphocytes upon testing with cyclic AMP and mitogens.

Since it has appeared from prior studies that cyclic AMP influences the immune system in the total organism as well as their lymphocytes in vitro, and, with the possibility that normal lymphocyte populations con-

Table 4. Lymphoproliferative Disease Lymphocyte (LPL) Profile[a]

Disease	cAMP 10^{-2} M	cAMP 10^{-7} M	PHA	PHA + cAMP 10^{-2} M	PHA + cAMP 10^{-7} M
Chronic lymphocytic leukemia	77	114	304	426	333
Chronic lymphocytic leukemia	105	85	938	172	1227
Lymphoproliferative disease	177	141	840	168	1034
Lymphosarcoma	101	—	882	415	2132
Hodgkin's disease	84	96	7034	1751	9296

[a]Expressed as % of control tritiated thymidine uptake, which in every case was 100%.

tinue to be responsive in the presence of abnormal clones, it was apparent that a study of cyclic AMP given to animals should be pursued with regard to the influence on tumor challenge. The effect on the immune system of cyclic AMP has been documented by others, including improvement of antibody response and increased cellular immune function (2). The influence of cyclic AMP on tumor cells in tissue culture and in animals has also been reported experimentally. These studies were, therefore, programmed to consider the effect of cyclic AMP on the immune system of animals in relation to growing tumors before and after immunization with dead tumor cells, and in relation to a possible effect of cyclic AMP on the tumor itself.

Prior studies from this laboratory have indicated the effect of cyclic AMP on one system involving tumor challenge (9). When cyclic AMP (or RNA) were given separately or together after live syngeneic tumor challenge, prolongation of life was obtained in the challenged mice after immunization with X-irradiated tumor cells (8,9). (See Table 5.)

The mice used for these studies were obtained from Flow Laboratories, Dublin, Virginia; both male and female mice weighing 22 to 25 g were employed. This syngeneic tumor was maintained in the inbred strain and given subcutaneously or ip as described. The live tumor dose for challenge was 5×10^5 tumor cells given. Immunizing doses were 5×10^5 tumor cells irradiated in vitro with 4000 r. The first immunization dose was given subcutaneously 14 days prior to challenge and the second dose intraperitoneally 7 days prior to challenge. Immunized animals not later challenged with live tumor all lived normal lives. In these studies cAMP was obtained from the Sigma Chemical Company, St. Louis, Missouri, and dissolved in 0.9% sodium chloride solution for injection. The animals were observed daily.

The results of immunization as previously reported indicated that immunization alone prolonged the survival time of animals minimally, ap-

Table 5. The Effect of cAMP and RNA on the Survival of Mice Given Tumors, With and Without Prior Immunization

					% surviving after No. of days shown				
		Treatment							
Exp.	XTC	LT	RNA (2 mg × 6)	cAMP (1 mg × 6)	15	20	25	30	35
A	√	√	—	—	100	90	30	10	0
B	—	√	—	—	90	50	10	0	0
C	√	√	√	—	100	100	70	70	50[a]
D	—	√	√	—	100	70	0	0	0
E	√	√	√	√	100	90	70	40	40[a]
F	√	√	—	√	100	90	80	70	70[a]
G	√	—	—	—	100	100	100	100	100[a]

[a]Significant prolonged survival (chi square).
XTC = irradiated tumor cells.
LT = live tumor.

proximately 2 days (8). This was considered a minimal immunizing procedure. Fig. 1 shows the survival of animals challenged with C3HED syngeneic tumor given to C3H-HEJ mice. The effect of cyclic AMP 1.0 mg given ip on days 1, 3, 6, 8, 10, 13 after live tumor showed little difference in the outcome. (See Fig. 2.) However, the injection of cyclic AMP on 6

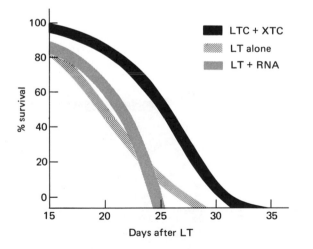

Fig. 1. Animal survival after receiving live tumor cells (LTC). XTC refers to irradiated tumor cells (used for preimmunization).

days over a two-week period after immunization and after live tumor challenge significantly prolonged the survival time of these animals, including the appearance of long survivors (8). These are illustrated in Fig. 3. The question as to why this should occur and whether it was related to the immunization effect prompted further study.

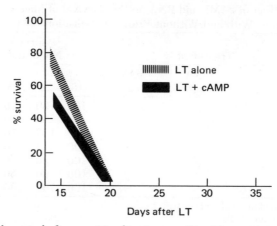

Fig. 2. Animal survival after receiving live tumor cells, with or without cAMP treatment.

The initial considerations concerned a study of (1) the type and timing of immunization, (2) the dose and timing of cyclic AMP, (3) the effect of the surface characteristics of the tumor cell used for immunization, and (4) what part of the immune system was responding. The type of immunization varying among procedures, including irradiated tumor cells,

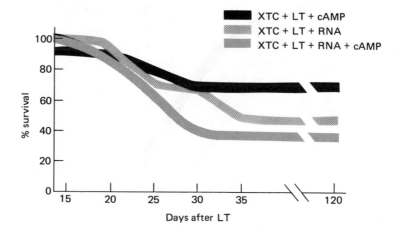

Fig. 3. Animal survival after immunization with X-irradiated tumor cells followed by live tumor (LT). Effect of treatment with RNA, cAMP, or both.

mitomycin C inhibited tumor cells, neuraminidase-treated tumor cells inhibited by mitomycin C, and some variation in timing, is described in detail in a separate paper. The dose of cyclic AMP did have an effect; specifically, the study indicated that the best time of administration was soon after live tumor challenge.

A main consideration for further study was the effect of cyclic AMP on the parts of the immune system. The following study was employed to indicate the response of T and B cells, compared with spleen cells, in mice upon challenge with cyclic AMP, killed tumor cells, live tumor cells, or combinations of these. The uptake of tritiated thymidine by populations containing a majority of T or B cells was measured to delineate the influence of cyclic AMP on immunologic function.

I. Method

Forty C_3H (female 15–20 g) were divided into two equal groups; 20 mice were retained as normal controls, and the other 20 were challenged with antigen, cAMP, antigen plus cAMP, or tumor, etc., as dictated in each of the separate studies. (See Fig. 4.)

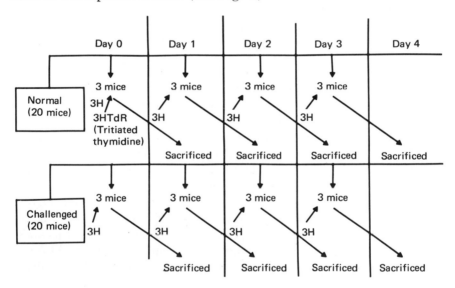

Fig. 4. Method for evaluating ^3H thymidine uptake in vivo in experimental and control groups of animals.

The group selected as normal controls received no challenge. Two animals were selected on each day of the study to receive 15 μCi of H^3 thymidine ip. After an elapsed period of 24 hr, the injected animals were sacrificed by cervical dislocation; the thymus, spleen, and bone marrow specimens were removed from each animal separately and cellularized in preparation for subsequent assay of H^3 uptake. (See Fig. 5.) Cellularization was accomplished, in a minimal volume of cold MEM, by chopping with scissors and passage through a 20 gauge needle. The cells were counted, and an aliquot containing 10^6 cells was precipitated on a Whatman GF/A glass filter. The filters were transferred to scintillation vials and counted in Omnifluor (New England Nuclear) using a Unilux II scintillation counter.

The experimental animals were challenged in various ways. Those mice challenged with antigen received one dose of 5×10^5 6C3HEDOG tumor cells ip, which had been rendered incapable of growth by incubation with mitomycin C for 1 hr at 37°C. The concentration of mitomycin C used was 25 μg per 10^6 tumor cells. (The cells were washed free of mitomycin C before injection.) Mice that received cAMP were given 1.0 mg of cAMP ip on two different occasions (days 0 and 2 of the study). Antigen and cAMP were combined in a third study; mice received the same antigen as before on day 0, while cAMP was again used at the 1.0 mg ip dosage on days 0 and 2. Finally, the animals that were injected with live tumor received 5×10^5 6C3HEDOG tumor cells subcutaneously in the left flank. This dosage gave a uniformly fatal outcome with a mean survival time of 19 days in a parallel control group.

After challenge, the animals in this group were handled the same as the control group; two animals received injections of H^3 thymidine on successive days of each study.

Counts were recorded for 10 min and were then converted to disintegrations per 10 min by use of a predetermined quenching curve. Data presented here are expressed in counts per 10 min per 10^6 cells for continuity from the beginning of the study to present.

The results of this study indicate differences related to type of challenge, specifically with cyclic AMP by itself and/or with live tumor. The findings in the following figures are changes compared to normal animals treated identically on the same day as controls.

The most significant changes were seen in the B cell catagory, although spleen cells showed somewhat similar responses. (See Fig. 6.) T cells did not show as great a response under these experimental conditions. The challenge with dead tumor cells as antigen showed a biphasic curve; an initial response by the second day was followed by a slight

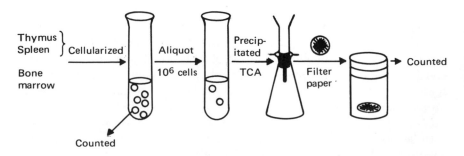

Fig. 5. Processing of tissues for ^{3}H thymidine counting.

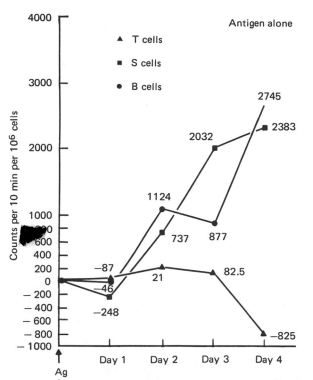

Fig. 6. Differences in ³H thymidine uptake in response to antigen (dead tumor cells) in different lymphocytic cell types in vivo. The O value is the response in animals not given antigen.

decrease (or plateau) and a higher peak on the fourth day. The B and spleen (S) cells appear somewhat similar, though the S cells did not show any diminution on the third day. The T cells were only slightly above normal and finished up lower than the normal cells under these circumstances.

Figure 7 shows the effect of two doses of cyclic AMP, one on the first day of the study and one two days later. A peak by B cells on the first day is higher and earlier than expected with antigen. There is a second peak after a second dose on the third day; this also diminishes toward normal by the fourth day. Relatively less activity is seen in T cells or spleen cells.

When antigen (dead tumor cells) plus cyclic AMP are combined, there is a peak on the first day similar to the cyclic AMP response and a second peak which is higher on day 3. (See Fig. 8.) This appears to be a combined effect between cyclic AMP and antigen. The spleen cells are somewhat in between, and relatively little effect was seen on T cells.

In Fig. 9, when live tumor cells were given, the responses were followed for 8 days instead of 4. We shall concentrate here on the first 4 days, since that is what relates to control studies thus far. There is a peak, as might be expected, on day 2 of both B cells and spleen cells, and some

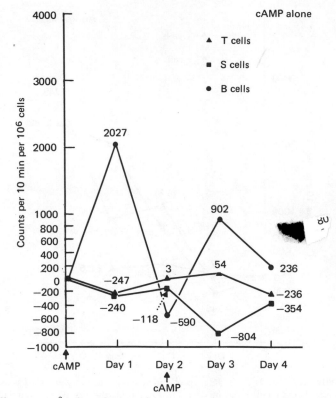

Fig. 7. Differences in ³H thymidine uptake in response to cAMP in different lymphocytic cell types in vivo.

increase in T cells. By day 4 these are all lower, and the second effect as seen with antigen is not apparent here. There is a second effect later as may be seen when the studies are followed further. This later recedes as anticipated.

When this study is repeated with live tumor cells and cyclic AMP given on day 0 and 2, the peak on day 1 is seen as in other cyclic AMP studies. A decrease in response is followed by an additional peak on day 4. This is somewhat delayed compared to antigen and cyclic AMP, but could be related to the cyclic AMP given as before.

These studies in intact animals after the administration of cyclic AMP, live and dead tumor cells have been initiated to consider the relationships that exist in the tumor-challenged host. Though the populations of T and B cells are undoubtedly impure to some extent, the indications are that there is activity in these immunologic cells upon stimulation by either antigen or cyclic AMP. The sequence of events appears to be different with live tumor as compared with tumor cells used as antigen. The influence of cyclic AMP on these events would appear to be a positive one indicating earlier and higher peaks of tritiated thymidine up-

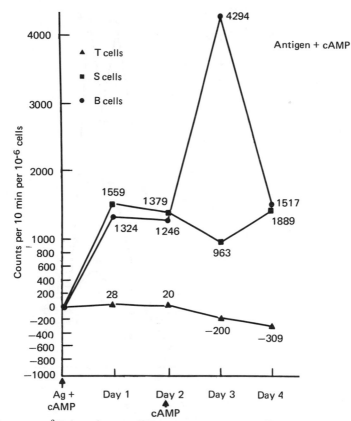

Fig. 8. Differences in ³H thymidine uptake in response to antigen (dead tumor cells) and cAMP in different lymphocytic cell types in vivo.

take. The interpretation of these studies await further experimentation, possibly utilizing variations in dosage, timing, and other more sophisticated measuring parameters.

Because low doses of cyclic AMP appear to influence the whole animal challenged with live tumor cells, and especially since this appears to relate to immunization or immunologic function, it would appear that a proper hypothesis on the outcome should include the effects of cyclic AMP on the immune response. The proper precautions should be observed in the interpretation of these results, though they may well provide a stimulus to look further into the effects of cyclic AMP on tumor immunity.

II. Summary

It has been observed from this laboratory and by others that small doses of cyclic AMP appear to stimulate normal human lymphocyte transformation, and large doses of cyclic AMP inhibit this phenomenon.

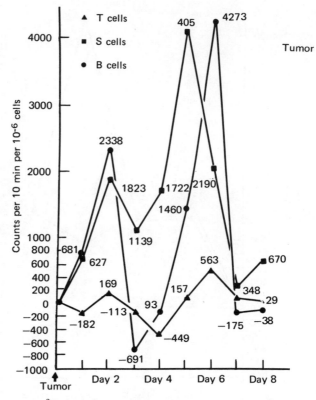

Fig. 9. Differences in ³H thymidine uptake in response to antigen (live tumor cells) in different lymphocytic cell types in vivo.

The addition of cyclic AMP in the presence of antigens or mitogens also appears to affect the magnitude of responses in normal lymphocytes in similar directions. The effect of added cyclic AMP on lymphocytes from patients with chronic disorders shows a somewhat different profile, though these cells are still able to respond apparently in one direction or the other. On the contrary, the addition of cyclic AMP to lymphocytes from patients with lymphoproliferative disorders shows little or no response. It is not known whether this is a difference in the surface of these cells, the permeability of cell membranes, the intracellular location of cyclic AMP, or an altered effector part of the cell.

Mice challenged with syngeneic tumors show prolonged survival when cyclic AMP is given in proper dose and timing, and is related to previous immunization. Cyclic AMP appears to have an effect on the B cells of intact mice as measured by tritiated thymidine uptake (as well as spleen cells), but a relatively smaller effect on T cells under the conditions of this experimental approach. These effects vary in the presence of antigen; when antigen and cAMP are combined, earlier and higher peaks are observed. The effect of live tumor on these parameters shows yet a

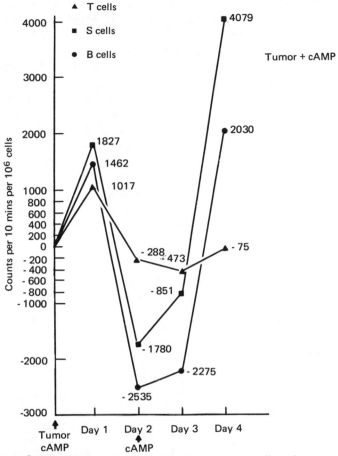

Fig. 10. Differences in ³H thymidine uptake in response to live tumor cells and cAMP in different lymphocytic cell types in vivo.

different pattern. This altered response appears sluggish on a similar time scale, since lower and later peaks occur. These are influenced by the addition of cyclic AMP, though they do not become similar to the pattern observed with cyclic AMP and antigen. Therefore, it may be postulated that the influence of the tumor (or antigen) on the immunologic system of the host may be altered by the addition of cAMP. Cyclic AMP itself changes some properties of the immunologic response.

Further Discussion of Dr. Sheppard's Paper

Dr. Morse: I would like to know if in cultures of normal cells that have reached saturation density and high levels of cyclic AMP, there are reagents which will unfreeze these cells and not affect the levels of cyclic AMP or the converse? I would also ask the opposite question with respect to transformed cells.

DR. WILLINGHAM: I know of no agents that relieve cells from contact inhibition that do not drop cyclic AMP levels. If someone else does, I would be interested.

DR. HIRSCHHORN: Dr. Willingham, can you tell me if inhibition of growth in normal 3T3 cells upon addition of dibutyryl cyclic AMP occurs on contact or prior to contact?

DR. WILLINGHAM: This is a difficult thing to say precisely because if you look at growth curves, the elevation in cyclic AMP does not occur until the cells are of sufficient density so that you could predict there is a good chance they are going to contact another cell. Whether the initial contact produces a sudden spike in the cyclic AMP inside the cell or not, we don't know, since we can't measure it.

DR. HIRSCHHORN: When you add dibutyryl cyclic AMP, do you see inhibition prior to contact or only after the contact occurs?

DR. WILLINGHAM: No, we see inhibition of growth whenever we elevate cyclic AMP levels, even when there is one cell on a dish, we can inhibit its growth.

References

1. Abdou, N. I. "Bone Marrow: The Bursa Equivalent in Man?" *Science* **172**:446–447 (1972).
2. Ishizuka, M., Gafni, M., and Braun, W. "Cyclic AMP Effects on Antibody Formation and Their Similarities to Hormone-Mediated Events," *Proc. Soc. Exp. Biol. Med.* **134**:963–967 (1970).
3. Kruger, J., and Gershon, R. K. "DNA Synthetic Response of Thymocytes to a Variety of Antigens," *J. Immunol.* **108**:581–585 (1972).
4. McCrery, J. E., Hall, J. C., and Rigby, P. G. "The Kinetics of Cyclic AMP and Human Lymphocyte Transformation (HLT)." In *Proceedings of the Sixth Leukocyte Culture Conference.* New York: Academic Press, pp. 153–163 (1972).
5. McCrery, J. E., and Rigby, P. G. "Lymphocyte Stimulation by Cyclic AMP, GMP and Related Compounds," *Proc. Soc. Exp. Biol. Med.* **140**:1456–1459 (1972).
6. MacManus, J. P., and Whitfield, J. F. "Cyclic AMP-Mediated Stimulation by Calcium of Thymocyte Proliferation," *Exp. Cell. Res.* **69**:281–288 (1971).
7. Rigby, P. G., and Ryan, W. L. "The Effect of Cyclic AMP and Related Compounds on Human Lymphocyte Transformation (HLT) Stimulated by Phytohemagglutinin (PHA)," *Eur. J. Clin. Biol. Res.* **15**:774–777 (1970).
8. Rigby, P. G. "The Effect of Exogenous RNA on the Improvement of Syngeneic Tumor Immunity," *Cancer Res.* **31**:4–6 (1971).
9. Rigby, P. G. "The Effect of Cyclic Adenosine 3'5'-Phosphate on Tumor Immunity," *Cancer Res.* **32**:455–457 (1972).
10. Rixon, R. H., Whitfield, J. F., and MacManus, J. P. "Stimulation of Mitotic Activity in Rat Bone Marrow and Thymus by Exogenous Adenosine 3'5'-Monophosphate (Cyclic AMP)," *Exp. Cell Res.* **63**:110–116 (1970).
11. Siskind, G. W., and Thorbecke, G. J. "Kinetics of the Proliferative Response to Antigen in Vitro of Rabbit Lymph Node Cells Taken at Various Times After Immunization," *Immunol.* **20**:151–160 (1971).

FINAL DISCUSSION

ALAN ROBISON, CHAIRMAN

DR. ROBISON: I thought I might start by mentioning an interesting experiment relating to the action of theophylline. I believe this is pertinent because a number of people who have done experiments with theophylline or caffeine noted an effect, and assumed that the effect was probably secondary to the inhibition of phosphodiesterase. We have recognized for a long time that theophylline and other phosphodiesterase inhibitors probably have a variety of other mechanisms of action, and Schwabe and Ebert recently did a very interesting experiment pointing to an additional mechanism involving cyclic nucleotides. They incubated cyclic AMP with isolated rat fat cells at different cell densities. When they incubated the cells at a high density, as I recall, about 100,000 rat cells per milliliter, and then added isoproterenol, a beta adrenergic agonist, they produced what a lot of people had seen before, a relatively small increase in the level of cyclic AMP. When they did the same experiment in the presence of theophylline, they saw a much more striking rise. But, when they reduced the density of the cells down to something like 20,000 per milliliter, then isoproterenol by itself produced a large effect, and theophylline did not increase the response much above that. They interpreted the experiment as suggesting that at high cell densities, theophylline was acting as an antagonist of some inhibitory substance produced by the cell, an inhibitor of adenyl cyclase. Thus, at a high cell density, there is a lot of inhibitor and, therefore, a large effect of theophylline. But when the cells are reduced in number, the inhibitor is not present in high concentration, and isoproterenol by itself produces a good response.

I mention this experiment for a couple of reasons: first, to make the general point that there are many actions of these drugs that can't possibly be blamed on phosphodiesterase inhibition, and second, to illustrate the importance of cell density as a variable. My impression is that many people in the fat cell area have not paid as much attention to this variable as they perhaps should. As to what that inhibitor might be, there are many possibilities. An interesting review was published in *Pharmacologi-*

391

cal Reviews not long ago by Burnstock, discussing the possibility that adenosine or other purine derivatives might act as neurotransmitters in the autonomic nervous system. Burnstock and others have accumulated evidence over the years that adenosine or adenosine derivatives can be released from nerve cells to produce various effects, completely independently of norepinephrine, acetylcholine, dopamine, and the other mediators. It is interesting to note that regardless of whether adenosine produces a so-called excitatory effect (for example, contraction of smooth muscle) or an inhibitory effect (relaxation of smooth muscle), the effect is antagonized by theophylline. Ted Rall showed in brain slices that adenosine is an effective agent in raising the level of cyclic AMP in brain cells, and this was inhibited by theophylline.

DR. BOURNE: Could you comment on the ability of drugs, like theophylline, to inhibit phosphodiesterase in broken cell or purified enzyme preparations versus the ability of theophylline to potentiate the effects of drugs that increase adenyl cyclase activity in intact cells? In leukocytes, it appears we need a high concentration of theophylline to do either one, and I wonder whether that might not be the case in many other tissues, which would perhaps make the effect of theophylline you have just described more prominent.

DR. ROBISON: Most of the early broken-cell experiments in which theophylline was studied as an inhibitor of phosphodiesterase were done with high substrate concentrations, on the order of 10^{-4} M, and under those circumstances, millimolar concentrations of theophylline produce about 50% inhibition. Most of the experiments with intact cells in which there is potentiation of hormones that stimulate adenylate cyclase have required concentrations in the same range. Increasingly, a number of people have reported effects of theophylline with concentrations of around 10^{-7} M or even lower, and the conclusion has often been that you couldn't possibly blame that on phosphodiesterase inhibition, but I wonder.

Dr. Hardman, would you comment on the effect of theophylline on the so-called high and low K_m phosphodiesterase enzymes?

DR. HARDMAN: I don't think this question has been fully answered. In our hands, the potency of methylxanthine seems to be about the same for the high and the low K_m enzymes. This seemed strange to us at the outset, but there is no reason why affinity for a competitive inhibitor should not be proportional to the affinity for the substrate. The inhibition by methylxanthines of the high K_m enzyme, at least the enzyme that Butcher prepared from bovine heart, is competitive. I believe the inhibition against the low K_m enzyme also is competitive. The problem is that the kinetics of the low K_m enzyme are usually anomalous, and it is very difficult to do a clean-cut kinetic analysis of inhibition of this enzyme.

DR. CHANDRA: I would like to make a comment about the theophylline effect observed by us and several other investigators. In view of the recent work of Schroder and Plagemann [*Cancer Res.* 32:1082 (1972)] which I described yesterday, the correlation between the antitumor activity and the phosphodiesterase inhibition by theophylline must be reconsidered. As an alternative possibility, we have found that theophylline is a strong inhibitor of dark repair of bacterial DNA indicating a possible direct action in the nucleus. In this connection, it is of interest that other investigators have demonstrated that caffeine and theophylline may intercalate to DNA.

DR. ROBISON: One of the problems we have had at this conference, in my opinion, is that there are a lot of controversial problems that did not get as well-aired as they might. I shall start off with a relatively un- complicated one. Actually, Dr. Parker mentioned this in his introductory talk. There is evidence, on the one hand, that the replication of some cells is stimulated by cyclic AMP. In addition to the thymic lymphocytes that Dr. MacManus talked about, I think we can include salivary acinar cells in this category: the hyperplasia of those cells in response to isoproterenol appears to be mediated by cyclic AMP. Yet there are a large number of cells, all of the fibroblastic cell lines that Drs. Sheppard, Willingham, and others have studied, for example, in which growth is clearly inhibited by cyclic AMP. I suppose it does not necessarily have to bother us that some cells are affected one way and other cells another, but it is not easy to see exactly how this would work. I asked Dr. Hadden if he would present his ideas on this subject.

DR. HADDEN: Yesterday afternoon, Dr. Sheppard, Dr. Willingham, Dr. MacManus, and I got together and discussed some of the apparent differences in our concepts. We came up with a working concept of what, in an hypothesized normal cell cycle, cyclic AMP might be doing and what manipulations to raise the cellular cyclic AMP level might do to affect the cycle. The so-called G_0 phase is slightly controversial, but I in- clude it in order to be able to localize an initiation phase at the end of G_0. The observations of what cellular cyclic AMP levels are doing during this cycle are based primarily on Dr. Sheppard's data. The control level was arbitrarily chosen as the value during the S phase. As Dr. Sheppard and others have shown, there is a dip in the cyclic AMP level in association with the induction of proliferation by trypsin or other substances which act in the 3T3 cell system. The cyclic AMP level subsequently rises and is high during G_1, and at some point it falls to the control level in association with entry into the S phase. Subsequently, a fall below control occurs in G_2 very close to M, and the level remains low during M. Subsequently, cyclic AMP rises back to control levels during the G_0 phase (the growth- inhibited phase).

We, and others, have shown that increasing the cyclic AMP level within the initiation phase has an inhibitory effect. Dr. Parker's data indi- cate that increases of cyclic AMP levels with adenylate cyclase active agents have a diminishing inhibition effect during the first 6 hr of lymphocyte transformation. At some time during late G_1, and I think we agree this is where the MacManus-Whitfield manipulations are being applied, there is a stimulatory effect of cyclic AMP possibility due to a premature induction or shortening of S. In any case, both DNA synthesis and the flow of cells into mitosis are enhanced during this period. The effect crosses over in late S. There is thought to be a period during S when there is not a great effect of cyclic AMP active agents. During G_2 in some systems, there is inhibition by cyclic AMP. As can be seen, there is something of a parallel between what cyclic AMP is doing in the cell during the division cycle, and what manipulations of it do to affect the replication processes.

Let me insert here a couple of comments about cyclic GMP, based on preliminary data. In some circumstances in which cyclic AMP has been shown to drop in association with the initiation of cellular proliferation, we have shown that cyclic GMP rises. Thus, we can talk about cyclic GMP as moving in an opposite direction to cyclic AMP during the initiation phase. Dr. MacManus has published data that indicate that exogenous cyclic GMP is inhibitory during late G_1 to M. This is another example of an action of cyclic GMP opposite to that of cyclic AMP.

DR. MACMANUS: Cyclic GMP (in the presence of calcium) added at the beginning of G_1 blocks during G_2. That is, the effect is not expressed until G_2.

DR. HIRSCHHORN: I hate to throw in contradictory data, but I believe there are two observations in the literature as well as our own work that exogenous cyclic GMP inhibits PHA-induced lymphocyte growth, as does exogenous cyclic AMP, when added at the beginning of G_1.

DR. PARKER: Our data are similar to Dr. Hirschhorn's: GMP added with or just after phytohemagglutinin inhibits lymphocyte transformation in the same way that cyclic AMP does. In fact, on the molar basis, it is a more potent inhibitor than cyclic AMP. I don't think this result necessarily obviates the possibility that cyclic GMP might be important in transformation, since we don't know how much is getting into the cell.

DR. MACMANUS: I would like to expand a little on what Dr. Hadden said. I think in thymic lymphocytes, we are dealing with a situation where cells are actively cycling, and in these cells we don't have an initiation phase. The data I presented yesterday suggested that a rise in cyclic AMP is associated with the initiation of DNA synthesis. In liver, during regeneration or in 3T3 cells that are being stimulated to proliferate by serum, the stimulus at time zero can be associated with a drop in cyclic AMP or a rise in cyclic AMP, or no change in cyclic AMP. My view is that the initial stimulus is a rise in cyclic GMP, and before the initiation in DNA synthesis, there is the rise in cyclic AMP.

DR. PLESCIA: I think the relative effects on activation, with respect to control, depends upon what that the basal level of cyclic AMP is. Perhaps each of the varying observations reported is due to the fact that we are trying to compare changes in cyclic AMP with respect to a rather elusive basal value. I think Dr. Braun underscored this very clearly in his own work, i.e., that these changes in cyclic AMP levels could either be inhibitory or stimulating depending upon the particular stage of the cell. Dr. Braun stated that if a cell is activated increasing further the cyclic nucleotides may not only push it into further activation, but perhaps into a inhibitory phase. Perhaps, as we look at all these data, we should pay more attention to the base line at the start of the experiment.

DR. ROBISON: To move on, the idea of altering sensitivity to the external environment, such as altered hormonal sensitivity, is something we have all become used to, and I was wondering whether altered sensitivity to cyclic nucleotides might not also be important. This came up, for example, in thinking about the data that Dr. Sheppard and Dr. Willingham mentioned suggesting that in fibroblasts, what they would want to do to get rapid replication would be to lower the level of cyclic AMP. Yet, Dr. Ryan showed that in certain human tumors, the level of cyclic AMP, if anything, is much higher than in normal cells. Obviously, there are ways to get tumorigenesis without altering the level of cyclic AMP—that, at least, should be obvious. But sticking to cyclic AMP, it seems to me that the possibility of reduced or diminished sensitivity to the action of cyclic nucleotides must also be considered. Dr. Bourne has some recent interesting observations that relate to this.

DR. BOURNE: I should like to present briefly the results of some recent experiments investigating the mechanism of action of cyclic AMP in a cultured line of lymphoma cells. Although these experiments were not immunological, they may be of interest to this group from two points of view: (1) They appear to shed some light on the role of cyclic AMP binding proteins and cAMP-dependent protein kinases in tumor growth; (2) they suggest a mechanism for several inhibitory effects of cyclic AMP on cultured lymphoid cells, which may eventually prove relevant to the widespread inhibitory effects of the nucleotide on normal lymphocytes.

The exposure of cultured lymphoid cells to glucocorticoids leads to inhibition of cell growth and ultimately to cytolysis. Using a steroid-sensitive mouse lymphoma cell line, Daniel, Litwack, and Tomkins found that high concentrations of dibutyryl cyclic AMP also cause slowing of cell multiplication, followed by cytolysis [*Proc. Nat. Acad. Sci.* (U.S.A.) 1973, in press]. When these cells were exposed first to low (noncytolytic) concentrations of dibutyryl cyclic AMP, and then to gradually increasing concentrations of the nucleotide (over a period of one month), these investigators obtained a population of lymphoma cells which could no longer be lysed by dibutyryl cyclic AMP, 1×10^{-3} M. These resistant (R) cells could be maintained in culture, in the absence of exogenous cyclic AMP, for more than six months without recovering their sensitivity to cyclic AMP.

Daniel and her colleagues next inquired whether cyclic AMP resistance could be due to impaired binding to a specific cytoplasmic receptor. They incubated radioactive cyclic AMP with cytosols from sensitive (S) or resistant lymphoma cells and subjected the mixtures to get filtration (Sephadex G-25). There was much greater (about fivefold) macromolecular binding by cytoplasmic extracts from S than from R cells. They also chromatographed extracts of S and R cells on DEAE Sephadex. In both cases the cyclic AMP binding migrated with cAMP-activated histone kinase, but both the binding and the cAMP-dependent kinase were diminished on R cells.

These results suggested that the binding protein, the "cyclic AMP receptor," is a regulatory subunit of cAMP-dependent histone kinase in these lymphoma cells, as in many other cell types. They also suggested that the resistance of R cells was due to deficiency of a cytoplasmic cyclic AMP receptor, thus rendering dibutyryl cyclic AMP biologically inactive.

The next step was to ask whether R cells were resistant to cyclic AMP generated within the cell, as they were to exogenous dibutyryl cyclic AMP. We found that prostaglandin E_1 stimulated accumulation of cyclic AMP in both R and S cells, but slowed the growth only of the S (parental) line (Daniel, Bourne, and Tomkins, in preparation). In contrast to dibutyryl cyclic AMP, PGE_1 caused only slowing of cell growth, but not cytolysis. One possible reason for the failure of S cells to die when exposed to PGE_1 could be that the rise in cyclic AMP in S cells was small and transient (barely detectable after 2 hr). The situation in the resistant cells was quite different: PGE_1 produced a 10-fold greater initial rise in cyclic AMP (1000 pmoles per 10^7 cells versus 100 in S cells), which persisted for hours. The inability of such large amounts of intracellular cyclic AMP, persisting over time, to slow the growth of R cells certainly suggests that the deficiency of cyclic AMP receptor in the cytosol of these cells made it relatively inactive. Parenthetically, it appears that the much greater accumulation of cyclic AMP in R cells is at least partly accounted for by a relative deficiency of cyclic AMP phosphodiesterase in these cells; this suggests that lymphoma cell phosphodiesterase may be regulated by the cyclic AMP receptor, a hypothesis presently under investigation.

These experiments suggest that the cyclic AMP receptor, or cytosol-binding protein, may play an essential role in the growth-slowing and cytolytic effects of endogenous and exogenous cyclic AMP. But whether the biochemical mechanism by which cyclic AMP bound to receptor protein kills these lymphoma cells or slows their growth is not known. We have found that the death of lymphoma cells exposed to dibutyryl cyclic AMP is preceded by inhibition of incorporation of radioactive leucine and uridine into macromolecules. This effect was not seen in the resistant (R) cells. It may be, then, that cyclic AMP slows cell growth and even-

tually causes cell death by inhibiting transport of macromolecular precursors, a mechanism similar to that proposed for the glucocorticoid-in-
duced cytolysis of lymphoid cells.

These cultured lymphoma cells may, in some respects, serve as a
model of events that occur in normal lymphoid cells (as with glucocorticoid sensitivity). It is possible that the inhibitory effects of exogenous
and endogenous cyclic AMP on many functions of lymphoid cells, including release of antibody, mitogenic response to phytohemagglutinin, and
cytolysis of target cells, occur through a common pathway. It appears
likely that this pathway will turn out to begin at the cyclic AMP receptor.

Dr. Robison: I think those data are of great interest from a variety
of points of view, but one thing they do, as least, is to underline the possible importance of changing the sensitivity of the cell to cyclic AMP,
which in many circumstances could be even more important than changing the level of cyclic AMP.

Dr. Hadden: Is there any obvious difference between the resistant
and the sensitive cell, particularly in their cycling time?

Dr. Bourne: No. Their doubling time is approximately 24 hr in both
sets. Morphologically, the sensitive and resistant cells are exactly similar,
and they cannot be distinguished by differences in protein or RNA synthesis. We have not cloned these cells or examined their chromosomes.
There is a lot of work to be done.

Dr. Parker: The fact that a cAMP-resistant cell line can be obtained in vitro underlines one of the limitations of attempting to raise in-
tratumor cyclic AMP concentrations as an approach to tumor therapy.
One of the great problems in cancer chemotherapy is cell variability and
the fact that cells that are resistant to the agents that are being used have
a selective growth advantage, utlimately results in a refractory cell
population.

Dr. Hardman: Dr. Bourne, you referred to the induction of
phosphodiesterase in the sensitive cells. Do you know whether or not
binding protein or cAMP-dependent protein kinase can be induced in the
sensitive cell?

Dr. Bourne: I don't know.

Dr. Robison: We already have generated a pretty good argument
about cyclic GMP, but in a sense the situation is not very satisfactory. I
am talking about Dr. Goldberg and Dr. Hadden thinking that
phytohemagglutinin causes cyclic GMP levels to be raised. Dr. Parker
tried to repeat those experiments under the same conditions and finds
that the cyclic GMP does not change. There is not much more we can do
about that here, other than to hope that these investigators will get
together and exchange samples and eventually get this problem
straightened out. As for how it will turn out, I would say we have a 50-50
situation here. But just for the sake of talking about it, let's assume that
cyclic GMP does go up and that it does do all the things that Dr.

Goldberg and Dr. Hadden think it does. So far we haven't talked about
how this might happen. Dr. Hadden, you showed some experiments with
isolated nuclei, and I was wondering if you could go over those again. The
general question is how do you think cyclic GMP does all the things that
you and Dr. Goldberg think it does?

DR. HADDEN: We showed yesterday the results of incubating isolated lymphocyte nuclei in the presence of the cyclic nucleotides using
the Mirsky incubation medium, which includes calcium as an essential
factor in the events observed. We found no effect of cyclic GMP on the
intact cell, but cyclic GMP at extremely low concentrations stimulated
the incorporation of tritiated uridine into acid-insoluble, ribonuclease-
sensitive nuclear material. We have done nothing further to characterize
the RNA fraction involved.

I can only speculate as to what is going on in the nucleus. Pogo et al.
(1966) observed that one of the first events in the nucleus following
stimulation of the intact cell by phytohemagglutinin was the acetylation
of the arginine-rich histone fraction, and this event was followed by
nuclear RNA synthesis within 30 min. We assume that cyclic GMP is
doing what PHA does to the nucleus and involves the modification of the
arginine-rich fraction of histones. This remains to be proved. In contrast
to these changes are the modifications of the lysine-rich histones, which
have been observed in replicating cells and which have been related to
cyclic AMP action by Langan. The modification in this case is
phosphorylation. The modifications of histones may relate to the cell cycle. Mirsky recently reported that with differential extraction of the
histone fractions, it can be estimated that lysine-rich histones are involved in masking perhaps 80% of the DNA, whereas arginine-rich
histones mask most of the remaining 20%. We assume that both classes of
histones are modified during the division cycle.

DR. ROBISON: You mentioned RNA synthesis in those experiments.
You are thinking in terms of phosphorylation via a protein kinase sensitive to cyclic GMP?

DR. HADDEN: The early modification of histones observed with PHA
was acetylation, so one might expect enzymes related to this process to
be involved.

DR. HIRSCHHORN: I think you will have to look a little bit more
closely at your RNA synthesizing system, because if you used uridine, in
order for it to be utilized in RNA synthesis, it first has to be phosphorylated. In other words, it is the triphosphate that is used by the enzyme to
make RNA, and therefore, in your system you are looking not only at the
synthesis of RNA, but also at all the enzymes taking you from uridine to
UTP. That implies that you have cytoplasm in your preparation. You are
not really working with clean nuclei.

DR. HADDEN: My knowledge of this area is that the triphosphate
forms of the nucleotides don't get into the nucleus. Of course, I am going

to see if I can't improve my success rate in these experiments using UTP, and indeed, if there is a difference between uridine and UTP incorporation, it would imply that cytoplasmic contamination might play a role.

DR. HIRSCHHORN: The important point is you are not necessarily looking at the action of cyclic GMP upon RNA synthesis.

DR. WEISSMANN: There has been a lot of talk in the literature about isolated nuclei. One of the ways of checking whether one has isolated nuclei is by doing an analysis of cytoplasmic enzymes. Fractionation studies that do not include appropriate monitoring of enzymes associated with specific subcellular fractions are meaningless. Nuclei, for example, will regularly adhere to components of the cytosol, and unless the monitoring is done, most of the conclusions about the nuclear localization of either cyclic nucleotides or enzymes are going to be absolutely useless.

DR. HADDEN: A good point.

DR. WEDNER: We have looked at the incorporation of C^{14} UTP into RNA in isolated nuclei, and we get a slight inhibition with 8-bromo cyclic GMP or with cyclic GMP. We don't have calcium in the medium, since the polymerase has been reported to be inhibited by calcium and stimulated by manganese and magnesium, so this may be the difference.

DR. HADDEN: The fact that you get inhibition would imply that you get activity in your control.

DR. WEDNER: Yes.

DR. HADDEN: In the presence of EGTA, we don't.

DR. WEDNER: We get inhibition with calcium. If we take the calcium out in the final step, after purifying the nuclei in calcium-containing solutions, and then put manganese and magnesium in, we get good incorporation of UTP into acid precipitable material.

DR. WEISSMANN: This experiment is almost impossible to do properly, because manganese also stimulates nuclear phosphatases roughly 50-fold. In other words, the conditions for stimulating endogenous RNA polymerase activity will also stimulate the phosphatases. Have you considered this?

DR. WEDNER: No, nevertheless, we get actinomycin C sensitive incorporation of uridine triphosphate radioactivity into acid precipitable material.

DR. ROBISON: Well, I think it's clear that we have a problem here that needs to be looked at further. To introduce the next problem, let me refer to a hypothetical figure, which suggests certain changes in cyclic AMP during development (Fig. 1). The reason behind postulating this early rise in cyclic AMP was partly that we observed it during late embryogenesis in rats. But it was also supposed to depict what I thought might be the in vivo analogue of contact inhibition of growth. This was based on reported work by Dr. Ryan, who is here, and also by Dr. Pastan's group, with whom Dr. Willingham is working now, suggesting very strongly that contact inhibition of growth in cultured cells was

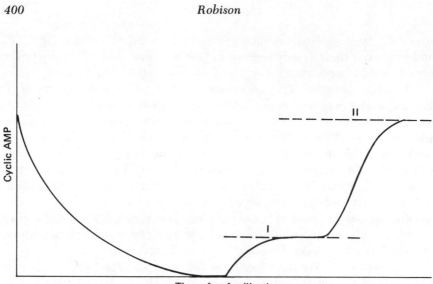

Fig. 1. Hypothetical changes in cyclic AMP (or in the sensitivity to cyclic AMP) during growth and development. Levels fall after formation of zygote but later rise after mechanisms for stimulating adenyly cyclase develop. Time scale of left-hand portion of diagram is greatly expanded relative to later stages. Baseline levels I and II will be reached at different times in different cells.

mediated by cyclic AMP. In other words, it seemed that cell-to-cell contact might somehow lead to the stimulation of membrane adenylate cyclase or by some other mechanism would raise the level of cyclic AMP and that this would inhibit cell replication. Then Dr. Sheppard obtained data showing that contact-inhibited cells had high levels of cyclic AMP before and after confluency, while transformed cells had low levels of cyclic AMP all the time.

Here is a very important discrepancy. This is the first conference I have ever attended where there is someone from Dr. Pastan's lab, together with both Dr. Ryan and Dr. Sheppard, and yet we have not gotten into an argument about it. Fortunately, I caught Dr. Willingham last night, and he agreed to defend the position that contact inhibition is mediated by cyclic AMP.

DR. WILLINGHAM: It makes me somewhat nervous to try defend the whole National Cancer Institute as Dr. Robison suggests. All I can say is that we are still perfectly confident that the hypothesis that cyclic AMP mediates contact inhibition is true, and is not some artifact of measurement, because we have done experiments in many normal cell lines. All of these cells show elevations of cyclic AMP levels when they show density-dependent inhibition of growth. I don't think that there is any conflict over the situation in transformed cells. These cells have low cAMP levels all the time, and in fact, at confluency the levels go down a little bit. But the original observation by Jacques Otten that cyclic AMP levels were inversly proportional to growth rate, that is, low cyclic AMP levels meant the cells had a rapid growth rate, still holds. We are also still certain that at confluency cyclic AMP levels rise precipitously.

DR. SHEPPARD: We are just as convinced there is no difference. We do agree with the observation that there is a direct correlation between the basal level of cyclic AMP and the rate of growth. We can see the same thing with regard to saturation density. That is, cells that grow faster and to higher densities have very low basal levels of cyclic AMP. But this has nothing to do with the contact-inhibition phenomenon. One of the points I raised yesterday is the question of what really is the biological stimulus that regulates the basal level of cyclic AMP? Our thoughts right now lie in the area of nutrient transport and/or nutrient availability. Cells that grow faster get nutrients such as glucose, carbon, amino acids, and phosphates into the cell, and metabolize them at a faster rate than the slower-growing contact-inhibited cells. Our idea is that a high level of cyclic AMP is the result of a deprivation of some basic nutrients. It is analogous to what happens in bacteria, where if you deplete carbon and glucose, the cyclic AMP levels rise.

There is quite a bit of evidence showing that transformed cells have much higher rates of transport of various nutrients. I am going to refer to Dr. Pasten's data with the line of chick cells that is infected by the mutant RSV virus. At 36°C the cells grow rapidly, at 41°C the cells exhibit contact inhibition. What Dr. Pasten has shown is that when you change the temperature of the permissive cell line to 41°C, there is a precipitous fall in cyclic AMP. There is also a very large increase in the rate of transport of glucose at this time. Other cell lines, which we have shown to have low levels of cyclic AMP, also have high rates of glucose transport. The correlation I am trying to make is that possibly cyclic AMP levels are responding in the cell to the availability of glucose, amino acids, or other basic nutrients that are necessary for the growth of the cell.

DR. ROBISON: Are you suggesting that if Pasten's group didn't have glucose in the medium, they would not have seen the fall in the level of cyclic AMP?

DR. WILLINGHAM: May I interject to point out that the change in cell growth in going from 36°C to 41°C can be blocked by elevating cyclic AMP levels artificially. So, we don't feel that the fall in cyclic AMP is purely a consequence of changes in the metabolism of these cells. We think it is a prerequisite before those changes occur.

DR. SHEPPARD: I would bet that if you added dibutyryl cyclic AMP, the transport of glucose, which is facilitated by this decrease, would be inhibited too. There is a very close relationship between transport and the cyclic AMP level.

DR. WILLINGHAM: That I don't know. All I know is that the cyclic AMP level falls, and the cells become transformed. Concomitantly, if you artificially raise the levels of cyclic AMP before transformation is allowed to occur at that temperature, you can block transformation.

Dr. Weissmann: You can test this hypothesis of glucose entry quite readily. Have you tried to inhibit it with cytochalasin, or with 2-deoxyglucose?

Dr. Sheppard: We have not looked at either one of those compounds. We have looked at the effect of a glucose-free medium on the growth of some transformed lines. Some very preliminary work suggests that in the absence of glucose, cyclic AMP levels are increased.

Dr. Willingham: I would agree that deprivation of a number of nutrients in the medium will raise cyclic AMP levels, but whether this has anything to do with transformation we don't know. It is the chicken and the egg phenomenon.

Dr. Robison: The only thing I can contribute to this discussion is that glucose does not alter the level of cyclic AMP in the perfused rat liver. Dr. Ryan studied the contact-inhibition problem, and I think he agrees more with the Pasten group than with Dr. Sheppard. As I recall, Dr. Ryan, you could not detect cyclic AMP before confluency, and could easily detect it after, which seems to be in line with the idea that cyclic AMP could mediate contact inhibition.

Dr. Ryan: We have data that agree with both Dr. Sheppard and Dr. Willingham. What occurs depends in part upon the cell and the nutrient conditions. I suspect that cells in culture really lack something that cells in vivo have. The reason I say this is that cyclic AMP levels in cells in culture are 20 to 200 times higher than levels in cells in vivo. So I suspect that there is something missing in the tissue culture system that produces a type of nutrient deprivation and elevates the cyclic AMP levels of cells in culture, no matter whether you are talking about confluence or elsewhere in the cell cycle.

Dr. Willingham: I am sure that is true, because one of the most potent regulators of cyclic AMP levels in randomly growing cells in culture is serum, and certainly in vivo cells are in a very different environment than they are in 10% serum. A connection between our tissue culture systems and what happens in the whole animal is very tenuous.

Dr. J. Hadden: I would like to make a comment concerning the relationship of cellular cyclic nucleotide levels to membrane transport. In collaboration with Dr. Dick Estensen, we have studied in lymphocytes the influence of two mitogenic agents on the transport of glucose. The two agents are phorbol myristate acetate and phytohemagglutinin, both potent stimulators of cellular levels of cyclic GMP and, in certain circumstances both have been shown to drop cyclic AMP. We find an increase in 2-deoxyglucose transport with both agents. The effect is immediate, which suggests there is a very close temporal relationship among the action of these agents on the surface of the cell, on changes in cellular cyclic nucleotide levels, and on transport.

DR. SHEPPARD: There is a possible reconciliation. If confluency inhibits transport, then in certain systems in which there is a significant density effect on the transport of some necessary nutrient, cyclic AMP levels might rise. This would be dependent on the cell line because, in fact, we have cells that are sensitive to glucose-free media and cells that are not sensitive to glucose-free media.

DR. WILLINGHAM: In response, all I can say is that all our culture conditions are stable. We don't change glucose levels; we use Delbecco's medium with 10% cat serum, and in continuously growing cells at the appropriate time, we get an elevation of cyclic AMP. What causes this—we are not sure. We are presuming it is caused by the phenomenon of contact inhibition. It may very well be possible that it is caused by lack of transport of a particular nutrient, but I don't think so, since the media can continuously be changed, at least in the components of the media we feel are necessary for cell growth.

DR. ROBISON: I am afraid that is all the time we have to talk about cyclic AMP and cell growth inhibition. At least we raised the argument. Dr. Lichtenstein raised an interesting question the other evening; how do relatively short-lived metabolic events early in time lead to delayed or persistent changes in cell function? This is a very interesting question from the standpoint of cyclic nucleotide metabolism. I really think it gets us into some very fundamental problems relating to the cell cycle, especially from the point of view of the things we are interested in at this conference.

DR. LICHTENSTEIN: We must interpret a whole series of observations. The most striking, perhaps, are in Dr. Braun's work, where the early introduction, in vivo, of agents that putatively cause a prompt and transient rise in cyclic AMP produces an effect later on immune function. The experiments Dr. Henney presented, where he showed that a sustained inhibition of cellular immunity in vivo can be produced by agents causing relatively transient changes in cyclic nucleotide levels, are another case in point. If mice were immunized with mastocytoma cells and given cholera enterotoxin two to four days later, a transient increase in the cyclic AMP level of liver or spleen cells was demonstrable. Then 10 days later, long after the cyclic AMP level had come back to normal, splenocytes from these animals were turned off in the sense that they were unable to kill target cells in vitro. We have not yet examined the specificity of this effect. So, if the stimulus comes at a time when the cyclic AMP level is increased, there are very profound and long-lasting effects.

Perhaps these events can be interpreted by reference to studies of immediate hypersensitivity. If you add isoproterenol to a leukocyte population, the cyclic AMP level goes up rapidly and comes down again in 10 to 20 min to normal. If you add antigen after this return to base line, you get a perfectly normal response. However, if antigen is added while the cyclic AMP level is elevated, you get no response. At that point in time, the cell is insensitive; in fact, it is what we call "desensitized." It will no longer recognize any antigenic stimulus. I feel that observations in this system may have some general application to the reception of many different kinds of signals. I was wondering whether there might not be, in cellular immunity, a phenomenon such as desensitization. We can produce a desensitized state in immediate hypersensitivity in a variety of ways. As far as I know, however, it has never been observed in vitro in a cellular immune system. When you think about it, moreover, we really don't know what desensitization is. There may be similar phenomena in nonimmune systems.

Desensitization might not be the explanation. We have tried to demonstrate it experimentally. For example, Dr. Henney exposed lymphocytes in vitro to antigen for several hours together with an agent that raises cyclic AMP. This does not lead to desensitization; the cells killed when washed free of drug. Maybe in his system, it takes more time to cause desensitization. On the other hand, it is possible that the kind of work that Dr. Melmon and Dr. Bourne have reported, where the effects noted were presumably dependent on a subpopulation of cells, which might get changed during the period of increased cyclic AMP and then modulate subsequent events, provides an explanation.

What I have tried to do is to raise the question and stimulate discussion as to how immune events that occur long after the pharmacologically induced changes in cyclic nucleotide levels have waned might be explained.

DR. HADDEN: Have you done any studies on binding of the antigen under these circumstances?

DR. LICHTENSTEIN: We have never formally demonstrated antigen binding to human basophils. When we add 10^{-6} μg of antigen and then wash the cells, we get it all back, $\pm$ 10 to 20%, which is as well as we can make the measurement.

DR. HADDEN: A comment from the cellular immune realm having to do with mitogen receptors: We know the lymphocyte stimulated with phytohemagglutinin loses not only sensitivity to PHA but its ability to bind it, suggesting that there is loss of the receptor. Whether it is lost into the cell and digested or liberated into the medium—we don't know. The kind of observation you showed us could involve a similar mechanism.

DR. HENNEY: Dr. Hadden's point is well taken, especially with regard to the cell-mediated immune response. We have not yet established whether in fact we get any cell differentiation in the presence of cholera enterotoxin, or whether we get differentiation of cells that are subsequently rendered inactive. What Dr. Bourne has shown in vitro is that plasma cells themselves can be prevented from secreting antibody by adenylate cyclase stimulation. Thus, we don't know if what we observe—reduced serum antibody titers in cholera enterotoxin treated animals—is due to constipated plasma cells or a lack of plasma cells. Similarly, we can shut off T effector cells in vitro very readily by adenylate cyclase stimulation. Again, we do not know whether in the presence of cholera enterotoxin we get in vivo differentiation into T cells capable of recognizing antigen, but not able to effect cytolysis or whether there is no T cell proliferation. We are currently investigating this more carefully.

DR. ROBISON: With reference to the argument that you and Dr. Strom were getting into the other day, do you suppose that alterations in the nature or sensitivity of surface receptors could account for some of your differences?

DR. HENNEY: I think that is possible. Dr. Strom uses effector cells seven days after antigenic stimulation. We use cells 11 to 12 days after antigen. It may be that cholinergic receptors on lymphocyte surfaces occur only transiently. As we showed yesterday, the proportion of effector T cells bearing histamine receptors does fluctuate during the immune response, and it is quite conceivable that other hormone receptors do too.

DR. SHEARER: I think we have to distinguish between the events that occur at the termination of immune responses and those concerned with the initiation of the immune response. I am more interested in what is happening at initiation. I would like to present some data that were obtained in Israel. These experiments were carried out by Moses, Weinstein, Bourne, Melmon, and myself at the Weizmann Institute. The experiments involve in vivo cell transfer into syngeneic irradiated recipients. The irradiated mice are injected with a constant number of spleen cells, and then 24 hr later given antigen. We measure the humoral antibody generated by the donor responses in the recipient at the peak of the antibody titer, namely, two weeks later. Prior to injection of the cells, we have preincubated the cells with soluble antigen for 15 to 30 min, at 37°C.

We found that if the cells were then injected into the recipients, irrespective of whether the soluble antigen was washed out of the cell suspension or not, there was a depressed response when compared with cells that were incubated in media above and then transferred. We don't know whether this is a tolerancelike phenomenon or not. It seems to be tolerancelike in the sense of antigen specificity; namely, if we used an unrelated antigen such as gamma globulin for the preincubation, this had no effect on the subsequent response of the transferred cells to a stimulation with homologous antigen.

What we did find that is interesting in the context of this meeting is that if we incubated the cells and antigen with histamine, or prostaglandins, or isoproterenol, or epinephrine, we abolished the tolerancelike phenomenon. That is, if we have drugs plus antigen, together with the cells used for transfer, the immunization is followed by a normal immune response. We don't know how to interpret this, but it seems that some of the effects that one might see, even for humoral responses, can occur early with respect to antigen, cells, and drug-mediated response.

DR. BOURNE: One way of thinking about these results would be that the initial contact with antigen, perhaps at a higher concentration than the cell would ever see in vivo, does something to the antigen receptors, but when cyclic AMP is raised, the response is blocked. It is interesting that in the same experiments, poly A:U had the same effect in reversing the tolerance that the cyclic AMP active drugs did, but it did not increase cyclic AMP. Our interpretation is that it is possible that a lot of the results that Werner Braun reported were true insofar as poly A:U is able to tremendously change immune responsiveness, but perhaps not true with regard to cyclic AMP being the mediating event.

DR. MORSE: I would like to know where those cells go? What I am concerned about is that you are observing an architectural sort of thing, yet talking about responsive cells: Where do those cells go after they are injected under the circumstances cited.

DR. SHEARER: A number of studies have been done in the in vivo cell transfer system, using radioactive cells or chromosomally marked cells, and tracing where the responding cells occur, using Jerne's placque assays for antibody-forming cells. Almost all of the antibody-producing cells during the first two weeks appear in the spleens of the recipients.

DR. MORSE: That is without treatment, and I am asking what happens when you do the manipulations you spoke of.

DR. SHEARER: That experiment has not been done.

DR. PERPER: We have treated lymphocytes and monocytes with prostaglandins, and they do circulate differently in that their life span in the blood is increased. Whether that relates to homing—I don't know.

DR. MELMON: I don't believe that the homing of the cells is dependent upon the direct effect of the free drug that we used to change the cells. It is important to consider the point that these drugs have cardiovascular effects independent of anything else, and if they are injected at the same time that cells are injected, they may alter the distribution of the cells. The question as to whether the cells get to the same places each time is very important, and should be followed up. The point that I tried to make is that it would be unlikely that a sufficient amount of free drug was the cause of the change in the half-life of the cells that Dr. Perper measured, because with and without washing we get the same effects on the transfer.

DR. BLOOM: Dr. Shearer is focusing on the very important early events about which I find myself still very puzzled. That is, if some of the drug effects last between 10 sec and 30 min and one looks four days later at the cells' ability to kill, or three days later at the cells' ability to incorporate thymidine, it is not at all clear to me why changes should occur. But when one looks at Conacavalin A stimulation of transformation or cytotoxicity, presumably the mediator effects occur in 10 sec to 30 min, and the chain starts. What is extraordinary is that if you add α methyl mannoside and take the Con A back off the cell as long as 6 to 12 hr later, there is no transformation or lymphocyte target cell-mediated cytotoxicity. I would bet that the cyclic AMP went up in the first 10 or 15 sec. So, I don't see the relationship between those instantaneous events and what is really happening later.

DR. MELMON: It is very common in pharmacology to pick up events long after a drug has been administered. Just consider drug toxicity where one is using, for example, antitumor agents, and the effects that they may have on cells other than the tumor, and the fact that maybe three or four generations of people or animals may be required before some abnormality is seen in the progeny. It may be difficult, then, to go back and implicate the drug, but that does not deny the possibility that long after a drug has been around, its phenotypic expression, if you will excuse the use of the term, becomes manifest.

DR. ROBISON: It is the kind of question that has been of interest and puzzling to endocrinologists for ages, and I guess equally so to immunologists. Perhaps the most striking example of all would be the small increase in testosterone that occurs a day or two before birth. This makes a lot of difference in what happens to us later on in life.

DR. PLESCIA: The problem of how these agents modify the initiation of the immune response is complicated by the fact that depending upon when these agents are added, relative to the antigen, changes the nature of the response. For example, if poly A:U is added two to four days prior to the administration of sheep red cell antigen (sRBC) in mice, one gets a suppression rather than potentiation of the response compared to base line. We have been able to find antigen-carrying cells in the spleen of mice within 4 hr after the administration of sRBC. These cells are not antibody producing, but when allowed to interact with B cells lead to the formation of antibody. The administration of poly A:U two to four days prior to the injection of sRBC leads to a substantial reduction in the number of these antigen-carrying cells 4 hr after the administration of sRBC, so it appears that these agents can interact with cells that are involved in immune response, either in a direct interaction or regulatory function, and thus one can modify the response in the initiation phase either up or down, depending upon whether these agents are present with the antigen or given before.

DR. PERPER: The work of Dr. Claman bears on these early events: If you treat an animal early in the course of immunization and look for a later event, which is measured by securing cells from a particular organ, you must know what that agent has done to the distribution of lymphocytes. Dr. Claman's data show that if one treats with corticosteroids you don't kill the cells, but rather redistribute them so that the circulating lymphocytes are now found in the bone marrow and no longer found in the peripheral lymphoid tissue. Thus, if you are going to take a killer cell from a lymph node of an animal treated early on with a compound which might have changed the distribution of the cell, the cells might now be in the bone marrow.

DR. SHEARER: I want to refer again to one of the points raised about the redistribution of cells in the recipients. Perhaps it was not clear in our remarks, but the recipients were bled, and the serum antibody titers were measured. We did not take cells from particular organs, so I think this is less likely to be due to redistribution. You can, of course, still argue that due to redistribution the antibody might not be released into the serum.

DR. PARKER: But if redistribution took place, obviously there would be a very different relationship between the transferred cell and resident cells in the various organs. This might markedly alter the ability of the cell to interact with antigen-containing macrophages and other cells.

DR. ROBISON: I think it is clear that we have some interesting phenomena even if we are not doing too well with them at the molecular biological level. Another interesting question Dr. Bloom raised in his talk, again at the physiological level, was how the immune response was stopped. Why don't these replicating T cells turn into a lymphoma, or B cells into a myeloma, for example. This is a very important problem, and I thought some of the experimental findings that Dr. Melmon discussed earlier might bear further comment. I asked Dr. Melmon if he would expand a little in this area and possibly even discuss the age-old question of what the physiological significance of histamine might be.

DR. MELMON: It happens that the amines that we used make cellular cyclic AMP increase, and this seems to have some eventual effect on immunologic processes. Whether the immunologic events are cause and effect related to cyclic AMP changes—I don't know. It is fascinating, however, that at least three separate types of observations seem to indicate that the end point (final expression) of an immunologic response may be modulated (inhibited) by the drugs that stimulate accumulation of cyclic AMP in cells in in vitro experiments. Thus, it could be that the immunologic response may be modulated by cellular substances not ordinarily thought of as participating in the immunologic response. I think the data we have presented might provide a stimulus to seriously study the potential of endogenous hormones other than corticosteroids to modulate the immune response. Our cell subtraction experiments support the hypothesis that the receptors for these amines exist on cells that are important in various aspects of the immune response. The receptors may be there for no purpose or may simply express themselves on every cell, but neither of these possibilities seems likely. We know that when we separate cells on the basis of their receptors for histamine, there is selectivity in cell separation, and depending on the time after immunization when one chromatographs the cell, there may be greater or lesser numbers of cells with varying functions with these receptors for the hormones on them.

I do not know whether histamine, isoproterenol, or any of the drugs we have been discussing has a role in vivo in modulating immunologic responses, but I do believe that the receptors on these cells are not likely to be artifacts, nor are they ubiquitous. They either are there to express themselves, or develop in some ill-defined but meaningful pattern as certain cell functions develop. It could be that the substance with a true modulating role of the immunologic response has not been identified, that is, that the receptors for amines may develop concordinantly with receptors for other substances. I think there has to be a serious look at the systematic interruptions or perturbations of various aspects ot the immune response by the drugs if we are to approach the understanding of both a physiologic role for the amines and mechanisms for turning off an immune process. There should be a systematic subtraction of cells with and without receptors that can be used to reconstitute an acceptor animal, and then a study of the response of this animal to the antigen. There is a chance that when this is done, one can finally answer Alan Robison's question: "What is the physiologic role of selected biogenic amines?"

In addition, it seems clinically reasonable to examine carefully those diseases that seem to progress from an inflammatory state to an abnormal immune process. For instance, in scleroderma there are symptoms highly reminiscent of an acute inflammatory reaction (why, we do not know) and other symptoms that are clearly immunologic abnormalities.

My bet is that the inflammatory process contributes to the immune response in some clinical settings. Our data certainly suggest such a connection. Perhaps cells with immunologic function are modulated by a variety of hormones released during the inflammatory process. Perhaps there are specific functions (e.g., antibody production, regulation of cell replication, regulation of overall antibody release) that may be modulated by the mediators of inflammation. It would be very exciting to be able to say that our in vitro observations may carry over to in vivo settings.

I do not think our data give more than a hint of what may come. A regulator cell function simply says that there is a population of cells in the animals we used that are probably not precursors of antibody-forming cells, which can be subtracted from the overall population of the spleen, which exist in relatively small numbers, and which probably work by somehow modulating the immune response at an early stage of the process. We know that the cell is there and that the substance it makes can create the same effect without the cell, but we do not know what the substance is. We do not know why that cell has a receptor for biogenic amines, nor do we know what the cell will do when stimulated by the agonists for which it has receptors.

DR. BLOOM: The excitement is that the cell is there. I find this particular area most exciting. Lymphocytes have been seized upon by people interested in genetics and molecular biology to study the generation of variability, and as an analogue to evolution, not through all times, but within one individual whom you can immunize with 8 million different antigens. What Dr. Melmon just said is that you can look at the lymphocyte in another way, that the lymphocyte series gives one a whole range of differentiation to study, from the stem cell to the activated cell to the memory cell to the terminal effector cell.

What I would hope we shall find after we have cleared the first hurdles of characterizing the binding properties of lymphoid cells for known and unknown pharmacologic agents is that these receptors come and go in an ebb and flow of differentiation. I am impressed with some of the data we have heard, that pretreatment of animals with drugs does very little, but treatment four days after cells have been selectively activated may do quite a lot five days later. The possibility exists that this is the way one exerts some form of control. I think first there is no possibility of there being only one or two mechanisms of control. The danger for self-destruction inherent in the immune system is too great to allow it to be so simple. Much of the more interesting speculation might be lumped under the words of feedback. We are now in a position to look at the receptors in inactive cells and determine at what stage new receptors appear. I would expect pluses and minuses in terms of differentiation of aging and senescence. We need to know whether these receptors persist at the same level or whether as a cell becomes committed to performing specialized functions, it becomes tougher and tougher to stop, because that is all it can do. It may well be that once a cell is fully committed, there is no worry about its becoming something else.

DR. PARKER: As a corollary to what Dr. Melmon has said, our own data indicate that there are real differences in the responsiveness of T and B cells to various hormonal agents, including histamine, corticosteroids, catecholamines, and PHA. As mentioned previously in this conference, interactions between T and B cells appear to be important in the development of immune responses. One view has been that the T cell plays a passive role in holding the antigen in a favored configuration for the B cell, but there is some evidence that when lymphocytes are stimulated by antigen in tissue culture, they release material that amplifies the stimulatory action of antigen. While the basis for the amplification is not known, it is entirely possible that cells responding to antigen secrete pharmacologically active substances which participate in the short-range cell communication among activated antigen-sensitive cells. In this event, small-molecular-weight hormones or hormonelike macromolecules would be obligatory participants in the synergistic interaction among sensitized cells in addition to their presumed role in helping regulate the overall tempo of immunological inflammation.

However, I wonder how important catecholamines and histamine really are in terms of possible direct regulatory effects on lymphocytes, as opposed to indirect effects through secondary changes in vascular permeability and other events affecting the local tissue environment. In some of the immunological systems studied, high concentrations of amine are required to obtain a well-defined metabolic effect, and the effect is not sustained despite the continuing presence of the amine. Moreover, it is remarkable how little effect repeated injections of moderately high doses of catecholamines usually have on antibody formation and cellular immunity in vivo. It may be worth keeping in mind the ability of lymphocytes to interact with insulin changes at different stages of the cell cycle, but there is no evidence that insulin levels per se have any very marked effects on immunological responsiveness.

SUBJECT INDEX